AF556889

FOCUS ON PLANT MOLECULAR BIOLOGY - 1

NITRIC OXIDE SIGNALING IN HIGHER PLANTS

The Editors

Jose Ronaldo Magalhaes was born in Uba, state of Minas Gerais, Brazil, in 1949. He achieved his BS in Agronomy Engineer at Universidade Federal de Vicosa (UFV), Brazil, in 1973 and He got his Ph.D. in Plant Physiology at Purdue University in 1983. He worked as a Post-Doctor fellow in Plant Biochemistry, at Purdue University in 1986/1987, at Tsukuba University, Japan, in 1989, and at University of California-Davis, in 1998/2000. He is currently a senior scientist at Embrapa (The Brazilian Agriculture Research Agency), CNPq (Brazilian National Research Council), Associate Editor of the International Journal Physiology and Molecular Biology of Plants, and advisor of graduate students in several Universities, such as Universidade de Sao Paulo (USP), UNICAMP, UFV, UFRRJ, and UEM. He is presently leading research on Nitric Oxide Signaling in Higher Plants, focusing on plant stress resistance. Dr. Magalhaes has over 200 publications, including 65 scientific papers in international journals, 4 books, and 9 book chapters. He has attended 77 international meetings in several countries, including USA, Australia, Japan, Russia, Netherlands, Portugal, and Latin America countries

Dr Rana Pratap Singh was born in 1959 in district Kushinagar in Uttar Pradesh (India) and is currently a senior reader in Dept. of Biosciences at MD University, Rohtak (India). He has a first division M.Sc. degree in Botany from Gorakhpur University and Ph.D. in life Sciences from DA University; Indore. He has published more than 70 original research papers ,review articles, book chapters and fourteen books (ed.) to his credit. he has been awarded R.D. Asana Medal of Indian Society of Plant Physiology, New Dehli and JEB Young Scientist Award of the Academy of Environ. Biol. India, for significant research contributions. He has visited many reputed research laboratories in India, Canada and China in various fellowships, visiting scientists and in scientific meetings etc. Dr. Singh is also working as Editor-in chief of an international Research Journal in Plant Science, "Physiology and Molecular Biology of Plants" and also placed on editorial board of 'Plant Cell Technology and Molecular Biology' ,'Plant Archieves' and 'Brassica'. His current research interest is understanding of plant relations to nitrogen metabolism especially under environmental stresses and to develop low cost technology for eco-friendly nutrients management using biotechnological approaches.

Leonidas Paixao Possos was born in 1954 in Tres Coracoes, Minas Gerais State, Brazil. He obtained his B.S. degree in Agronomy at the Vicosa Federal University ,Brazil, in 1975, and his Ph.D degree in Plant Physiology at the University of Arizona, USA, in 1989. He has worked as a researcher in the Brazilian Agricultural Research Agency (Embrapa) since 1976 .Dr Passos is a leader in investigating biotechnological and physiological profiles of forage species aiming at enhancing the efficiency of genetic breeding programs for subtropical and tropical ecosystems. His team developed new approaches on in vitro embryogenic tissue formation and plant regeneration for miniaturized evaluation purposes, and also novel techniques for assessing soluble carbohydrate responses of forage grasses to abiotic stresses. He has published 26 scientific papers and reports in several countries, and authored a book in plant physiology laboratorial procedures. Currently, he manages the Biotechnology and Plant Physiology Lab of Embrapa's National Dairy Cattle Research Center and is an acting member of the Brazilian Society of Animal Science, American Society of Agronomy and Soil Science Society of America.

FOCUS ON PLANT MOLECULAR BIOLOGY - 1

NITRIC OXIDE SIGNALING IN HIGHER PLANTS

EDITORS

Jose R. Magalhaes
EMBRAPA-Gado de Leite
36038-330, Juiz de Fora
MG, Brazil

Rana P. Singh
Department of Biosciences
MD University, Rohtak - 124 001, India

Leonidas P. Passos
EMBRAPA-Gado de Leite
36038-330, Juiz de Fora
MG, Brazil

STUDIUM PRESS, LLC
HOUSTON, USA

ISBN: 0-9761849-2-3

First Published 2005

Published by :
STUDIUM PRESS LLC
P.O. Box 722200, Houston,
Texas 77072 U.S.A.
Tel. : (713) 541-9400 Fax : (713) 541-9401
E mail : studiumpress@studiumpress.com

Printed by :
Salasar Imaging Systems
Delhi 110035 (India)

PREFACE

The book is a timely, unique and comprehensive updated compendium on production and utilization of nitric oxide (NO) in higher plants. Written by the most outstanding experts from all over the world, *Nitric Oxide Signaling in Higher Plants*, this extraordinarily detailed reference, provides the latest advances in the explosive growth of NO research covering nitric oxide in plant biology from its past to the future prospects in the biology.

It covers spatial and temporal aspects of NO production, visualization of NO in cells suspensions, leaves, and roots, measurement of NO by chemiluminescence detection, NO signaling through the hypersensitive disease resistance response, plant programmed cell death, peroxisomes as a source of NO in plant, nitric oxide as an essential component of biotic and abiotic stresses-induced signaling pathways, the crucial role of this molecule in the adaptation to environmental constraints, NO bridging the gap between environmental stimuli and endogenous signals in plants, nitric oxide produced by nitrate reductase in plants contrasting with NO synthase, nitric oxide as a mediator of abscisic acid signaling in guard cells, NO as a potential second messenger in cytokinin signaling, Nitric oxide and hormone signal networks, NO involvement in cellular redox homeostasis, NO acting at the plant mitochondrial level by binding to cytochrome oxidase, NO in perception and action of light in seed germination; bio-mimetic NO-synthesizing peroxidase-mediated pathways and lignification, and much more. It considers the historical background of plant NO research followed by the current state of art, the recent advances and the future prospects, emphasizing the necessity for an integrated knowledge of the plant sciences.

Nitric oxide is one of the simplest and most ubiquitous molecules in the atmosphere and in soils. The molecule has been detected beyond the earth in Venus and Mars, and in interstellar space too.

Joseph Priestley (1733-1804), a familiar name for the plant biologists, in particular for photosynthetic researchers, has been credited in the science history with the discovery of "oxygen (O_2)" and "photosynthesis" in 1774. Prior to this famous discovery, he had already found "NO", termed as "nitrous air", the first documented gas in science. He prepared NO by the action of "spirit" of HNO_3 on a number of different metals including iron, copper, silver and mercury.

For two hundred years since the discovery of NO gas, there was a long silence in plant NO research. Suddenly, NO was highlighted as a toxic molecule. NO become known as a pollutant byproduct from industry, produced by the combustion of fuels or the manufacture of nitrogenous fertilizers through industrial activities. The controversy as to whether NO is phytotoxic represents the first chaotic period in NO research in plants.

In fact, little more than two decades ago the idea that NO could serve in any physiological process was simply dismissed. Now NO is a broad discipline covering almost every field of molecular biology, physiology, biochemistry and medicine.

Until recently, we had believed that NO was just an exogenous biologically irrelevant molecule. After the discovery of NO-producing enzymes and the cGMP signal transduction systems, our appreciation of NO drastically changed, as seen in the awards of 'Molecule of the Year' by Science magazine in 1992 and the Nobel Prize for physiology and medicine in 1998. The discovery of functions for NO in mammalian cells was an exciting breakthrough for science. Nitroglycerin, a material of dynamite invented by Alfred Nobel, has been used clinically over 100 years to treat angina pectoris but its mechanism of action has only recently been elucidated in terms of NO. There is no doubt that NO research is now one of the most important and active fields of science in which more than ten thousand papers were published last year (based on the ISI data base).

Today, Nitric oxide is considered to be a "magic" gas molecule, which is critical to numerous biological processes, including vasodilator, neurotransmission (Viagra reverses impotence by enhancing a NO-stimulated pathway) and macrophage-mediated microorganism and tumor killing. The discovery that NO is a biologically produced free radical has revolutionized our thinking about physiological and pathological process. This discovery has ignited enormous interest in the scientific community.

Nitric oxide is emerging as a new chemical messenger in plants that functions in a broad spectrum of physiological and developmental processes. An intriguing cross talk between the environmental stimuli, hormone actions and NO functions seems to be the key for the understanding how plant responses are synchronized for the adaptation to a continuous changing environment. NO is a wide spread intracellular and intercellular signaling molecule involved in the regulation of a notable spectrum of diverse cellular functions.

This book is a summary of the most recent discoveries in the field of NO biology, biochemistry, and signaling molecule in plants. It is hoped that the book will further fuel the interest in the rapidly expanding field of nitric oxide biology and encourage additional research into better understanding of the role of NO in plant development. For creating a consistent and beautiful "NO world" in science, it seems obvious that we need to be open to incorporating and integrating comprehensive knowledge just like plants did in evolution.

The contents have been carefully organized in a planned sequence to give a detailed insight of the subject to beginners as well as to the experts.

These chapters critically evaluate the state of art, the current status and future prospects of the knowledge in NO research which are very useful for the scientists, academicians, researchers, students, planners and industrialists etc. working in the area of biotechnology, plant agriculture, horticulture, agronomy, plant physiology, molecular biology, plant pathology and forestry. The book reveals the updated knowledge on the nitrogen relations to plants which includes a historical perspective, recent discoveries and future directions at one place to focus for the creating worthful knowledge for the human beings by their efforts.

The reader will immediately recognize that each chapter in this volume has been written by a world outstanding expert in NO research. A special effort has been made to concentrate on new findings in the nitric oxide field. It is through this vision that new leads will develop to more complete understanding of the cell and molecular biology of nitric oxide.

In bringing this volume to fruition, credit must be given to the scientists in various specialized fields of NO research. Our great appreciation goes to the contributors to this volume. We extend our sincere thanks and most gratitude to all these colleagues. We extend our warm appreciation and thanks to Studium Publishing LLC for their keen interest in bringing out this title with quality work. We are thankful to our research scholars and family members for their understanding and patience during preparation of this title and efforts made by Ms. Usha, Department of Biosciences, M.D. University and Mr. Ashok Datta and Ms. Reema of LaPrints, New Delhi, India for preparing camera ready version. Finally, acknowledgement is expressed to Brazilian Agriculture Research Agency (Embrapa - Gado de Leite), Brazilian Research Council (CNPq) and M.D. University, Rohtak, India for support.

May, 2004 **The Editors**

CONTENTS

CONTRIBUTORS

Barceló, A. Ros
Department of Plant Biology, University of Murcia, E-30100 Murcia, Spain
E-mail : rosbarce@um.es

Barroso, Juan B.
Departamento de Bioquímica y Biología Molecular, Facultad de Ciencias Experimentales, Universidad de Jaén, Spain

Bright, Jo
Centre for Research in Plant Science, Faculty of Applied Sciences, University of the West of England, Bristol, Coldharbour Lane, Bristol BS16 1QY, UK

Butt, Yoki Kwok-Chu
Dept. of Applied Biology and Chemical Technology, The Hong Kong Polytechnic University, Kowloon, Hong Kong SAR
Fax : (852)-23649932

Carreras, Alfonso
Departamento de Bioquímica y Biología Molecular, Facultad de Ciencias Experimentales, Universidad de Jaén, Spain

Corpas Francisco J.
Departamento de Bioquímica, Biología Celular y Molecular de Plantas, Estación Experimental del Zaidín, CSIC, Granada, Spain
E-mail : javier.corpas@eez.csic.es; Fax : 34 958 129600

del Río, Luis A.
Departamento de Bioquímica, Biología Celular y Molecular de Plantas, Estación Experimental del Zaidín, CSIC, Granada, Spain
Fax : 34 958 129600

Delledonne, Massimo
Dipartimento Scientifico e Tecnologico, Università degli Studi di Verona, Verona, Italy
E-mail : massimo.delledonne@univr.it; Fax : +39-045-8027929

Desikan, Radhika
Centre for Research in Plant Science, Faculty of Applied Sciences, University of the West of England, Bristol, Coldharbour Lane, Bristol BS16 1QY, UK

Ferrarese-Filho, Osvaldo
Dep. Bioquímica - UEM - 87020-900 Maringá, PR - Brazil

Gabaldón, C.
Department of Plant Biology, University of Murcia, E-30100 Murcia, Spain

García-Mata, Carlos
Instituto de Investigaciones Biológicas, Facultad de Ciencias Exactas y Naturales, Universidad Nacional de Mar del Plata, CC1245, 7600 Mar del Plata, Argentina
Fax : (+54)-223-4753150

Giba, Zlatko
Institute of Botany, Faculty of Biology, University of Belgrade, Belgrade, Yugoslavia
Fax : 381-11-769903

Gould, Kevin S.
Plant Sciences Group, School of Biological Sciences, University of Auckland, Private Bag 92019, Auckland, New Zealand.

Graziano, Magdalena
Instituto de Investigaciones Biológicas, Facultad de Ciencias Exactas y Naturales, Universidad Nacional de Mar del Plata, CC1245, 7600 Mar del Plata, Argentina
Fax : (+54)-223-4753150

Grubišic, Dragoljub
Institute for Biological Research "Siniša Stankovic", 29, novembra 142, 11060, Belgrade, Yugoslavia

Hancock, John
Centre for Research in Plant Science, Faculty of Applied Sciences, University of the West of England, Bristol, Coldharbour Lane, Bristol BS16 1QY, UK

Kaiser, Werner M.
Julius von Sachs Institut für Biowissenschaften, Lehrstuhl für Molekulare Pflanzenphysiologie und Biophysik, Julius von Sachs Platz 2, D-97082, Würzburg -Germany
E-mail : kaiser@botanik.uni-wuerzburg.de; Fax : (0)-931-888-6158

Klinguer, Agnès
UMR 1088 INRA/Université de Bourgogne et FRE/CNRS 2625, Plante-Microbe-Environnement, INRA, 17 rue Sully, BP 86510, Dijon 21065 cedex, France

Konjevic, Radomir
Institute of Botany, Faculty of Biology, University of Belgrade, Belgrade, Yugoslavia
E-mail : konjevic@bfbot.bg.ac.yu; Fax : 381-11-769903

Lamattina, Lorenzo
Instituto de Investigaciones Biológicas, Facultad de Ciencias Exactas y Naturales, Universidad Nacional de Mar del Plata, CC1245, 7600 Mar del Plata, Argentina
E-mail : lolama@mdp.edu.ar; Fax : (+54)-223-4753150

Lamotte, Olivier
UMR 1088 INRA/Université de Bourgogne et FRE/CNRS 2625, Plante-Microbe-Environnement, INRA, 17 rue Sully, BP 86510, Dijon 21065 cedex, France

León, Ana M.
Departamento de Bioquímica, Biología Celular y Molecular de Plantas, Estación Experimental del Zaidín, CSIC, Granada, Spain
Fax : 34 958 129600

Lo, Samuel Chun-Lap
Dept. of Applied Biology and Chemical Technology, The Hong Kong Polytechnic University, Kowloon, Hong Kong SAR
E-mail : bcsamlo@inet.polyu.edu.hk; Tel. : (852)-27666686; Fax : (852)-23649932

Lum, John Hon-Kei
Dept. of Applied Biology and Chemical Technology, The Hong Kong Polytechnic University, Kowloon, Hong Kong SAR
Fax : (852)-23649932

Magalhaes, Jose R.
EMBRAPA-Gado de Leite, 36038-330, Juiz de Fora, MG, Brazil.
E-mail : josemag@cnpgl.embrapa.br; Fax : (55-32)-3249-4852

Modolo, Luzia V.
Departamento de Bioquímica, Instituto de Biologia, Universidade Estadual de Campinas (UNICAMP), Campinas, SP, 13083-970, Brazil
Fax : 55-19-3788-6129

Neill, Steven
Centre for Research in Plant Science, Faculty of Applied Sciences, University of the West of England, Bristol, Coldharbour Lane, Bristol BS16 1QY, UK
E-mail : Steven.Neill@uwe.ac.uk

Pagnussat, Gabriela C.
Instituto de Investigaciones Biológicas, Facultad de Ciencias Exactas y Naturales,, Universidad Nacional de Mar del Plata, CC1245, 7600 Mar del Plata, Argentina
Fax : (+54)-223-4753150

Palma, José M.
Departamento de Bioquímica, Biología Celular y Molecular de Plantas, Estación Experimental del Zaidín, CSIC, Granada, Spain
Fax : 34 958 129600

Planchet, Elisabeth
Julius-von-Sachs-Institut für Biowissenschaften, Lehrstuhl für Molekulare Pflanzenphysiologie und Biophysik, University of Würzburg, Julius-von-Sachs-Platz 2, D-97082 Würzburg, Germany
Fax : (0)931-888-6158

Pomar, F.
Department of Plant Biology, University of Murcia, E-30100 Murcia, Spain

Pugin, Alain
UMR 1088 INRA/Université de Bourgogne et FRE/CNRS 2625, Plante-Microbe-Environnement, INRA, 17 rue Sully, BP 86510, Dijon 21065 cedex, France

Quiros, Miguel
Departamento de Química Iuorgánica, Facultad de Gieucias, Universidad de Granada, Spain

Rockel, Peter
Forschungszentrum Jülich, Institut für Chemie der belasteten Atmosphäre, D-52425 Jülich, Germany

Romero-Puertas, María C.
Dipartimento Scientifico e Tecnologico, Università degli Studi di Verona, Verona, Italy
Fax : +39-045-8027929

Salgado, Ione
Departamento de Bioquímica, Instituto de Biologia, Universidade Estadual de Campinas (UNICAMP), Campinas, SP, 13083-970, Brazil
E-mail : ionesm@unicamp.br; Fax : 55-19-3788-6129

Sandalio, Luisa M.
Departamento de Bioquímica, Biología Celular y Molecular de Plantas, Estación Experimental del Zaidín, CSIC, Granada, Spain
Fax : 34 958 129600

Saviani, Elzira E.
Departamento de Bioquímica, Instituto de Biologia, Universidade Estadual de Campinas (UNICAMP), Campinas, SP, 13083-970, Brazil
Fax : 55-19-3788-6129

Scherer, Günther F.E.
Universität Hannover, Institut für Zierpflanzenbau, Baumschule und Pflanzenzüchtung, Abt. Spezielle Ertragsphysiologie, Herrenhäuser Str. 2, D-30419 Hannover, Germany
E-mail : scherer@zier.uni-hannover.de; Fax : +49-511-762-3608

Selvi-Srinivas, Manickam
UMR 1088 INRA/Université de Bourgogne et FRE/CNRS 2625, Plante-Microbe-Environnement, INRA, 17 rue Sully, BP 86510, Dijon 21065 cedex, France

Silva, Filomena, L.I.M.
EMBRAPA - Gado de Leite – 36038-330 Juiz de Fora, MG, Brazil
Fax : (55-32)-3249-4852

Sonoda, Masatoshi
Julius-von-Sachs-Institut für Biowissenschaften, Lehrstuhl für Molekulare Pflanzenphysiologie und Biophysik, University of Würzburg, Julius-von-Sachs-Platz 2, D-97082 Würzburg, Germany
Fax : (0)931-888-6158

Stoimenova, Maria
Julius-von-Sachs-Institut für Biowissenschaften, Lehrstuhl für Molekulare Pflanzenphysiologie und Biophysik, University of Würzburg, Julius-von-Sachs-Platz 2, D-97082 Würzburg, Germany
Fax : (0)931-888-6158

Wendehenne, David
UMR 1088 INRA/Université de Bourgogne et FRE/CNRS 2625, Plante-Microbe-Environnement, INRA, 17 rue Sully, BP 86510, Dijon 21065 cedex, France
E-mail : wendehen@dijon.inra.fr
Tel. : +33-3-80-69-37-21; Fax : +33-3-80-69-32-26

Yamasaki, Hideo
Integrative Biology Group, Center of Molecular Biosciences (COMB), University of the Ryukyus, Nishihara, Okinawa 903-0213, Japan
Fax : +81-98-895-8944
E-mail : yamasaki@comb.u-ryukyu.ac.jp

Zottini, Michela
Dipartimento di Biologia Università degli Studi di Padova - Via U. Bassi, 58/B - I-35131 Padova - ITALY
Tel. : +39-049-8276247; Fax : +39-049-8276300
E-mail : mzottini@civ.bio.unipd.it

Chapter 1

NITRIC OXIDE RESEARCH IN PLANT BIOLOGY : ITS PAST AND FUTURE

Hideo Yamasaki

Integrative Biology Group, Center of Molecular Biosciences (COMB), University of the Ryukyus, Nishihara, Okinawa 903-0213, Japan
Fax : +81-98-895-8944
E-mail : yamasaki@comb.u-ryukyu.ac.jp

Summary

Nitric oxide (NO), a gaseous free radical, had been known for many years simply as a toxic air pollutant. For plant biologists, therefore, phytotoxicity of NO was a major subject of interest until recently. After the discovery of the enzyme NO synthase (NOS) in vertebrate animals and pivotal NO signaling functions postulated in late 1980s, many researchers began to consider that plants may posses a NO signaling system similar to mammals. In fact, many biochemical studies had suggested the presence of mammalian-type NO synthase (NOS) in plants. Despite extensive reserch, however, no plant homologue of such NOS has been found in plant genomes to date. Alternatively, assimilatory nitrate reductase (NR) has been found to produce NO through one electron reduction of nitrite using NAD(P)H as an electron donor. In addition to NR-mediated NO production, novel NO synthases that produce NO with L-arginine as the substrate have recently been found but these enzymes showed no sequence similarity to mammalian-type NOSs. Thus, it has become clear that plant NO systems are considerably different from those of animals

In : Nitric Oxide Signaling in Higher Plants, 2004
(Eds Jose R. Magalhaes, Rana P. Singh and Leonidas P. Passos)
Studium Press, LLC, Houston, USA, pp 1-23

in many ways. In this review I wish to present some prospects for the future plant NO research by looking back its history and looking out over the horizon to where we should go.

Keywords : Arginine pathway, integrated plant biology, nitrate reductase, nitration, nitrite pathway, Priestly, RNS, ROS

1. INTRODUCTION

Nitric oxide (NO) is one of the simplest and most ubiquitous molecules in the atmosphere and in soils. The molecule has been detected beyond the earth in Venus and Mars, and in interstellar space (Liszt and Turner, 1978). Until recently, we had believed that NO is just an exogenous biologically irrelevant molecule. After the discovery of NO producing enzymes and the cGMP signal transudation systems in mammals, our appreciation of NO drastically changed, as seen in the awards of 'Molecule of the Year' by Science magazine in 1992 and the Nobel Prize for physiology and medicine in 1998. The discovery of functions for NO in mammalian cells was an exciting breakthrough for science. Nitroglycerin, a material of dynamite invented by Alfred Nobel, has been used clinically over 100 years to treat angina pectoris but its mechanism of action has only recently been elucidated in terms of NO. There is no doubt that NO research is now one of the most important and active fields of science in which more than ten thousand papers are published annually (based on the ISI data base in 2002).

In contrast to our knowledge of NO biology in mammals, however, NO production mechanisms in plants and the corresponding signaling pathways are largely not understood. Until very recently, most plant researchers had considered that a mammalian-type NO synthase (mammalian NOS) is involved in the NO production mechanism of plants. Despite many papers suggesting the presence of NOS-like activity in plant cells, no such homologue has been found in the *Arabidopsis* genome nor in others (Guo *et al.*, 2003). This may imply that NO production and signaling mechanisms are considerably different from mammalian systems. In fact, there is an increasing number of papers reporting that assimilatory nitrate reductase (NR) produces NO in plant cells through a different mechanism from the

NOS system. If plant NO systems are distinct from animal's, it follows that we can not readily apply the knowledge obtained from mammals to plants.To help to draw a consistent and clear picture of plant NO systems, here I review the historical background of plant NO research followed by the future prospects, emphasizing the necessity for an integrated knowledge of plant sciences.

2. THE BEGINNING : BEHIND THE DISCOVERY OF PHOTOSYNTHESIS

Joseph Priestley (1733-1804) is a familiar name for plant biologists, in particular for photosynthetic researchers (Fig. 1). Priestley has been credited in the science history with the discovery of "oxygen (O_2)" and "photosynthesis" in 1774. Prior to this famous discovery, he had already found "NO", termed as "nitrous air", the first documented gas in science.

He prepared NO by the action of spirit of HNO_3 on a number of different metals including iron, copper, silver and mercury. In his studies on the phenomena of combustion and respiration, NO was applied to quantify the amount of O_2 that was designated "dephlogisticated air " at that time by Priestley. Using two different ways, he demonstrated that a decrease in O_2 concentration within a closed glass vessel can be observed after burning (Fig. 1). One is a "biological assay" to measure remaining O_2 by observing mouse mortality. The other way is a "chemical assay" in which NO is applied to quantify the oxygen level inside the vessel. NO forms deep red fumes of NO_2 in atmospheric conditions:

$$2NO + O_2 \rightarrow 2NO_2 \qquad \text{....... (1)}$$

Because NO_2 is reactive with water, the gas formed is absorbed by the water decreasing the gas phase volume. He seemed to prefer the chemical assay using NO over the biological assay because of its higher accurateness (Ainscough and Brodie, 1995). Thus, NO research was begun by the photosynthetic founder and the inter-relationship of NO was demonstrated even at this early stage of experimentation.

3. AFTER A LONG SILENCE: AIR POLLUTANT STUDIES

For two hundred years since the discovery of NO gas by Priestley, there was a long silence in plant NO research. Suddenly, NO was highlighted as a toxic molecule. The Industrial Revolution that started in Europe had stimulated the consumption of fossil fuels for industrial

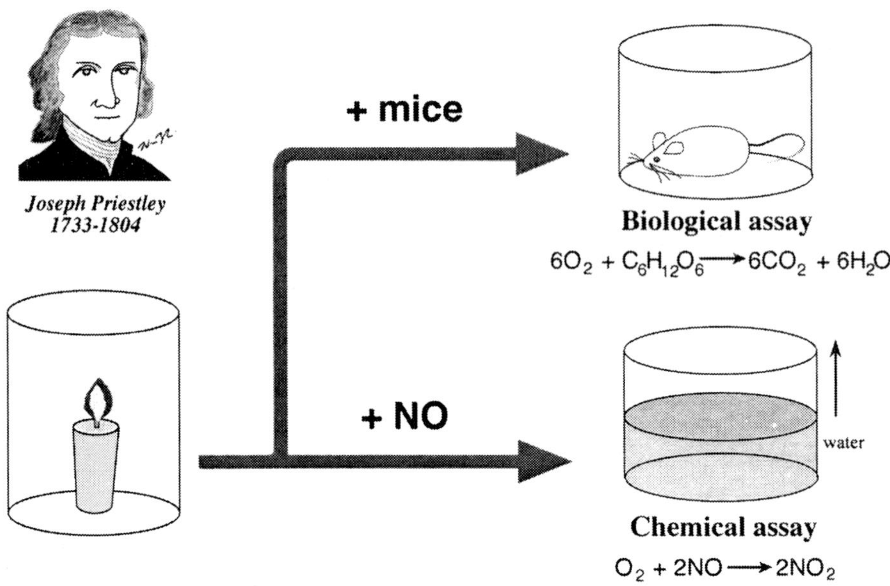

Figure 1 : The beginning : the founder of photosynthesis discovered NO. Joseph Priestley is renowned for his discoveries in 1774: photosynthetic activity of plants and respiration phenomena in biology and oxygen gas in chemistry. Two years prior to this famous experiment, he found NO, the first gas to be discovered in the history of science. He applied the chemical reaction of NO in the air to quantify the amount O_2. Because of high water solubility of NO_2 formed after the reaction, the volume inside the vessel will be decreased when NO is provided into a sealed glass vessel dipped onto the water. By monitoring the volume of the air, he quantified the amount of O_2 that remained within the vessel, the chemical assay that was necessary for his experiments.

activities. As the result of the rapid industrialization of urban cities, air pollution problems arose. NO become known as a pollutant byproduct from industry.

NO_x ($NO_x = NO + NO_2$) is produced by the combustion of fuels or the manufacture of nitrogenous fertilizers through industrial activities. NO_x concentrations may range up to 100 ppb in the air over industrialized areas (Wildt *et al.*, 1997) and it is a cytotoxic agent to both plants and animals (Wellburn *et al.*, 1980). However, in contrast with other air pollutant gases, such as sulfur dioxide (SO_2) and O_3, the phytotoxicity of NO_x has been considered to be relatively week. Fumigation with NO_x sometimes does not show deleterious effects but it often stimulats plant growth most likely by acting as a nitrogen fertilizer (Yoneyama *et al.*, 1979). The controversy as to whether NO_x is phytoxic (Wellburn, 1990) represents the first chaotic period in plant NO research. Because of many contradictory results, probably due to differences in experimental conditions and in plant species choosen,it was difficult at that time to deduce the generalized effects of NO on plants (Wellburn, 1990). Nevertheless, air pollutant studies provided important clues for the future plant NO research since they demonstrated various effects of NO through out the plant body in many species.

4. THE DAWN OF PLANT NO RESEARCH: PHANTOMS OF MAMMALIAN NOS

On endogenous NO production and its physiological roles in vertebrates, Furchgott (Furchgott, 1988), Ignarro *et al.* (Ignarro *et al.*, 1987) and the Moncada laboratory (Palmer *et al.*, 1987) independently revealed that endothelium–derived relaxing factor (EDRF) was NO, a new field exploded on to the scientific scene. Shortly, Bredt and Snyder (1990) reported the first evidence for the enzyme nitric oxide synthase (NOS) by isolating and purifying the protein (nNOS) from rat brain. Untill the late 1990s, however, these scientific breakthroughs had not influenced plant NO reseach. In 1996, Evidence for the presence of NOS activities in plants were reported from different laboratories (Cueto *et al.*, 1996; Leshem and Haramaty, 1996; Ninnemann and Maier, 1996). At the same time, physiologically positive functions for NO in plants were proposed for the establishment of symbiosis (Cueto *et al.*, 1996), induction of phytoalexin production (Noritake *et al.*, 1996) and regulation of growth and development (Leshem and Haramaty, 1996). Unfortunately, these pioneering works did not attract the general interest of plant researchers, particularly those working on molecular biology.

Two landmark papers (Delledonne *et al.*, 1998; Durner *et al.*, 1998) sparked researchers to consider that plants may posses a mammalian-type NO signaling system comprising cGMP and NOS. Inhibition of L-arginine to citrulline conversion activity by mammalian NOS inhibitors suggested the presence of mammalian-type NOS in plants (Cueto *et al.*, 1996; Delledonne *et al.*, 1998). Although a protein of 166 kDa which crossreacted with antibodies raised against mouse macrophage or rabbit brain NOS was identified in maize cells (Ribeiro Jr *et al.*, 1999), no protein nor gene homologues to mammalian NOS have been isolated to date (Chandok *et al.*, 2003; Guo *et al.*, 2003) and neither has proteomics analysis using mammalian NOS antibodies found such proteins in plants (Butt *et al.*, 2003). Taking together all available information, we must conclude that plants do not posses an NO-producing enzyme structurally identical to mammalian NOSs (eNOS, nNOS, iNOS) as previously presumed.

5. NITRITE PATHWAY: THE SIMPLEST MECHANISM FOR NO PRODUCTION

It is not well appreciated that endogenous NO producing activities had been found in plants prior to the discovery of NOS in animals. The presence of a nonenzymatic pathway for NO production has been known for many years, *i.e.* an inorganic nitrogen pathway (Evans and McAuliffe, 1956). The slow and spontaneous liberation of NO can be observed with nitrite (NO_2^-) at neutral pH. Since the amount of NO released by this reaction is usually below the limit of detection, spontaneous NO liberation from nitrite has been considered to be chemically possible but physiologically irrelevant. However, a large amount of NO can be produced from nitrite when a mild reductant is present. Ascorbate (AsA), a physiologically important reductant in plants, has also been reported to induce chemical NO production from nitrite (Yamasaki, 2000):

$$2HNO_2 + 2AsA \rightarrow 2NO + 2MDA + 2H_2O \qquad \text{.......} (2)$$

$$2MDA \rightarrow AsA + DHA \qquad \text{.......} (3)$$

$$2HNO_2 + AsA \rightarrow 2NO + DHA + 2H_2O \qquad \text{.......} (4)$$

One electron oxidation of ascorbate produces monodehydroascorbate (MDA), an oxidized form of ascorbate. Two molecules of MDA spontaneously disproportionate to ascorbate and dehydroascorbate (DHA). As the result of these reactions, two molecules of NO can be produced by ascorbate (reaction 4). Because the reaction requires HNO_2 but not NO_2^-, it is virtually negligible at neutral and alkaline pHs (Yamasaki, 2000). Presumably, chemical NO production via the nitrite pathway is pronounced only in acidic compartments or tissues under healthy conditions (Lundberg *et al.*, 1997; Weitzberg and Lundberg, 1998). However, wounding by physical as well as biotic factors that results in physical destruction of cell compartments can induce NO production by mixing nitrite (cytosol), acid (vacuole) and reductants (chloroplast).

6. LEGUME NITRATE REDUCTASE: THE FIRST NO PRODUCING ENZYME

In 1980s, enzymatic NO production with nitrite had been reported with legume plants. Nitrate reductase (NR), a well-known enzyme catalyzing reduction of nitrate (NO_3^-) to nitrite (NO_2^-) in nitrate assimilation pathway, was the first plant enzyme conclusively shown to have NO producing ability (Yamasaki, 2000). The first clue was provided by Harper in 1981 who demonstrated the effects of gas purging on the *in vivo* NR assay (Harper, 1981). The *in vivo* NR assay is a technique used to determine the nitrate reducing activity of tissues. He found that the level of NO_x (NO + NO_2) evolved from soybean leaves while purging with anaerobic gas (N_2 or argon) was greater than that with aerobic gas purging (air or O_2). Since no NO_x evolution occurred in boiled (heat denatured) leaf disks, Harper presumed an enzymatic NO_x evolution pathway from NO_2^- in leaves (Harper, 1981). Using a chlorate screening procedure, Harper and co-workers isolated soybean mutants lacking the constitutive NR activity in leaves but retaining inducible NR activity in response to NO_3^- (Nelson *et al.*, 1983; Ryan *et al.*, 1983). They demonstrated that a soybean mutant lacking constitutive NR activity did not evolve NO_x during *in vivo* NR assays (Nelson *et al.*, 1983), the first evidence that NO_x evolution was associated with the activity of constitutive NR in the leaves. It was later determined that there were two forms of constitutive NR in wild-type soybean leaves, which were designated C_1NR and C_2NR (Streit *et al.*, 1985). Because the C_1NAD(P)H:NR (EC 1.6.6.2) is unique to Phaseoleae tribe of the family Leguminosae (Dean and Harper, 1988), the evolution of NO_x through the NR-dependent pathway was thought to be limited to legume species among higher plants (Dean and Harper, 1988). These excellent pioneering works done in legume plants did not receive much attention from plant researchers, presumably due to the absence of information on physiological role(s) of enzymatic NO production. After a silence, the findings were revived and renewed 10 years later.

7. RESETTING KNOWLEDGE: REVIVAL OF NR

In 1997, Wildt and coworkers reported NO emissions from a several plant species other than Leguminosae including sunflower, sugarcane, corn, rape, spruce, spinach and tobacco (Wildt *et al.*, 1997). This observation raised a possibility that NO-producing ability might not be a specific phenomenon of legume plants but could be more generally distributed among higher plants. In 1999, Yamasaki and coworkers demonstrated that a purified maize NR (EC 1.6.6.1) is capable of producing NO (Yamasaki *et al.*, 1999). A similar nitrite dependent NO production was found in NAD(P)H:NR (EC 1.6.6.2) of the fungus *Aspergillus* (Yamasaki, 2000). Furthermore, NR of *Escherichia coli* was reported to produce NO in the presence of nitrite (Ji and Hollocher, 1988). It became evident that the NO-producing activity of NR is a more general feature than previously thought; the nitrite pathway was reborn (Sakihama *et al.*, 2002b).

NR has long been known as a plant enzyme extensively studied to improve crop productivity. In addition to photosynthesis (carbon assimilation), nitrate assimilation is an ecologically important process in terms of the biospheric N cycle. Most plant biologists had focused only on the beneficial role of NR. In addition to NO producing capability, NR was found to produce peroxynitrite ($ONOO^-$), the reaction product of NO with superoxide (O_2^-). $ONOO^-$ has been considered to be a major cytotoxic agent of RNS derived from NO (Yamasaki, 2000). Yamasaki and Sakihama (2000) showed the *in vitro* formation of $ONOO^-$ by maize NR, evidence demonstrating that NR potentially produces three types of toxic molecules (i.e. NO, O_2^-, $ONOO^-$), an unknown harmful feature of an old enzyme (Yamasaki and Sakihama, 2000). These new aspects of NR were welcome to NR researchers (Mallick *et al.*, 2000a; Mallick *et al.*, 2000b; Morot-Gaudry-Talarmain *et al.*, 2002; Rockel *et al.*, 2002).

8. TWO NO PRODUCTION PATHWAYS IN PLANTS

Figure 2 summarizes enzymatic NO producing pathways in plants. Until now, we have understood that there are, at least, two distinct pathways in plants (Sakihama *et al.*, 2002b): the arginine pathway and the nitrite pathway (Fig. 2). The former can be mediated by NO synthase (NOS) with the substrates L-arginine, O_2, and NADPH. As the result of this reaction, the product L-citrulline is produced along with NO. Competitive inhibition should occur when analogues of L-

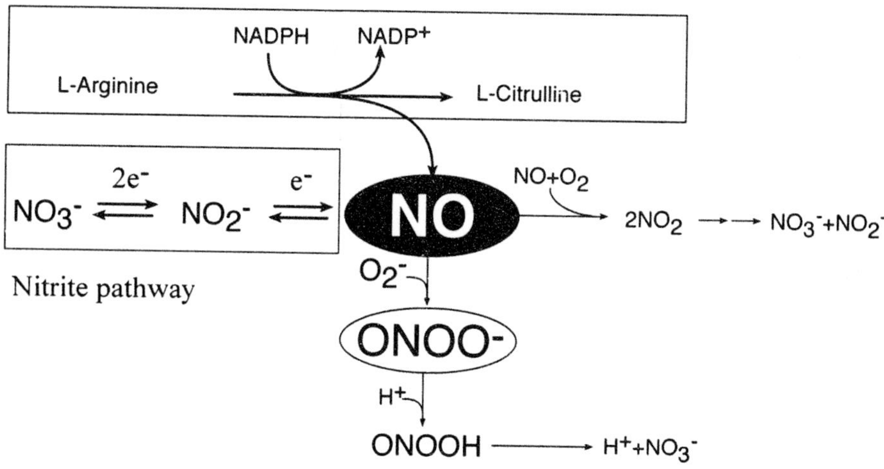

Figure 2 : Two enzymatic NO producing pathways in plants
Plants have two distinct pathways for NO production. One is the arginine pathway in which NO is synthesized from the amino acid L-arginine, O_2 and NADPH. The products L-citrulline and NO are formed as a result. Mammalian-type NO synthase had been supposed to catalyze this reaction but there is no evidence to show the presence of such an enzyme. Recently, structurally different proteins have been reported to mediate the pathway. The other pathway is the nitrite pathway in which NO is produced through the one electron reduction of nitrite. This can occur in an acidic environment but at neutral pH assimilatory nitrate reductase (NR) catalyzes the reducing reaction. It is very important to note that NO and reactive nitrogen species (RNS) are eventually converted to nitrite or nitrate.

arginine (such as L-NAME) are present. Although there is little possibility for the presence of mammalian-type NOS in plants, a plant NOS structurally different from the animal one could exist. In fact, two types of enzymes that show NOS activity have been found to date (Chandok *et al.*, 2003; Guo *et al.*, 2003). The findings of NOS unique to plants may provide explanations for previous studies that reported NOS-like activity in plants (Guo *et al.*, 2003). As described above, the nitrite pathway can proceed even in the absence of an enzyme, the fundamental difference between the arginine pathway and the nitrite pathway. Furthermore, in the nitrite pathway, assimilatory NR is not the only enzyme involved in biological NO production; many enzymes including redox domain(s) potentially reduce nitrite to form NO. Bacterial denitrification processes can be attributed to the nitrite pathway. On either pathway, nitrite or nitrate should eventually be produced as the final degradation product of NO. The complexity of the relationship between substrates and products in NO metabolism brings difficulties in identifying the molecules directly involved in the signaling systems; its degradation product (nitrite or nitrate) but not NO itself may display a signaling activity or its *vice versa*.

9. LIGHT AND SHADOW OF NO

Looking back at the history of how NO come to be recognized in science, it is interesting to note that there have always exited two extremes: hero or villain; protective or toxic; friend or foe. These apparent diametrically opposed roles for NO sometimes lead to confusion even for current researchers. Why does NO show such "Jekyll and Hyde"-like behavior? Figure 3 intends to be of help for showing how we can interpret the dualism of NO. The "Yin-Yang" symbol represents simply "polarities in unity". According to oriental philosophies (*e.g.*, in China, Japan or Korea), nature includes polarities, such as light and shadow, good and evil, male and female,

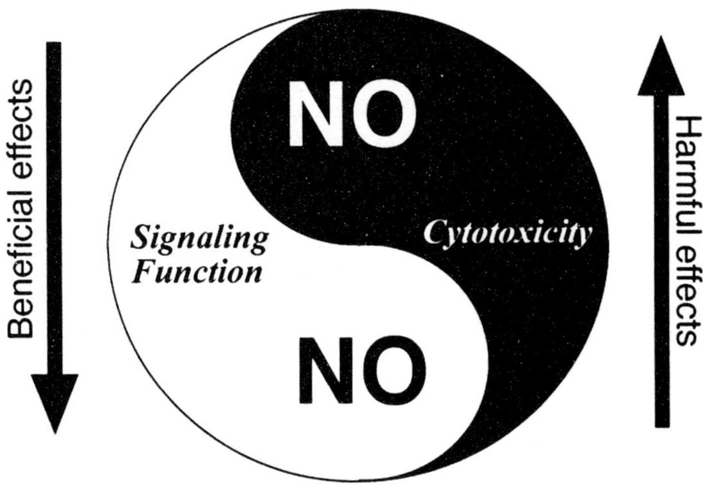

Figure 3 : The Yin-Yang symbol for NO
The Yin-Yang (means shadow-light) symbol represent the polarities of reality (*e.g.*, shadow and light; male and female, evil and good). NO also includes this dualism in nature: beneficial and harmful effects on plants. To obtain maximal signaling function of NO, cytotoxicity should be minimized by appropriate mechanisms. Without protecting mechanisms, harmful effects may exceed beneficial signaling functions.

but makes beautiful unity. We can find this dogma in characteristics of many biomolecules including amino acids (zwiterion, negative and positive charges in a molecule) or lipids (amphiphilic nature, hyrdophilic and hydrophobic groups in a molecule). In plant biology, Barry Osmond (1994) has favored applying the Yin-Yang symbol to address the nature of photoinhibition: light is energy or an inhibitor for photosynthesis? Sakihama *et al.* (2002a) have also applied this concept to account for the dualism of plant phenolic compounds: antioxidant or prooxidant? For understanding the nature of ubiquitous and essential biomolecules, knowing the balance is more important than giving a cast; there should be no ultimate answer for the cast of NO.

10. EVOLUTIONAL ASPECTS: BEHIND THE DUALISM

The dualism of NO in effects may be better explained by introducing evolutional aspects rather than a philosophical argument. Figure 4 illustrates how biological relationships change in the process of evolution. This principle can be seen in many biological interactions, *i.e.* the process of establishing symbiosis. In an early stage, hostileinvaders (element A) just cause harmful effects to living organisms (stage 1). Later, living organisms (element B) evolve protecting mechanisms against harmful invaders (stage 2). After the acquisition of the protecting mechanism, invaders become friends and may have a new function for mutual benefits via tolerance stage (stage 3). Finally, living organisms may actively produce the element inside for new functions (stage 4). We can apply this sequential change in relationship to interactions between living organisms and inorganic molecules, *i.e.*, O_2 or NO.

Until the discovery of the enzyme superoxide dismutase (SOD) by McCord and Fridovich (McCord and Fridovich, 1969), no one (except Priestley) had considered that the oxygen molecule might show toxicity for plants and animals. The finding of SOD, which reduces oxygen toxicity through the scavenging of O_2^- produced in cells in anaerobic conditions, was the begging of ROS research in life science. It is now a consensus that the concentration of atmospheric oxygen started to increase after the evolutional successes of oxygenic photosynthetic organisms such as cyanobacteria three billion years ago (Asada, 2000). At that time, O_2 was just a byproduct produced during photosynthetic electron transport. As seen in the

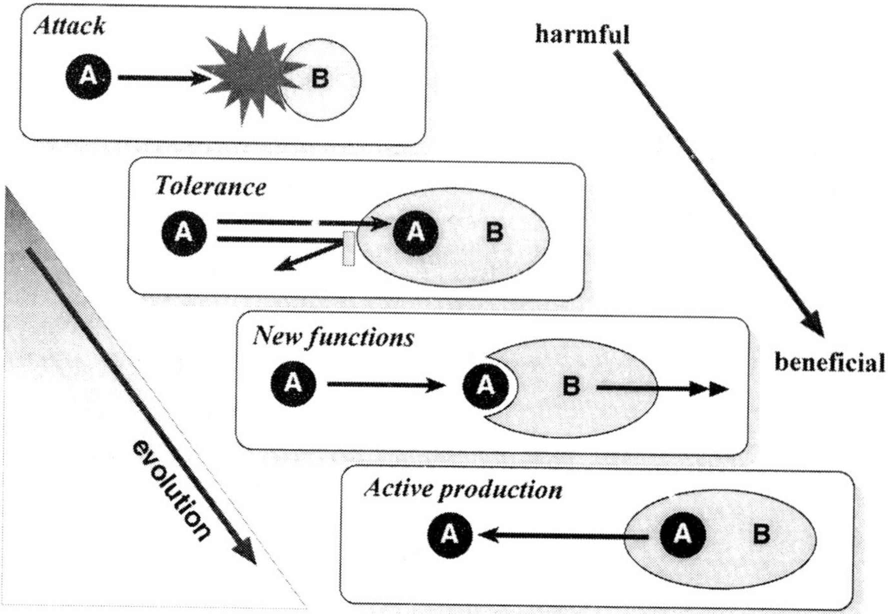

Figure 4 : Evolution of relationship between two elements
This scheme represents conceptual steps how living organisms have evolved to cope with unfavorable outer element(s). Suppose the element *A* is harmful to the element *B*. At the beginning, *B* suffers attack by *A* (stage 1). Then, *B* develops protection mechanism and further evolves tolerance mechanisms (stage 2). Finally, harmful elements play new functions for the element *B* (stage 3). Eventually, *B* is able to actively produce *A* to utilize it. This concept can account for the dualism of NO in its effects on plants: harmful and beneficial in the same cells (see text). Cytotoxicity of NO may be a reflection of the relationship between NO and living organisms at an early stage (stage 1) and signaling functions would be the most evolved and advanced relationship (stage 4). This concept can be also applied to the establishment of symbiosis in biological interactions.

oxygen susceptibility of many anaerobic bacteria, most of living organisms were not able to live in a high O_2 concentration (stage 1). After the acquisition of SOD and other antioxidant systems, the present life forms including plants and animals evolved to dominate by coping with oxygen toxicity. Thinking of our human body for example, we can see the old fashion relationship (stage 1) between oxygen and living organism in the innate immune response; the ROS H_2O_2 produced by macrophages via O_2^- formation is used to kill pathogenic invaders. Peroxisomes include many types of H_2O_2

producing enzymes (Corpas *et al.*, 2001). Because the compartment also contains abundant catalase to destroy the byproduct H_2O_2, cells are tolerant against H_2O_2 produced during the metabolisms (stage 3). H_2O_2 is a membrane permeable mild oxidant that can oxidize functional SH-groups of enzymes, a characteristic ideal for metabolic switching mechanisms (new function, stage 4). Like H_2O_2, NO also shows versatility that can also be explained well by the above concept.

Heme is well known to bind NO to form the nitrosyl-Fe complex and this chemical nature sometimes disturbs cellular functions (Takahashi and Yamasaki, 2002; Yamasaki and Sakihama, 2000; Yamasaki *et al.*, 2001) as well as metabolic switching mechanism such as guanyl cyclase activation (Neill *et al.*, 2003; Wendehenne *et al.*, 2001). Recently, hemoglobin genes have been found in many organisms other than vertebrates including plants, fungi and bacteria (Arredondo-Peter *et al.*, 1998; Watts *et al.*, 2001; Wittenberg *et al.*, 2002) raising the large question of what role hemoglobins might have other than oxygen delivery? A detoxification role against NO toxicity by hemoglobins has been proposed in bacteria (Wittenberg *et al.*, 2002). In plants non-symbiotic hemoglobin has been suggested to play important roles in roots under hypoxic conditions (Dordas *et al.*, 2003). It is interesting to consider the possibility that the hemoglobin family originally evolved to reduce NO toxicity and subsequently evolved as O_2^- delivery purpose. Ancient life forms may had struggled with nitrogen toxicity mediated by nitrogen oxides prior to oxygen toxicity because NO_x was abundant in ancient earth atmosphere (Feelisch and Martin, 1995). Inside the cells, therefore, evolutionally advanced organisms like vertebrate animals and vascular plants may include such a long history of changes in the relationship to NO.

11. CANDIDATES FOR NO SIGNALING MOLECULES

In vertebrate animals, the regions of NO production are limited and isolated by abundant heme proteins such as hemoglobin in blood or myoglobin and other heme proteins. In turn, plant tissues, except legume root nodules, do not contain such abundant heme proteins compared to animals. This fundamental difference raises the question as to whether the NO signaling system is identical between plants and animals, echoing the case of plant NOS. In the future, many types of proteins will be reported to produce NO. However, it is essential to identify signal receptor systems as well as NO production systems. Also, it should be noted that NO itself does not necessarily act as signaling molecule.

The lifetime of NO *in vivo* is less than 10 sec (Wink *et al.*, 1996). It strongly depends on local environmental factors such as O_2 concentration or temperature. However, it should be emphasized that NO is a radical molecule that can not stand for a long time; in an aqueous solution NO may diffuse only less 500 μm from the production sites (Wink *et al.*, 1996). Therefore, to transmit the signal initiated by NO, there should be more chemically stable agents that display messenger roles in plant NO signaling systems. NO is the predominant RNS produced in biological systems. $ONOO^-$ is a secondarily produced RNS. Nitrotyrosine (NO_2-Tyr) is formed through chemical nitration of the amino acid tyrosine by $ONOO^-$ (Yamasaki, 2000). Plant phenolics including flavonoids and hydroxycinnamic acid (HCAs) can also be nitrated by RNS (Sakihama *et al.*, 2003b), implying that in addition to nitrosolglutathione (GSNO), plants may contain abundant nitrated or nitrosylated compounds in the tissues. It is plausible that nitrated or nitrosylated molecules and related compounds, rather than the NO molecule itself, may exhibit signaling functions or hormonal actions. Our knowledge on the structures of plant secondary metabolites are quite limited. In this context, the number of candidates would be more than 10^4 (Fig. 5). Because the composition and structure of secondary metabolites show species-specific manners (Cohen *et al.*, 2002b; Sakihama *et al.*, 2002), the diversity would be profitable to elicit a specific system while avoiding interference of other plants and organisms.

12. A COMMON LANGUAGE IN BIOSPHERE: SENSING NO IN THE ENVIRONMENT

Figure 6 illustrates how plants receive NO in nature. There are several sources that produce NO in the field conditions. For many years, bacterial denitrification and nitrification had been thought as the only biological NO producing mechanism (Berks *et al.*, 1995), but L-arginine-dependent NOS has been recently found in many bacterial species (Cohen and Yamasaki, 2003). Legume plants produce root nodules through establishing endosymbiosis between *Rhizobium* to fix N_2. Nodules contain abundant legume hemoglobin (leghemoglobin) that has been considered to protect the oxygen labile nitrogenase by

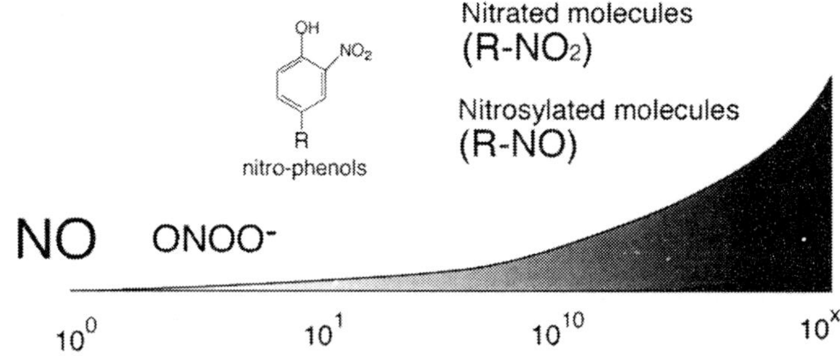

Figure 5 : Candidates of NO-induced signaling molecules
In NO-induced signaling systems of plants, it is necessary to consider not only NO itself but also its derivatives as candidates of signaling molecules. The first derived RNS is peroxynitrite ($ONOO^-$). This RNS potentially forms nitrated molecules such as nitrotyrosine or nitro-phenol for example. Because plants contain abundant phenolic compounds with wide structural varieties, an astronomic number of nitro or nitosylphenolics may be produced. For a specific elicitation of metabolism in a specific species, it is worthwhile to consider these nitrogen-contaning molecules as the candidates of signaling agents.

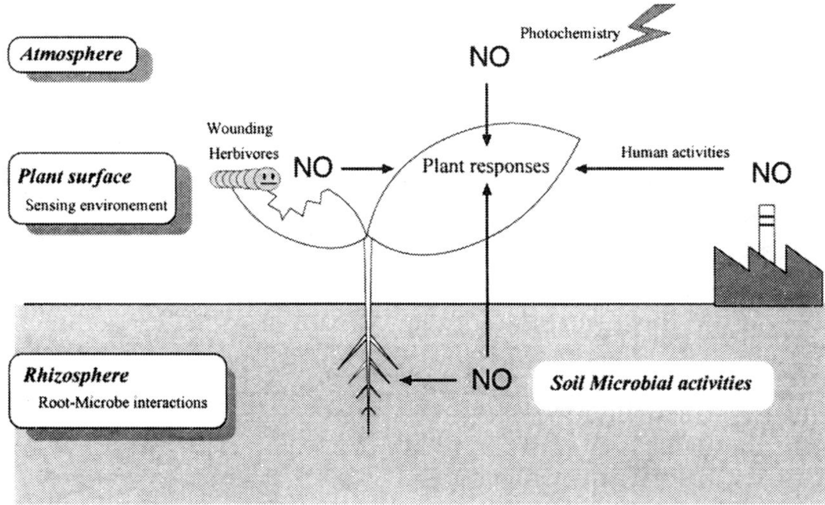

Figure 6 : Plants sense environmental changes through NO
Plants produce NO endogenously through nonenzymatic and enzymatic mechanisms. This endogenous NO production could function as plant hormone which induces signaling cascades. An important feature of NO is that the molecule can be also produced in a range of sources in the field. For plants without neural system to sense environment, change in NO level of surrounding environment would be a simple way to sense many factors necessary to acquire as information to survive.

removing O_2 from the inside of the nodule. An ESR study has shown the evidence for the NO-leghemoglobin complex in nodule (Mathieu *et al.*, 1998), suggesting that NO would be also produced as a byproduct during N_2 fixation as the case of nitrate assimilation by NR. Thus, NO can be produced by a variety of bacterial activities in soils. It is, therefore, very likely that plant roots sense these bacterial activities through the level of NO (and O_2 as well) in surrounding environment (Stöhr and Ullrich, 2002).

Unlike animal tissues, plant tissues in general, are not easily degraded because of high amounts of phenolic compounds such as lignin. Interestingly, plant tissues seem to contain sizable amount of organic nitrogen compounds whose structures remain unknown (Morikawa *et al.*, 2002). Nitrated aromatics have been recognized as explosive agents such as nitroglycerin or TNT (2,4,6-trinitrotoluene). Phyllosphere bacteria as well as Rhizosphere bacteria have been reported to degrade such nitrated aromatics. These bacterial degradation activities are expected to contribute to bioremediation of soils (Hannink *et al.*, 2001). The degradation activity may have originally evolved for degrading plant tissues that contain a high amount of nitrated phenolics produced by RNS actions (Cohen *et al.*, 2002c). It may be possible to draw a picture of a “NO world” in soil where NO is a common language for living organisms. Simultaneous germination is frequently observed after bush fire. This is probably due to the fact that NO is required for seed germination (Giba *et al.*, 1992; Giba *et al.*, 1999). The slash-and-burn method of agriculture in tropics seems to be a practical way to apply such nature of NO to germination of crops species. Thus, plants grown in the field seem to be always exposed to environmental NO as well as endogenous overproduction (Yamasaki *et al.*, 2001).

Active leaves contain high amounts of ascorbate in choloroplasts for detoxification of ROS (Asada, 2000). Chloroplasts under illumination also include nitrite that can be chemically converted to NO in the presence of ascorbate under acidic conditions. This chemical NO production should not be observed in normal favorable conditions but does occur when the tissue becomes acidic by some reason such as the grazing of leaves by herbivores. Because vacuoles usually contain abundant organic acids, destruction of compartmentation by the grazing mixes all solutes including nitrite, ascorbate and other organic acids, a condition where NO is produced in the absence of enzymes. NO released during grazing has been suggested to determine the mouth movement behavior of snails and their food preferences (Cohen *et al.*, 2002a). Mechanical damage of leaves is not only found in the grazing event but also in wounding phenomena. Like jasmonate (Thaler, 1999), chemical NO production may act as a signal to elicit anti-herbivore or anti-wounding response of plants.

The local concentration of atomospheric NO is affected by many factors. UV irradiation may facilitate photolysis of nitrite to produce NO. Availability of oxygen (O_2) and local concentration of CO_2 determines the lifetime of NO and its nitrating reaction (Yamasaki, 2000). The reaction between NO and O_2 is strongly temperature-dependent. In this context, NO is an ideal messenger molecule for plants to sense a wide spectrum of environmental changes including biotic and abiotic factors (Sakihama *et al.*, 2003a; Yamasaki, 2000). Therefore, plant NO signaling systems should be responsible for exogenous NO emission from the environment as well as endogenous NO production by NOS and NR. This trait provides a fundamental difference that distinguishes plant from animal, i.e. an "open system" versus "closed system" for NO signaling mechanisms. In contrast to plants, NO signaling systems of animals are localized in specific tissues, and are little influenced by environmental NO (closed NO signaling system). In addition to exploring the universality of NO signaling mechanisms among bacteria, fungi, invertebrates, vertebrates and plants, investigating the differences should be taken into consideration for the future NO research.

CONCLUSIONS AND FUTURE PROSPECTS

In addition to NO gas emission, legume plants have been reported to emit the nitrous oxide (N_2O), laughing gas (Zhang *et al.*, 2000). Since a tobacco mutant lacking NiR shows a higher N_2O emission (Goshima *et al.*, 1999), N_2O seems to be produced from NO by an unknown mechanism within the plants. In the context of global warming, N_2O is known as a strong greenhouse gas that is 300-fold more effective than CO_2. Recently, N_2O emission has been found not only in legume plants but also in many plant species, implying that the phenomenon may be common (Hakata *et al.*, 2003). Generally, a global scale of greenhouse gas emission, in particular CO_2, is ascribed to human activities through the combustion of fossil fuels. The emission of N_2O from plants suggests that plant or vegetation is a potential source of greenhouse gas in a global scale (Mosier, 2001). Looking back at the history of plant NO research, we can see repetition of surprise and confusion. To establish a straightforward direction for the future of plant NO research, I would like to emphasize the necessity of integrating our knowledge to cover interdisciplinary areas: botany, cell biology, molecular biology, biochemistry, radical chemistry, microbiology, stress physiology pathology, symbiosis biology, ecology, soil science, medical life science and atmospheric science. Figure 7 represents a web of plant NO research (NO web) that requires our integrated knowledge to understand physiological roles of NO in plants. It should be emphasized again that plants are closely associated with surrounding environment and they have to sense a range of stimuli and transmit the signal. Characteristics of NO (abundant atoms in the atmosphere, reactive to many types of molecules and membrane-permeable) would satisfy plant signaling systems without having a complex nerve system

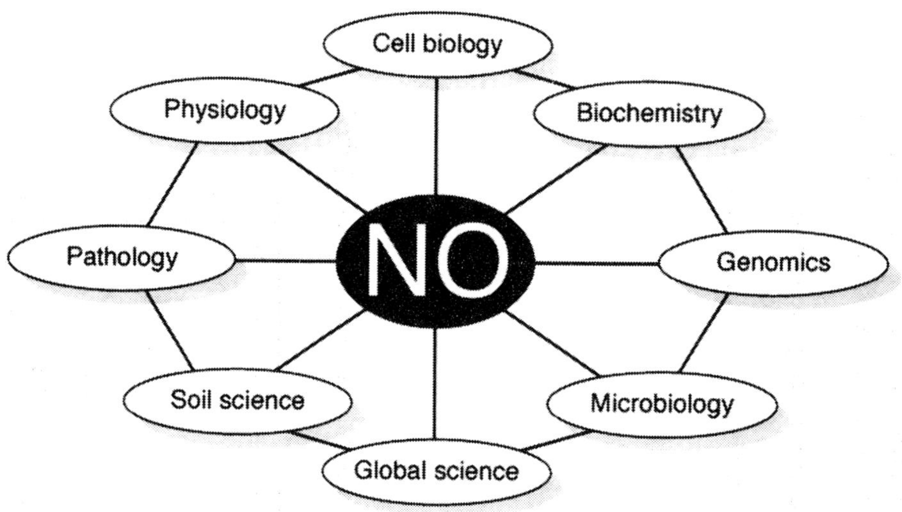

Figure 7 : Web of NO research
Because NO exists almost everywhere on the earth, NO research in plants needs comprehensive knowledge in multidisciplinary areas of science: *e.g.*, soil science, atmospheric science, pathology, microbiology, physiology, biochemistry, chemistry. For a better understanding of signaling feature of NO, integration of our knowledge is essential.

as in animals. For creating a consistent and beautiful "NO world" in science, it seems obvious that we need to be open to incorporating and integrating comprehensive knowledge just like plants did in evolution.

ACKNOWLEDGMENTS

I am grateful to Dr. M.F. Cohen of the USDA-Agricultural Research Service for a critical reading of the manuscript. Thanks are also due to many colleagues, particularly to Drs.Yasuko Sakihama, Shunichi Takahashi, Hisashi Shimoji.

REFERENCES

Ainscough, E.W. and Brodie, A.M. (1995). Nitric oxide - some old and new perspectives. *J. Chem. Edu.*, **72** : 686-692.

Arredondo-Peter, R., Hargrove, M.S., Moran, J.F., Sarath, G. and Klucas, R.V. (1998). Plant hemoglobins. *Plant Phsyiol.*, **118** : 1121-1125.

Asada, K. (2000). The water-water cycle as alternative photon and electron sinks. *Roy. Soc. Lond. Phil. Trans. B*, **355** : 1419-1431.

Berks, B.C., Ferguson, S.J., Moir, J.W.B. and Richardson, D.J. (1995). Enzymes and associated electron transport systems that catalyse the respiratory reduction of nitrogen oxides and oxyanions. *Biochim. Biophys. Acta*, **1232** : 97-173.

Bredt, D.S. and Snyder, S.H. (1990). Isolation of nitric oxide synthase, a calmodulin requiring enzyme. *Proc. Natl. Acad. Sci. U.S.A.*, **87** : 682-685.

Butt, T.K.C., Lum, J.H.K. and Lo, S.C.-L. (2003). Proteomic identification of plant proteins probed by mammalian nitric oxide synthase antibodies. *Planta,* **216** : 762-771.

Chandok, M.R., Ytterberg, A.J., van Wijk, K.J. and Klessig, D.F. (2003). The pathogen-27 induced nitric oxide synthase (iNOS) in plants is a variant of the P protein of the glycine decarboxylase complex. *Cell*, **113** : 469-482.

Cohen, M.F., Meziane, T., Tsuchiya, M. and Yamasaki, H. (2002a). Feeding deterrence of *Aozolla* in relation to deoxyanthocyanin and fatty acid composition. *Aqua. Bot.*, **74** : 181-187.

Cohen, M.F., Sakihama, Y., Takagi, Y.C., Ichiba, T. and Yamasaki, H. (2002b). Synergistic effect of deoxyanthocyanins from symbiotic fern *Azolla* spp. on *hrmA* gene induction in the cyanobacterium *Nostoc punctiforme. Mol. Plant-Microbe Intereact.*, **15** : 875-882.

Cohen, M.F., Williams, J. and Yamasaki, H. (2002c). Biodegradation of diesel fuel by an *Azolla*-derived bacterial consortium. *J. Environ. Sci. Health Part A*, **9** : 1593 -1606.

Cohen, M.F. and Yamasaki, H. (2003). Involvement of nitric oxide synthase in sucroseenhanced hydrogen peroxide tolerance of *Rhodococcus* sp. APG1, a plant-colonizing bacterium. *Nitric Oxide*, **9** : 1-9.

Corpas, F.J., Barroso, J.B. and del Rio, L.A. (2001). Preoxisomes as a source of reactive oxygen species and nitric oxide signal molecules in plant cells. *Trends Plant Sci.*, **6 :** 145-150.

Cueto, M., Hernández-Perera, O., Martín, R., Bentura, M.L., Rodrigo, J., Lamas, S. and Golvano, M.P. (1996). Presence of nitric oxide synthase activity in roots and nodules of *Lupinus* albus. *FEBS Lett.*, **398** : 159-164.

Dean, J.V. and Harper, J.E. (1988). The conversion of nitrite to nitrogen oxide(s) by the constitutive NAD(P)H-nitrate reductase enzyme from soybean. *Plant Physiol.*, **88** : 389-395.

Delledonne, M., Xia, Y., Dixon, R.A. and Lamb, C. (1998). Nitric oxide functions as a signal in plant disease resistance. *Nature*, **394** : 585-588.

Dordas, C., Rivoal, J. and Hill, R.D. (2003). Plant hemoglobins, nitric oxide and hypoxic stress. *Anal. Bot.*, **91** : 173-178.

Durner , J., Wendehenne, D. and Klessig, D. F. (1998). Defense gene induction in tobacco by nitric oxide, cyclic GMP, and cyclic ADP-ribose. *Proc. Natl. Acad. Sci. U.S.A.* **98** : 10328-10333.

Evans, H.J. and McAuliffe, C. (1956) Identification of NO, N_2O, and N_2 as products of the nonenzymatic reduction of nitrite by ascorbate or reduced diphosphopyridine nucleotide. In: *Inorganic Nitrogen Metabolism* (Eds. McElroy, W.D.and Glass B.) John Hopkins Press, Baltimore, pp. 189-197.

Feelisch, M.F. and Martin, J.F. (1995). The early role of nitric oxide in evolution. *Trends Ecol. Evol.*, **10** : 496-499.

Furchgott, R.F. (1988) Studies on relaxation of rabbit aorta by sodium nitrite: The basis for the proposal that the acid-activatable inhibitory factor from bovine retractor penis is inorganic nitrite and the endothelium-derived relaxing factor is nitric oxide. In: *Vasodilatation: Vascular Smooth Muscle, Peptides, Autonomic Nerves and Endothelium* (Ed. Vanhoutte, P.M.) Raven, New York, pp. 401-414.

Giba, Z., Grubisic, D. and Konjevic, R. (1992). Sodium nitroprusside-stimulated germination of common chick weed (*Stellaria media L.)* seeds. *Arch. Biol. Sci. Belgrade*, **44** : 17P-18P.

Giba, Z., Grubisic, D. and Konjevic, R. (1999). How do seeds sense environmental changes? The role of nitrogenous compounds. *Arch. Biol. Sci. Belgrade*, **51** : 121-129.

Goshima, N., Mukai, T., Suemori, M., Takahashi, M., Caboche, M. and Morikawa, H. (1999). Emission of nitrous oxide (N_2O) from transgenic tobacco expressing antisense NiR mRNA. *Plant J.*, **19** : 75-80.

Guo, F.-Q., Okamoto, M. and Crawford, N.M. (2003). Identification of a plant nitric oxide synthase gene involved in hormonal signaling. *Science*, **302** : 100-103.

Hakata, M., Takahashi, M., Zumft, W., Sakamoto, A. and Morikawa, H. (2003). Conversion of the nitrate nitrogen and nitrogen dioxide to nitrous oxides in plants. *Acta Biotechnol.*, **23** : 249-257.

Hannink, N., Rosser, S.J., Christopher, E.F., Basran, A., Murray, J.A.H., Nicklin, S. and Bruce, N.C. (2001). Phytodetoxification of TNT by transgenic plants expressing a bacterial nitroreductase. *Nature Biotech.*, **19** : 1168-1172.

Harper, J.E. (1981). Evolution of nitrogen oxide(s) during *in vivo* nitrate reductase assay of soybean leaves. *Plant Physiol.*, **68** : 1488-1493.

Ignarro, L.J., Buga, G.M., Wood, K.S., Byrns, R.E. and Chaudhuri, G. (1987). Endothelium-derived relaxing factor poduced and released from artery and vein is nitric oxide. *Proc. Natl. Acad. Sci. U.S.A.*, **84** : 9265-9269.

Ji, X.B. and Hollocher, T.C. (1988). Reduction of nitrite to nitric oxide by enteric bacteria. *Biochem. Biophys. Res. Commun.*, **157** : 106-108.

Leshem, Y.Y. and Haramaty, E. (1996). The characterization and contrasting effects of the nitric oxide free radicals in vegetative stress and senescence of *Pisum sativum* L. foliage. *J. Plant Physiol.*, **148** : 258-263.

Liszt, H.S. and Turner, B.E. (1978). Microwave detection of interstellar NO. *Astrophys. J.*, L73-78.

Lundberg, J.O.N., Carlson, S., Engstrand, L., Morcos, E., Wiklund, N.P. and Weitzberg, E. (1997). Urinary nitrite :more than a marker of infection. *Adult Urology*, **50** : 189-191.

Mallick, N., Mohn, F.H., Rai, L.C. and Soeder, C.J. (2000a). Evidence for the noninvolvement of nitric oxide synthase in nitric oxide production by the green alga *Scenedesmus obliquus*. *J. Plant Physiol.*, **156** : 423-426.

Mallick, N., Mohn, F.H., Rai, L.C. and Soeder, C.J. (2000b). Impact of physiological stresses on nitric oxide formation by green alga, *Scenedesmus obliquus*. *J. Microbiol. Biotechnol.*, **10** : 300-306.

Mathieu, B., Moreau, S., Frendo, P., Puppo, A. and Michael J. Davies, M.J. (1998). Direct detection of radicals in intact soybean nodules: presence of nitric oxide-leghemoglobin complexes. *Free Radic. Biol. Med.*, **24** : 1242-1249.

McCord, J.M. and Fridovich, I. (1969). Superoxide dismutase. An enzymic function for erythrocuprein (hemocuprein). *J. Biol. Chem.*, **244** : 6049-6055.

Morikawa, H., Fukunaga, K., Takahashi, M., Kawamura, Y. and Sakamoto, A. (2002). Formation of unidentified nitrogen (UN) in plants - UN from nitrates. *Plant Cell Physiol.*, **43** : S128.

Mosier, A.R. (2001). Exchange of gaseous nitrogen compounds between agricultural systems and the atmosphere. *Plant Soil*, **228** : 17-27.

Neill, S.J., Desikan, R. and Hancock, J.T. (2003). Nitric oxide signaling in plants. *New Phytol.*, **159 :** 11-35.

Nelson, R.S., Ryan, S.A. and Harper, J.E. (1983). Soybean mutants lacking constitutive nitrate reductase activity. I. selection and initial plant characterization. *Plant Physiol.*, **72** : 503-509.

Ninnemann, H. and Maier, J. (1996). Indications for the occurrence of nitric oxide synthases in fungi and plants and the involvement in photoconidiation of *Neurospora crassa. Photochem. Photobiol.*, **64 :** 393-398.

Noritake, T., Kawakita, K. and Doke, N. (1996). Nitric oxide induces phytoalexin accumulation in potato tuber tissues. *Plant Cell Physiol.*, **37** : 113-116.

Osmond, C.B. (1994) What is photoinhibition? Some insights from comparison of shade and sum plants. In: *Photoinhibition of Photosynthesis: from molecular mechanisms to field.* (Eds. Baker, N.R.and Bowyer, J.R) BIOS Science Publishers, Oxford, pp. 1-24.

Palmer, R.M.J., Ferrige, A.G. and Moncada, S. (1987). Nitric oxide release accounts for the biological activity of endothelium-derived relaxing factor *Nature*, **327** : 524-526.

Ribeiro Jr, E.A., Cunha, F.Q., Tamashiro, W.M.S.C. and Martins, I.S. (1999). Growth phase-dependent subcellular localization of nitric oxide synthase in maize cells. *FEBS Lett.*, **445** : 283-286.

Rockel, P., Strube, F., Rockel, A., Wildt, J. and Kaiser, W.M. (2002). Regulation of nitric oxide (NO) production by plant nitrate reductase in vivo and in vitro. *J. Exp. Bot.*, **53 :** 103-110.

Ryan, S.A., Nelson, R.S. and Harper, J.E. (1983). Soybean mutants lacking constitutive nitrate reductase activity. II. Nitrogen assimilation, chlorate resistance, and inheritance. *Plant Physiol.*, **72** : 510-514.

Sakihama, Y., Cohen, M.F., Grace, S.C. and Yamasaki, H. (2002a). Plant phenolic antioxidant and prooxidant actvities: phenlics-induced oxidative damage mediated by metals in plants. *Toxicology*, **177** : 67-80.

Sakihama, Y., Nakamura, S. and Yamasaki, H. (2002b). Nitric oxide production mediated by nitrate reductase in the green alga *Chlamydomonas reinhardtii*: an alternative NO production pathway in photosynthetic organisms. *Plant Cell Physiol.*, **43** : 290-297.

Sakihama, Y., Murakami, S. and Yamasaki, H. (2003a) Involvement of nitric oxide in the mechanism for stomatal opening in *Vicia faba* leaves. *Biol. Plant.*, **46** : 117-119.

Sakihama, Y., Tamaki, R., Shimoji, H., Ichiba, T., Fukushi, Y., Tahara, S. and Yamasaki, H. (2003b). Enzymatic nitration of phytophenolics: evidence for peroxynitrite-independent nitration of plant secondary metabolites. *FEBS Lett.*, **533** : 377-380.

Stöhr, C. and Ullrich, W.R. (2002). Generation and possible roles of NO in plant roots and their apoplastic space. *J. Exp. Bot.*, **53** : 2293-2303.

Streit, L., Nelson, R.S. and Harper, J.E. (1985). Nitrate reductase from wild-type and nr1-mutant soybean (*Glycine max* [L.] Merr.) leaves. I. Purification, kinetics, and physical properties. *Plant Physiol.*, **78** : 80-84.

Takahashi, S. and Yamasaki, H. (2002). Reversible inhibition of photophosphorylation in chloroplasts by nitric oxide. *FEBS Lett.*, **512** : 145-148.

Thaler, J.S. (1999). Jasmonate-inducible plant defenses cause increased parasitism of herbivores. *Nature*, **399** : 686-688.

Watts, R.A., Hunt, P.W., N., H.A., Hargrove, M.S., Peacock, W.J. and Dennis, E.S. (2001). A hemoglobin from plants homologous to truncated hemoglobins of microorganisms. *Proc. Natl. Acad. Sci. U.S.A.*, **98** : 10119-10124.

Weitzberg, E. and Lundberg, J.O.N. (1998). Nonenzymatic nitric oxide production in humans. *Nitric Oxide*, **2** : 1-7.

Wellburn, A.R. (1990). Why are atmospheric oxides of nitrogen usually phytotoxic and not alternative fertilizers? *New Phytol.*, **115** : 395-429.

Wellburn, A.R., Wilson, J. and Aldridge, P.H. (1980). Biochemical responses of plants to nitric oxide polluted atmospheres. *Environ. Pollut.*, **22** : 219-228.

Wendehenne, D., Pugin, A., Klessig, D.F. and Durner, J. (2001). Nitric oxide: comparative synthesis and signaling in animal and plant cells. *Trends Plant Sci.*, **6** : 177-183.

Wildt, J., Kley, D., Rockel, A., Rockel, P. and Segschneider, H.J. (1997). Emission of NO from several higher plant species. *J. Geo. Res.*, **102** : 5919-5927.

Wink, D.A., Grisham, M.B., Mitchell, J.B. and Ford, P.C. (1996). Direct and indirect effects of nitric oxide in chemical reactions relevant to biology. *Methods Enzymol*, **268** : 12-31.

Wittenberg, J.B., Bolognesi, M., Wittenberg, B.A. and Guertin, M. (2002). Truncated hemoglobins: a new family of hemoglobins widely distributed in bacteria, unicellular eukaryotes, and plants. *J. Biol. Chem.*, **277** : 871-874.

Yamasaki, H. (2000). Nitric oxide produced by nitrate reductase: implications for involvement of active nitrogen species in photoinhibition *in vivo*. *Roy. Soc. Lond. Phil. Trans. B*, **355** : 1477-1488.

Yamasaki, H. and Sakihama, Y. (2000). Simultaneous production of nitric oxide and peroxynitrite by plant nitrate reductase: in vitro evidence for the NR-dependent formation of active nitrogen species. *FEBS Lett.*, **468** : 89-92.

Yamasaki, H., Sakihama, Y. and Takahashi, S. (1999). An alternative pathway for nitric oxide production in plants: new features of an old enzyme. *Trends Plant Sci.*, **4** : 128-129.

Yamasaki, H., Shimoji, H., Ohshiro, Y. and Sakihama, Y. (2001). Inhibitory effects of nitric oxide on oxidative phosphorylation in plant mitochondria. *Nitric Oxide*, **5** : 261-270.

Yoneyama, T., Sasakawa, H., Ishizuka, S. and Totsuka, T. (1979). Absorption of atmospheric NO_2 by plants and soils. (II) Nitrite accumulation, nitrite reductase activity and diurnal changes of NO_2 absorption in leaves. *Soil Sci. Plant Nutr.*, **25** : 267-275.

Zhang, L., Boeckx, P., Chen, G. and van Cleemput, O. (2000). Nitrous oxide emission from herbicide-treated soybean. *Biol. Fertil. Soils*, **32** : 173-176.

Chapter 2

NITRIC OXIDE IS AN ESSENTIAL COMPONENT OF BIOTIC AND ABIOTIC STRESSES-INDUCED SIGNALING PATHWAYS IN PLANTS

David Wendehenne*•, Kevin S. Gould•#, Olivier Lamotte, Manickam Selvi-Srinivas, Agnès Klinguer and Alain Pugin

UMR 1088 INRA/Université de Bourgogne et FRE/CNRS 2625, Plante-Microbe-Environnement, INRA, 17 rue Sully, BP 86510, Dijon 21065 cedex, France;
#Plant Sciences Group, School of Biological Sciences, University of Auckland, Private Bag 92019, Auckland, New Zealand.
*Corresponding author : E-mail : wendehen@dijon.inra.fr
Tel. : +33-3-80-69-37-21; Fax : +33-3-80-69-32-26
•Joint first authors

Summary

The identification of nitric oxide (NO) as a biological signaling molecule in animals in the 1980s has had an extraordinary impact on scientific research. Nowadays, NO impinges on almost all areas of biology, including vascular biology, neuroscience, and immunology. In recent years, it has become evident that NO also plays a significant role in plants, especially in the responses to pathogens and abiotic stress. There is now clear evidence that NO is produced in plants by the activation of two enzymes: nitric oxide synthase (NOS), and nitrate reductase. A pathogen-inducible NOS-like enzyme has been identified as a variant of the P protein of the mitochondrial glycine decarboxylase complex. Once synthesized, NO influences cellular physiology by modifying the activities of numerous

In : Nitric Oxide Signaling in Higher Plants, 2004
(Eds Jose R. Magalhaes, Rana P. Singh and Leonidas P. Passos)
Studium Press, LLC, Houston, USA, pp 25-64

proteins through redox-based reactions. The subcellular confinement of NOS, and its associations with specific effector molecules, are an important regulatory mechanism for NO signaling in defence pathways. Components of defence-related NO signaling include MAPK, cyclic GMP, cyclic ADP ribose, and salicylic acid. Consistent with the reports of NO activities in animal cells, NO in plants appears to mobilize Ca^{2+} from internal stores through the direct modulation of Ca^{2+}-permeable channels. Diverse abiotic stressors also generate NO in plants, and can lead either to enhanced stress tolerance or symptoms of cytotoxicity, depending on the nature of applied stress, the concentration of resulting NO, and plant species. The demonstration that plants produce NO in response to biotic and abiotic stress highlights the crucial role of this molecule in the adaptation to environmental constraints.

Keywords : abiotic stress; calcium, cGMP; cADPR; MAPK; nitrate reductase; nitric oxide; nitric oxide synthase; plant defence; stress tolerance.

1. INTRODUCTION

Nitric oxide (NO) is a hydrophobic, gaseous molecule and highly diffusible free radical. Over the past 20 years, opinions derived from biomedical research on the role of NO have evolved from its being considered as an environmental pollutant to that of a key mediator of endogenous signals involved in a wide range of physiological and pathophysiological processes. Many insights into the regulation of NO production in animal cells have arisen from the identification of the enzyme responsible for producing NO, nitric oxide synthase (NOS) (Nathan and Xie, 1994). These studies have been paralleled by an enhanced understanding of the biological reactivity of NO, dictated primarily by its reaction with cysteine residues and the transition metal centres of a broad spectrum of functional proteins (Stamler *et al.*, 1992). These are redox-based, post-translational modifications, which modulate protein function to transduce cellular signals with a remarkable spatial and temporal resolution (Stamler *et al.*, 2001).

As sessile organisms, the ability of plants to respond to local environmental signals through highly sensitive perception/

transduction systems assumes particular significance. Studies of the *Arabidospis thaliana* genome indicate that the plants devote at least 10% of their genes to proteins involved in signal transduction, including those for receptors, kinases, phosphatases and transcription factors (The *Arabidopsis* genome initiative, 2000). Moreover, technological advances in the last decade have facilitated an understanding of how plant cells integrate and process external signals. Ca^{2+}, cGMP, cyclic ADP ribose (cADPR), reactive oxygen species (ROS; mainly the superoxyde anion $O_2{\cdot}^-$, H_2O_2 and the hydroxyl radical HO·), inositol triphosphate (IP_3) and plant hormones have all been implicated as important signalling molecules. Interest in NO as an endogenous signalling molecule in plants, however, did not gain impetus until the discovery that NO is produced in tobacco and *Arabidopsis thaliana* plants resisting pathogen infection (Delledonne *et al.*, 1998; Durner *et al.*, 1998). NO has since been shown to play an important role in diverse plant signalling processes, ranging from abscisic acid-mediated stomatal closure to programmed cell death (Desikan *et al.*, 2002; Pedroso *et al.*, 2000a).

This review focuses on the involvement of NO in biotic and abiotic stress responses in plants, with particular emphasis on the function of NO in plant defence. The mechanisms through which NO can be generated, as well as recent advances in the identification of putative targets for NO, are detailed.

2. NITRIC OXIDE SIGNALING ACTIVITIES IN RESPONSE TO BIOTIC STRESSES

2.1. Plant-Pathogen Interactions as a Model to Study NO-Signaling in Plants

2.1.1. Physiology of Resistance Responses

Over the past 60 years, mechanisms for the resistance of plants to disease have received considerable scientific attention. It is now widely accepted that plants defend themselves against attack from potential pathogenic microorganisms by activating sophisticated and integrated defence mechanisms. Plant defence responses normally include the production of ROS, the reinforcement of cell walls, and the transcriptional activation of defence-related genes. Some of these genes encode enzymes of the phenylpropanoid biosynthetic pathway which regulates the production of various secondary compounds, including lignin and low molecular weight antimicrobial compounds known as phytoalexins (Dixon *et al.*, 2001). Others belong to the family of pathogenesis-related (PR) genes encoding for a large and diverse collection of unrelated proteins (Fritig *et al.*, 1998). Although the function of some PR proteins remains unclear, many of them show antimicrobial activity *in vivo* and/or *in vitro*. In addition to these genes, changes in the expression pattern of genes encoding antioxidant enzymes, signalling proteins such as kinases, phosphatases and transcription factors, DNA repair proteins, heat shock proteins, hormone biosynthesis enzymes, and metabolic enzymes are commonly observed in response to microorganisms or to plant defence inducers, as demonstrated using cDNA microarray strategies (for example see Maleck *et al.*, 2000; Schenk *et al.*, 2000).

Plant disease resistance is sometimes associated with the development of the hypersensitive response (HR). HR is characterized by the rapid, localized death of plant cells in and around the initial infection site, and is interpreted as an attempt to prevent the spread of pathogens to non-infected parts of the plant. According to recent pharmacological and molecular studies, HR may represent a type of programmed cell death (Lam *et al.*, 2001). The plant growth regulators salicylic acid (SA), jasmonic acid and ethylene are apparently key regulators of this process through their capacities to regulate the production of ROS (Hoeberichts and Woltering, 2002). In addition, in many plant species, pathogen-induced defence responses lead to a systemic resistance, an important component of the disease resistance repertoire. Systemic resistance is characterized by the induction of long-term resistance that is effective against not only the initial pathogen, but also against subsequent infection by a broad spectrum of avirulent and virulent pathogens (Ryals *et al.*, 1996). To date, two types of systemic resistance have been studied: systemic acquired resistance (SAR) and induced systemic resistance (ISR). SA is a crucial signaling compounds for SAR (Ryals *et al.*, 1996), whereas ISR induction is SA-independent but instead requires ethylene and jasmonic acid (Pieterse *et al.*, 1998). Central to our understanding of systemic resistance has been the identification and isolation of the *Arabidopsis NPR-1* gene (Cao *et al.*, 1997). Both SA and jasmonic acid/ethylene systemic signaling occur through the activation of NPR-1. In turn, NPR1 acts as a transcription co-activator, differentially regulating gene expression according to the signaling pathway that is activated upstream of it (Cao *et al.*, 1997; Pieterse *et al.*, 1998).

2.1.2. Elicitor Signaling in Plant Defence Responses

Plant defence responses are initiated by direct or indirect recognition of elicitors, molecules derived from the microorganism. Elicitors of different structure have been isolated from microbial culture filtrates and plant and phytopathogenic fungi cell walls (Pugin and Guern, 1996). By studying the effects of purified elicitors on intact plants and/or plant cell suspensions, it has been possible to characterize some of the mechanisms underlying the activation of inducible defences. The simplest model requires that elicitors interact with plant receptors that are either located on the plasma membrane or in the cytosol. Several classes of putative elicitor receptors, whose role in disease resistance has been inferred from mutants phenotypes, have been identified (review by Dangl and Jones, 2001). In addition, receptor-like high-affinity binding proteins have been described for various elicitors (for example see Bourque *et al.*, 1999). The initial interaction between pathogen-derived ligands and plant receptors triggers a signaling cascade, most commonly investigated using pharmacological and biochemical approaches (review by Scheel, 1998).

A number of signal transduction pathways has been proposed to mediate the defence responses in host cells (for example see Blumwald *et al.*, 1998; Lebrun-Garcia *et al.*, 1999; Scheel *et al.*, 1998). The processes normally include ion channels-mediated changes in plasma membrane permeability, modulation of kinases (including mitogen activated protein kinases (MAPK)) and phosphatases, production of NADPH oxidase-dependent ROS, and induction of G proteins. These pathways are controlled by negatively-acting proteins such as phosphatases, implying that the defence response to potential pathogens may be mediated, to some extent, by the release from repression mechanisms (Bowler and Chua, 1994). How these events interrelate, and when and where they operate independently, appear to vary according to the plant-elicitor model under study, and are currently areas of intensive resesarch. Ca^{2+} is invariably involved as a second messenger, modulating most of the responses associated with the defence mechanisms including ROS production, defence gene expression, phytoalexin biosynthesis and HR (Blume *et al.*,

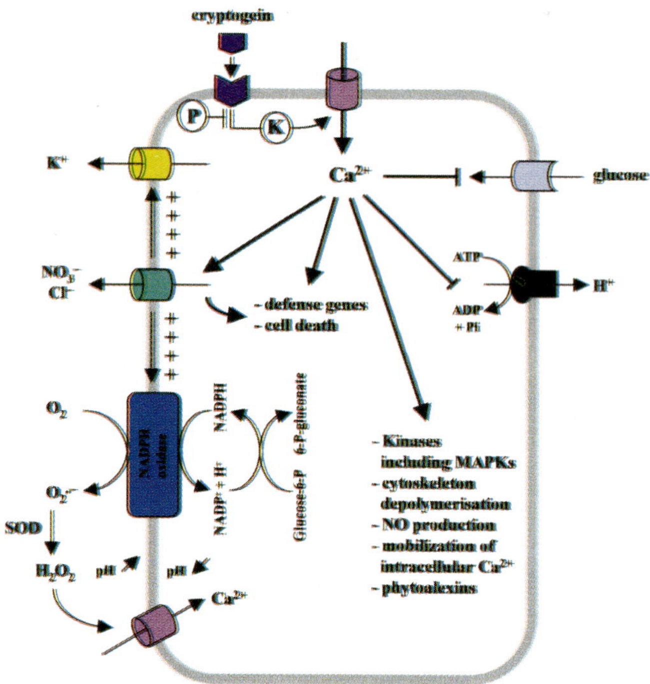

Figure 1 : Hypothetical model of the cryptogein transduction pathway in tobacco cells.

Cryptogein binding to high affinity binding sites is followed by kinase-mediated phosphorylation events leading to a large and sustained Ca^{2+} influx. In turn, the resulting increase in cytosolic Ca^{2+} concentration controls several cellular responses including the activation of the MAPKs salicylic acid-induced protein kinase (SIPK) and wound-induced protein kinase (WIPK), the release of Ca^{2+} from internal stores, an inhibition of glucose transport, an inhibition of H^+-ATPase, extracellular medium alkalinization, a disruption of the microtubular cytoskeleton, a production of NO, the synthesis of phytoalexins and the induction of various defence genes. In addition, the elevation of cytosolic Ca^{2+} concentration is required for the activation of anion channel-mediated NO_3^- and Cl^- efflux acting upstream of the NADPH oxidase-induced production of ROS, which, in turn, may activate Ca^{2+}-permeable channels from the plasma membrane. The oxidation of NADPH by NADPH oxidase leads to the activation of the pentose phosphate and therefore, to a decrease in glucose 6-phosphate concentration. In conjunction with Ca^{2+}, anion channels activity may also contribute to cryptogein-induced plasma membrane depolarisation, K^+ efflux, defence genes activation and ultimately cell death. All the cryptogein cascade seems negatively regulated by at least one phosphatase. For further details see Binet *et al.* (2001), Bourque *et al.* (1999), Bourque *et al.* (2002), Lebrun-Garcia *et al.* (1998), Lecourieux *et al.* (2002), Pugin *et al.* (1997), Simon-Plas *et al.* (2002); Wendehenne *et al.* (2002). K: kinase, P: phosphatase, SOD: superoxide dismutase.

2000; Lecourieux *et al.*, 2002). As an example, figure 1 summarizes the signaling cascade triggers in tobacco cells by cryptogein, a 10 kDa proteinaceous elicitor purified from the culture filtrates of the oomycete *Phytophthora cryptogea*, an avirulent pathogen of tobacco (Ricci, 1997).

One theme that emerges from these studies is the similarity between how animals and plants perceive potential pathogens and transduce the corresponding information to an appropriate response. For example, molecular analysis indicates that the putative receptor for the bacterial elicitor flagellin shares significant homology to a Toll-like receptor involved in innate immunity in animals (Gómez-Gómez and Boller, 2002). Moreover, many lines of enquiry have demonstrated the similarity between the NADPH oxidase in plant cells and in phagocytic animal cells (Pugin *et al.*, 1997; Simon-Plas *et al.*, 2002). These remarkable similarities, coupled with the fact that NO plays a crucial role in animal host defence, have prompted several research teams to address the potential involvement of NO in plant immune responses. Such studies have not only enhanced our understanding of the role of NO in plant defence, they also serve to illustrate the potential benefits of using plant-pathogen interactions as a model to investigate NO signaling in plants. Recent advances in our knowledge of the mechanisms of NO production and signalling functions are presented below. Because the NO involvement in HR induction is covered elsewhere in this volume (see Delledonne and coworkers), it is not discussed here.

2.2. NO Production in Plant Defence Responses

2.2.1. Enzymatic Sources of Nitric Oxide

Although the production of NO in plant cells is undisputed, many uncertainties remain regarding a possible enzyme involved in NO-synthesis. It has long been recognized that nitrate reductase (NR) catalyses both *in vitro* and *in vivo* the NAD(P)H-dependent reduction of nitrite to NO over a physiological pH range (Kaiser and Huber, 2001; Morot-Gaudry-Talarmain *et al.*, 2002; Yamasaki, 2000). Significantly, Desikan *et al.* (2002) recently demonstrated that NR-mediated NO synthesis is required for abscisic acid-mediated closure of stomata in *Arabidopsis thaliana*, thus providing the first demonstration of NR-dependent NO production in plants. There is also a growing body of experimental evidence implicating the existence of a mammalian-type NOS. In mammals, NOS are members of a family of cytochrome P450-like reductases linked to an NADPH oxidase enzyme (Nathan and Xie, 1994). NOS are active as homodimers, require several cofactors including NADPH, FAD, FMN and calmodulin (CaM) and oxidize L-arginine into NO and L-citrulline. Evidence for NOS functional activities has been reported for various plant tissues using biochemical and pharmacological strategies (Barroso *et al.*, 1999; Caro and Puntarulo, 1999; Delledonne *et al.*, 1998; Durner *et al.*, 1998; Lum *et al.*, 2001; Pedroso *et al.*, 2000a; Song and Goodman, 2001). In addition, Western blot analyses using antibodies raised against mammalian NOS have enabled the detection of immunoreactive proteins in plant extracts (Barroso *et al.*, 1999; Modolo *et al.*, 2002; Ribeiro *et al.*, 1999). However, although these studies indicate that a NOS-like protein is likely to be present in plant cells, there are no clear homologues of the mammalian NOS genes in *Arabidopsis thaliana* genome (The *Arabidopsis* genome initiative, 2000). Consistent with this, and using a convincing proteomic approach, Butt *et al.* (2003) recently demonstrated that proteins from maize embryonic axes which cross-react with mammalian NOS antibodies are not directly related to NOS. Collectively, therefore, these studies strongly indicate that plant NOS-like activities are catalysed by an enzyme that is structurally different from the mammalian type NOS.

The enzyme responsible for plant NOS-like activities was recently purified from tobacco leaves and identified as a variant form of the P protein of glycine decarboxylase (GDC), a multi-enzyme complex present in the mitochondrial matrix which catalyses the conversion of glycine into serine, CO_2 and NH_3 (Chandok *et al.*, 2003; Wendehenne *et al.*, 2003). Sequence analysis of the variant form of the P protein (variant P/iNOS) identified several motifs that are similar in function to those found in mammals NOS, including putative CaM binding site, flavin binding site, NADPH binding site and heam binding site. Variant P/iNOS clearly exhibits NOS activity and was shown to be sensitive to mammalian NOS inhibitors. Conversely, inhibitors of the P protein of GDC blocked variant P/iNOS ability to synthesize NO. Several experimental and structural arguments suggest that variant P/iNOS does not appear to be part of the GDC complex and may not be targeted to mitochondria. The surprising finding that plant iNOS is a variant P protein of the GDC complex confirms previous hypothesis that plant NOS may be structurally unrelated to its animal counterpart and explains the difficulties encountered during attempts to isolate this enzyme.

Nothing is known about the possible occurrence of NR-dependent production of NO from pathogenic elicitors. In contrast, involvement of a NOS-like enzyme in plant defence responses to avirulent pathogens or derived elicitors has been reported in several studies. Variant P/iNOS was reported to catalyse NO production in tobacco leaves challenged with tobacco mosaic virus (Chandok *et al.*, 2003). In this pathosystem, variant P/iNOS appears to be regulated primarily at the transcriptional level as shown by Northern blotting experiments. Moreover, NOS-like activities have been detected in tobacco challenged with a spontaneous avirulent mutant of *Ralstonia solanacearum* (Huang and Knopp, 1998), in soybean cell suspensions inoculated with an avirulent strain of *Pseudomonas syringae* pv. *glycinea* (Delledonne *et al.*, 1998), in soybean tissue treated by a fungal elicitor purified from *Diaporthe phaseolorum* f.sp. *meridionalis*, the causal agent of stem canker disease (Modolo *et al.*, 2002), and more recently in *Arabidopsis thaliana* plants inoculated with turnip crinkle virus (Chandok *et al.*, 2003). Consistently, mammalian NOS inhibitors were shown to partly inhibit pathogen- or elicitor-mediated NO production in various plant species (reviewed by Wendehenne *et al.*, 2001, 2002). For example, cryptogein-induced NO production could be reduced by more than 50% by the NOS inhibitors N^G-mono-methyl-arginine monoacetate (L-NMMA) and Nw-nitro-L-arginine methyl ester (L-NAME) both in tobacco leaves tissue and cell suspensions (Foissner *et al.*, 2000, Lamotte *et al.*, 2003). At present, it is not known whether the NOS-like activities reported in these studies are catalysed by variant P/iNOS. However, regarding their pharmacological/biochemical properties, this hypothesis seems likely.

In contrast to the attention that has been given to NO synthesis in plants challenged by pathogen or derived elicitors, the intermediary signalling events regulating its production remain poorly understood. Several lines of evidence indicate that intracellular Ca^{2+} levels may represent a critical point of control of the NOS-like enzyme. First, *in vitro* activities of plant NOS have been shown to be Ca^{2+} and CaM dependent (Chandok *et al.*, 2003; Delledonne *et al.*, 1998; Modolo *et al.*, 2002). Second, treatment with Ca^{2+} chelating agents or the Ca^{2+} surrogate lanthanum suppressed the cryptogein-induced NO production in tobacco cell suspensions (Lamotte *et al.*, 2003). Strikingly, addition of lanthanum in the mid-course of the cryptogein response prevents any further NO synthesis, indicating that the activity of the elicitor-induced NO-generating enzyme strictly depends on sustained Ca^{2+} influx from extracellular medium and therefore on the resulting increase of intracellular free Ca^{2+} concentration ($[Ca^{2+}]$). Apart from Ca^{2+}, the potential role of phosphorylation/dephosphorylation events in NO synthesis has also received some attention. When the kinase inhibitors staurosporine or K252a were added to tobacco cell suspensions along with cryptogein, a dramatic inhibition of the elicitor-induced NO production was observed (Lamotte *et al.*, 2003). This suggests that at least one upstream phosphorylation event is required for activation of the elicitor-induced NO-generating enzyme. The importance of phosphorylation/dephosphorylation has been further underscored by the demonstration that calyculin A, an inhibitor of Ser/Thr phosphatase type 1 and 2A, triggers a rapid NO burst in tobacco cells in the absence of the elicitor. Collectively, these data suggest that cryptogein signal transduction leading to NO synthesis may proceed by activation of protein kinase(s) and inhibition of a calyculin A-sensitive phosphatase(s). Such antagonistic modulation by kinases and phosphatases should allow fine tuning of NO in response to elicitors but also to other stimuli. The question of whether plant NO-generating enzymes are directly regulated by phosphorylation or dephosphorylation requires further investigation.

2.2.2. Spatial and Temporal Aspects of Nitric Oxide Production

Recent progress in the area of NO signaling in mammals has demonstrated the requirement for temporal and spatial resolution of NO synthesis if it is to function efficiently in signal transduction (Stamler *et al.*, 2001). NO-dependent responses are governed by the subcellular compartment of NO production and by the frequency/duration of its synthesis. The multiple-component signalling module N-methyl-D-aspartate (NMDA) receptor/NOS provides a noteworthy example of NO spatiotemporal requirements (Stamler *et al.*, 2001). In this module, the cofactors and second messenger which regulate NOS and NO targets are discretely colocalized, thereby ensuring the efficiency and specificity of the signal propagation.

For plants, possibly the best indication of the spatiotemporal aspect of elicitor-mediated NO signalling has been achieved from experiments based on the use of NO fluorescence indicators. The newly developed NO fluorophore 4,5 diaminofluoresceine diacetate (DAF-2DA), has permitted the real-time imaging of NO production in epidermal tobacco cells treated with cryptogein (Foissner *et al.*, 2000). Figure 2 shows the time course of cryptogein-induced NO accumulation in tobacco cell suspensions. Addition of cryptogein results in a rapid enhancement of fluorescence within 5 min in distinct cellular compartments, most probably the chloroplasts. The level of

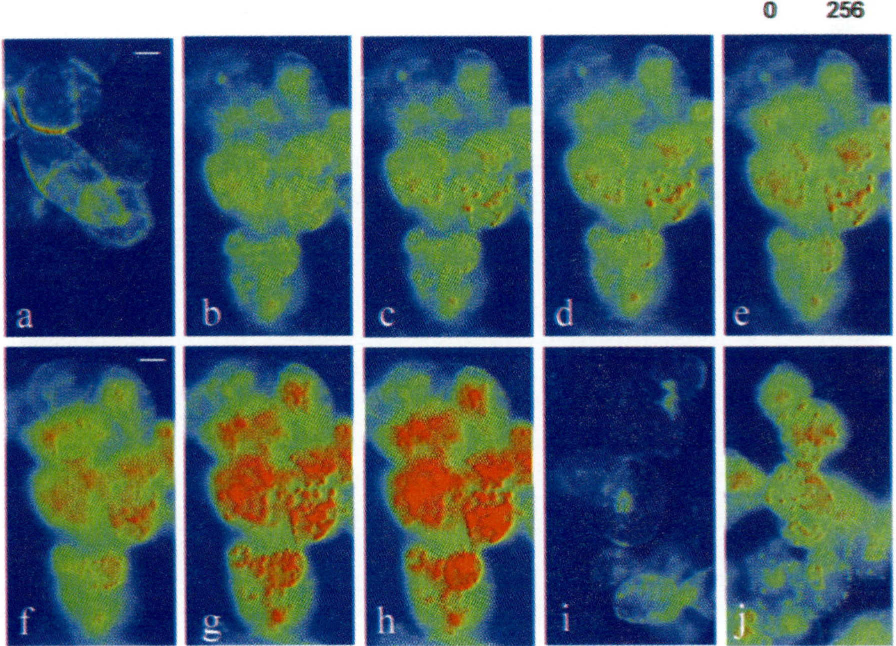

Figure 2 : Spatiotemporal aspect of cryptogein-induced Nitric oxide production in tobacco cell suspensions.
Cryptogein production was visualized using the NO specific fluorophore 4,5 diaminofluoresceine diacetate (DAF-2DA). Images were colour indexed such no fluorescence appear in blue, maximum DAF-2DA fluorescence appear in red. (a) Background fluorescence of cells without DAF-2DA. (b-h) Time course of intracellular NO production in cells treated with 50 nM cryptogein. (b): 0 min; (c): 5 min; (d): 10 min; (e): 15 min; (f): 30 min; (g): 45 min; (h): 60 min. (i) Effect of 500 µM cPTIO on 50 nM cryptogein-induced NO production monitored after 60 min of treatment. (j) Effect of 10 mM L-NAME on 50 nM cryptogein-induced NO production monitored after 60 min of treatment. Bars = 50 µm.

fluorescence increases with time, and after 15 minutes, the cells become almost entirely fluorescent, illustrating the high diffusivity of NO. This process is abolished by the membrane-permeable NO-scavenger, carboxy-2-phenyl-4,4,5,5-tetramethylimidazolinone-3 oxide-1-oxyl (carboxy-PTIO), and is significantly reduced by the mammalian NOS inhibitor L-NAME. Collectively, these data suggest that in response to cryptogein, NO is first produced in the chloroplasts through a NOS-like activity and then diffuses and/or is produced in other cellular compartments. The demonstration that NO generated in response to abiotic stress also occurs predominantly in the chloroplast provides further evidence that this organelle is an important source of NO in plants (see below; Garcês *et al.*, 2001; Gould *et al.*, 2003). The possibility that variant P/iNOS may be located in the chloroplasts needs to be addressed in future experiments.

The fact that NO production in plants is restricted to certain subcellular compartments may be critical both for specificity of targeting and for propagation of signals. In this regard, the notion that NO donors treatment of plants and cell suspensions recapitulate *in vivo* NO-dependent processes seems questionable. Although potentially highly informative, these strategies do not necessarily consider the spatio-temporal aspect of NO signalling. NO treatment of *Arabidopsis thaliana* cell cultures, for example, resulted in glutathione-S-transferase (*GST*) mRNA accumulation (Huang *et al.*, 2002a) whereas NOS inhibitors or NO scavenger had no inhibitory effect on the induction of *GST* gene in tobacco and soybean cells treated by cryptogein or an avirulent strain of *Pseudomonas syringae* pv. *glycinea*, respectively (Delledonne *et al.*, 1998; Lamotte *et al.*, 2003). Therefore and of importance, caution in the interpretation of data based on the use of NO donors need to be exerted.

2.3. Nitric Oxide Signaling in Plant Defence Responses

2.3.1. cGMP-dependent pathways

In mammals, NO-mediated cellular responses are in part triggered by an increase in the level of the second messenger cGMP. Structural studies confirm that NO binds to the heam prosthetic group of soluble gunylate cyclase, initiating a conformational change that increases the catalysis of cGMP synthesis several hundred fold (Lucas *et al.*, 2000). Consequently, cGMP alters the activity of three main target proteins: (i) cGMP-dependent protein kinases (PKG); (ii) cyclic nucleotide gated channels (CNGC); and (iii) cGMP-regulated phosphodiesterases. The NO/cGMP-mediated signaling cascade plays a central role in the regulation of diverse (patho)physiological processes, including smooth muscle relaxation, retinal phototransduction, apoptosis and intestinal fluid and electrolyte homeostasis (Beck *et al.*, 1999).

In plants, cGMP has also been established as a secondary signalling molecule (Walden *et al.*, 1998). For example, cGMP is required for gene expression and phytochrome phototransduction induced by gibberellic acid (Bowler *et al.*, 1994; Neuhaus *et al.*, 1997; Penson *et al.*, 1996). Interestingly, as reported for plant NOS, analysis of the *Arabidopsis* genome has not identified any genes with significant similarity to mammalian guanylate cyclase. Nevertheless, in addition to biochemical and pharmacological approaches, molecular studies strongly suggest that a protein with this functional capacity is likely to be present in plant cells. Indeed, cGMP-binding domains were found in various plant proteins, including putative CNGC and potassium channels (The *Arabidopsis* genome initiative, 2000; Leng *et al.*, 1999; Sentenac *et al.*, 1992). Moreover, a protein with guanylate cyclase activity but structurally unrelated to its animal counterpart was recently identified in *Arabidopsis* (Ludidi and Gehring, 2003). Because of its unusual structural organisation, this protein, termed AtGC1, defines a new class of guanylate cyclase.

Through the use of biochemical and pharmacological approaches, it has been demonstrated that cGMP may serve as a second messenger for NO-signalling in plants. NO treatment of spruce needles and tobacco leaves or suspension cells leads to a transient surge in endogenous cGMP concentration (Durner *et al.*, 1998; Pfeiffer *et al.*, 1994). Moreover, NO-mediated cell death and accumulation of mRNA encoding Phenylalanine ammonia lyase (PAL), a key enzyme of the phenylpopanoid synthetic pathway, were moderated by mammalian guanylate cyclase inhibitors in *Arabidopsis thaliana* and in tobacco suspension cultures, respectively (Durner *et al.*, 1998; Clarke *et al.*, 2000). Conversely, both *PAL* expression and activity were activated by the cell-permeable cGMP analogue 8-Br-cGMP (Durner *et al.*, 1998). By analogy with mammalian processes, these results imply that NO produced in response to elicitors of plant defence responses could partly initiate its biological effect through activation of guanylate cyclase. Unexpectedly, however, NO seems not to be a key regulator of the *Arabidopsis* guanylate cyclase AtGC1 (Ludidi and Gehring, 2003). Therefore, whether guanylate cyclase is indeed a target for NO in plants remains to be resolved. Identification of other members of the AtGC guanylate cyclase family as well as other putative plant guanylate cyclases should further helps our understanding of NO/cGMP-mediated signalling.

Recently, genetic analysis of the *Arabidopsis* mutants HLM1 and DND1 allowed the identification of CNGC-encoding genes which misfunction are responsible for the mutant phenotypes (Balagué *et al.*, 2003, Clough *et al.*, 2001). Because these mutants are impaired in their ability to produce HR in response to avirulent pathogens, it is postulated that the CNGC proteins HLM1 and DND1 might constitute upsteam components of the signalling pathways leading to HR. Expression of *HLM1* and *DND1* in heterologous systems followed by patch-clamp analysis has shown that these putative channels are permeable to K^+, Na^+ and Ca^{2+} and activated by cGMP. These results raise the possibility that HLM1 and DND1 might be regulated by cGMP in response to pathogens and derived elicitors. In this scenario, it may be of considerable interest to examine the functional interaction between NO and cGMP.

2.3.2. Ca^{2+}: A Second Messenger Mediating Nitric Oxide Effects

For mammalian systems, considerable evidence supports the hypothesis that NO regulates a broad spectrum of proteins by S-nitrosylation. In this process, NO modifies critical cysteine residue of target proteins without the assistance of enzymes, leading to either a decline or and increase in protein activity (fig. 3). Functional regulation by S-nitrosylation has been described both *in vivo* and *in vitro,* and to date almost 100 S-nitrosylated proteins have been identified (Stamler *et al.*, 2001; Jaffrey *et al.*, 2001). NO also targets tyrosine residue of proteins through peroxynitrite ($ONOO^-$), which is formed from the reaction of NO with superoxyde anion (figure 3; Koppenol *et al.* 1998; Stamler *et al.*, 1992). The biological significance of tyrosine nitrosation has been the subject of considerable scientific interest. Tyrosine nitrosation causes inactivation of several enzymes, influences tyrosine phosphorylation-mediated signal transduction, and is associated with many inflammatory and neurodegenerative diseases (Hanafy *et al.*, 2001).

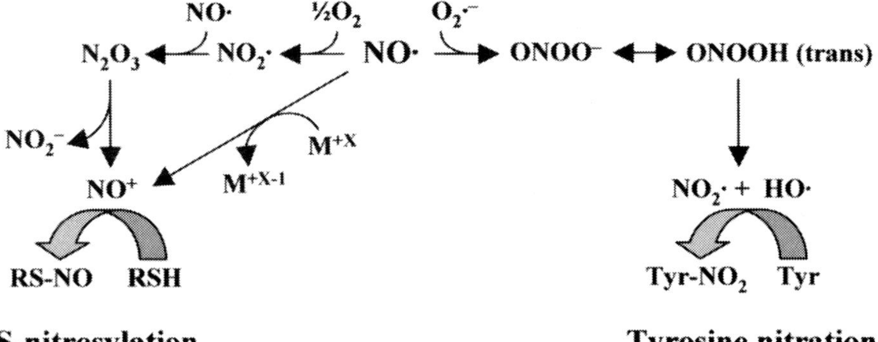

Figure 3 : S-nitrosylation and tyrosine nitration.
S-nitrosylation: in biological environments, NO reacts with O_2 to yield nitrogen dioxide radical ($NO_2\cdot$) which then oxidizes NO to from dinitrogen trioxide (N_2O_3). N_2O_3 decomposes into nitrite (NO_2^-) and a nitrosonium cation (NO^+), which reacts with critical cysteine residues to form S-nitrosothiols. Alternatively, NO^+ may be directly formed by the one-electron oxidation of NO. In this latter reaction, metalloenzymes (M^{+X}) act as the electron acceptor. **Tyrosine nitrosation**: the reaction of NO with superoxide ($O_2\cdot^-$) results in the generation of peroxynitrite ($ONOO^-$). At physiological pH values, $ONOO^-$ equilibrates with peroxynitrous acid (ONOOH) (pKa = 6.8). The trans configuration allows decomposition into $NO_2\cdot$ and the hydroxyl radical (HO·) which is thought to oxidize tyrosine into tyrosine radical. Next, a bond is formed between the tyrosine radical and nitrogen dioxide, forming 3-nitrotyrosine. For further details see Stamler *et al.* (1992).

S-nitrosylation and tyrosine nitration are emerging as major mechanisms for the post-translational modification of proteins in animals. An important class of proteins that constitutes key targets of NO is that of channels/transporters (Stamler *et al.*, 2001). Of these, Ca^{2+} channels including cardiac and skeletal ryanodine receptors (RYR), voltage-gated L-type, N-type and P/Q type Ca^{2+} channels, store-operated Ca^{2+} channels, CNGCs and the NMDA receptor have all been shown to be up- or down- regulated by NO, both *in vivo* and *in vitro* (Table 1). Consistently, many studies have reported that NO can induce an increase in intracellular [Ca^{2+}], the result of an influx of Ca^{2+} over the plasma membrane, the release of Ca^{2+} from intracellular stores, or a combination of the two. For example, in neuronal cells, NO mediates intracellular [Ca^{2+}] elevation by promoting S-nitrosylation of voltage-gated P/Q type channels and CNGC (Broillet *et al.*, 2000; Chen *et al.*, 2002). Similarly, NO produced by contracting muscle has been postulated to activate the

Table 1. Nitric Oxide-regulated Ca^{2+} channels in animal cells.
See the corresponding text for further details. cGMP-dependent process refers to unknown cGMP-dependent mechanisms; (?) refers to putative mechanisms. This table is modified from Stamler *et al.* (2001).

Ca^{2+} Channel	Regulation	Functional effect	Reference
NMDA receptor	S-nitrosylation	inhibition	Choi *et al.* (2000)
L-type	cGMP-dependent PKG	activation	Wang *et al.* (2000)
	cGMP-dependent process	inhibition	Campbell *et al.* (1996)
	peroxynitrite-induced tyrosine nitration (?)	activation	Campbell *et al.* (1996)
	membrane depolarisation	activation	Ohkuma *et al.* (1998); Willmott *et al.* (2000)
	S-nitrosylation	activation/ inhibition	Campbell *et al.* (1996); Poteser *et al.* (2001)
N-type	cGMP-dependent process	inhibition	Yoshimura *et al.* (2001)
P/Q-type	S-nitrosylation (?)	activation	Chen *et al.* (2002)
CNGC	S-nitrosylation	activation	Broillet *et al.* (2000)
Ryanodine receptor	S-nitrosylation	activation	Sun *et al.* (2001); Xu *et al.* (1998)
	cGMP/cADPR pathway	activation	Reyes-Harde *et al.* (1999); Willmott *et al.* (1996)
IP_3 receptor	cGMP-dependent process	activation	Tritsaris *et al.* (2000)

cardiac and skeletal RYR by S-nitrosylation of critical cysteine residues, thereby contributing to an increase of cytosolic [Ca^{2+}] (Xu *et al.*, 1998). In addition to S-nitrosylation and tyrosine nitration, NO regulates Ca^{2+} channels indirectly in two distinct ways (Table

1). First, NO opens L-type voltage-gated channels via depolarisation of the plasma membrane (Ohkuma *et al.*, 1998; Willmott *et al.*, 2000). Second, NO modulates Ca^{2+} channels through cGMP-mediated PKGs activation. Several scenarios for this process have been described: i) PKGs phosphorylate voltage-dependent Ca^{2+} channels or associated proteins on the plasma membrane, leading to activation of the channels (Wang *et al.*, 2000); (ii), PKGs induce production of IP_3, which in turn activates IP_3-sensitive stores, amounting to NO-induced release of Ca^{2+} (Tritsaris *et al.*, 2000); iii), PKGs activate ADP-ribosyl cyclase, the enzyme responsible for the production of cADPR, a specific modulator of RYR (Reyes-Harde *et al.*, 1999; Willmott *et al.*, 1996). This third pathway, involving cGMP, PKG and cADPR-mediated release of Ca^{2+} through RYR activation, has been implicated in various cell types and may constitute an important mechanism by which NO controls Ca^{2+} homeostasis in animals. It is interesting to note that several studies in which NO inhibits Ca^{2+} channels through a cGMP-dependent process have been also reported. However, this mechanism has not been studied in detail and is poorly understood.

In plants, the first evidence that NO operates via Ca^{2+} mobilisation originated from the study of Durner *et al.* (1998). It was found that cADPR induces *PAL* and *PR-1* mRNA accumulation in tobacco. This process was blocked by the RYR inhibitor ruthenium red, suggesting that cADPR mediates its effect through a RYR-dependent mechanism for Ca^{2+} release. In support of this model, patch clamp analysis confirmed that cADPR can induce Ca^{2+} release with the hallmark characteristics of mediation by RYR in plant cells (Allen *et al.*, 1995, 1998). The possibility that NO acts through cADPR to induce defence genes was confirmed in a series of experiments showing that NO-induced accumulation of the *PR-1* transcript was suppressed by the cADPR antagonist, 8-bromo-cADPR (Klessig *et al.*, 2000). Moreover, as it has also been reported for Ca^{2+}-dependent gene activation in animal cells (Peunova and Enikolopov, 1993), cGMP and cADPR appeared to act synergistically in inducing *PR-1* gene. However, although these original data argue for the possibility of a cross-communication between NO, cGMP and cADPR in plants, definitive proof linking cADPR to NO and cGMP is lacking at present. Measurements of changes in cADPR concentrations following NO stimulation of plant cells could provide an unequivocal answer to this question.

The most conclusive evidence for a fundamental role of Ca^{2+} in mediating NO effects in response to biotic stress arose from studies based on the recombinant aequorin technology. Aequorin is a photoprotein containing three EF-hands Ca^{2+}-binding sites, EF-hand designing a conserved helix-loop-helix structure that can bind a single Ca^{2+} ion. When these sites are occupied by Ca^{2+}, aequorin undergoes a conformational change and emits luminescence (for further details see Mithöfer and Mazars, 2002). By expressing aequorin at a specific subcellular compartment of plant cells, it has been possible to measure Ca^{2+} fluctuations occurring in response to biotic and abiotic stimuli, to second messengers including cGMP and cAMP, and to applied chemicals (Blume *et al.*, 2000; Knight *et al.*, 1991; Pauly *et al.*, 2000; Volotovski *et al.*, 1998). Using transgenic *Nicotiana plumbaginifolia* cell suspensions that constitutively express aequorin in the cytosol, Lecourieux *et al.* (2002) investigated the changes in cytosolic $[Ca^{2+}]$ ($[Ca^{2+}]_{cyt}$) triggered by cryptogein. The authors reported that cytosolic Ca^{2+} signature attributable to cryptogein is characterized by a first transient $[Ca^{2+}]_{cyt}$ increase resulting from a Ca^{2+} influx, which in turn leads to Ca^{2+} release from internal stores. The first peak is followed immediately by a second, sustained $[Ca^{2+}]_{cyt}$ attributable to the influx of extracellular Ca^{2+}. Of interest, the NO-scavenger cPTIO was shown to reduce by 50% the intensity of the first peak of $[Ca^{2+}]_{cyt}$ without significantly affecting the second peak (Lamotte *et al.*, 2003). Taken together, these data highlight a specific function for NO in the intracellular release of Ca^{2+} by tobacco cells elicited by cryptogein. Moreover, because cPTIO displayed a similar inhibition profile to those of RYR inhibitors, it was postulated that NO may specifically mobilize Ca^{2+} from internal stores through the activation of RYR. This process seems specific to NO rather than peroxynitrite since urate, an efficient and specific peroxynirite scavenger both in animal and plant cells (Alamillo and Garci-Olmedo, 2001; Hooper *et al.*, 1998) did not affect the cryptogein-induced Ca^{2+} signature.

Further support for an association between NO signalling and Ca^{2+} arose from the observation that stimulation of aequorin-transformed *Nicotiana plumbaginifolia* cell suspensions with the non-thiol NO-donor, diethylamine NONOate, resulted in transient $[Ca^{2+}]_{cyt}$ increase (Lamotte *et al.*, 2003). Biochemical and pharmacological analysis showed that the NO-evoked cytosolic Ca^{2+} signature was attributable to Ca^{2+} released from internal stores sensitive to RYR inhibitors. Taking account the NO/Ca^{2+} crosstalk observed in cryptogein signalling, this finding confirms the capacity for NO to mobilize intracellular Ca^{2+} through Ca^{2+}-permeable channels which are pharmacologically related to RYR.

One question central to our understanding of NO/Ca^{2+} signalling is how NO regulates Ca^{2+} release. Based on mammalian studies and the data discussed here (Durner *et al.*, 1998; Klessig *et al.*, 2000; Lamotte *et al.*, 2003), several scenarios can be envisioned (Figure 4). As indicated above, it is possible that NO exerts its effect indirectly through cGMP and/or cADPR. With respect to their involvement in plant defence responses, the cGMP-dependent proteins HLM1 and DND1 may be good candidates for such a pathway. On the other hand, NO may directly regulate Ca^{2+}-permeable channel through S-nitrosylation or tyrosine nitration. This hypothesis might well explain the finding that the cytosolic Ca^{2+} signature induced by NO donors is strongly diminished by reducing compounds which are known to compete with endogenous thiols for S-nitrosylation in mammals (unpublished data). Further work directed towards the electrophysiological characterization of putative NO-regulated Ca^{2+}-permeable channels should assist to clarify this question.

2.3.3. Nitric Oxide and MAP Kinases

In mammals, one of the major emphasis in recent years has concerned the functional relationship between NO and MAPK. It is now well-established that NO and related molecular species modulate the three MAPK signalling pathways (Lander *et al.*, 1996). Mechanistically, this process may occur by activation of an upstream low molecular weight G-protein, such as p21Ras, Cdc42 or Rac1 (reviewed by Beck *et al.*, 1999). The observations that the NO/MAPK pathways lead to the inactivation or suppression of pro-apoptotic protein such as p53, Bad and Bax, and/or to the upregulation of antioxidant enzymes lend themselves to the possibility that these signalling cascades are possible anti-apoptotic mechanisms (Schroeter *et al.*, 2002). Alternatively, NO might also directly modulate members of the different MAP kinase cascades by S-nitrosylation, causing either activation or inhibition of the enzymes. In turn, MAPK appear to be involved in the regulation of *NOS* gene expression. For example, the induction of endothelial *NOS* gene by epidermal growth factors

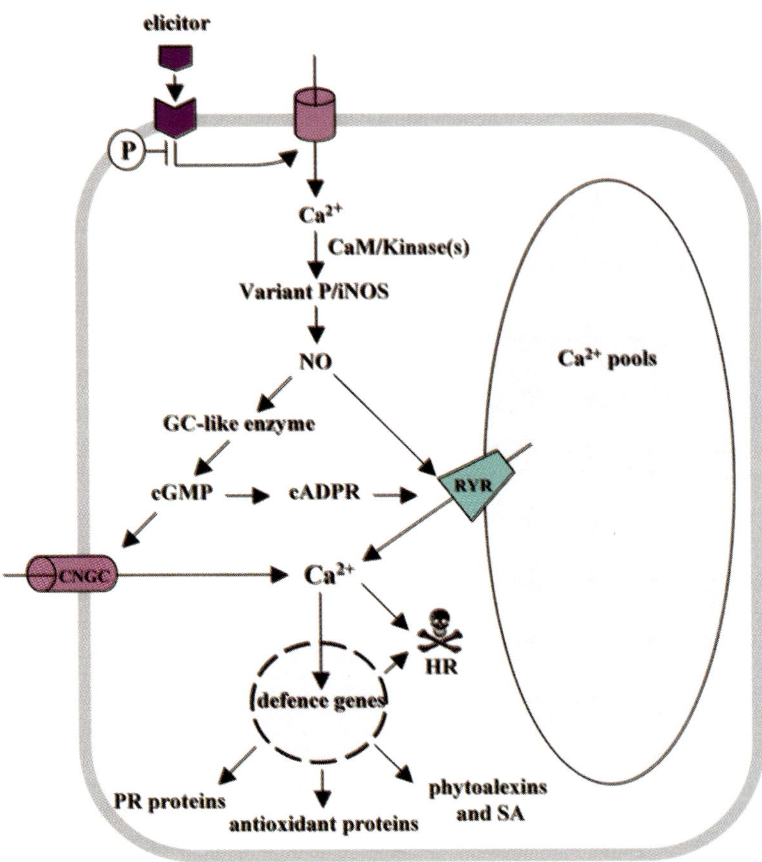

Figure 4 : Model for nitric oxide and Ca^{2+} relationships in plant defence.
In response to elicitors of plant defence responses, NO is produced by variant P/iNOS, the activation of which is strictly dependent on an increase of intracellular free Ca^{2+} concentration, CaM and phosphorylation events. This biosynthetic pathway is negatively regulated by one or several phosphatase(s) suggesting that NO production in response to potential pathogens may be mediated by release of repressive mechanisms. Once produced, NO activates RYR channels through S-nitrosylation accounting for Ca^{2+} release from internal Ca^{2+} pools. Moreover, NO may induce a guanylate cyclase-like enzyme resulting in cGMP synthesis. Consequently, cGMP promotes the synthesis of cADPR which, upon binding to RYR, leads to Ca^{2+} release from Ca^{2+} internal stores. cGMP may also activate plasma membrane CNGCs such as DND1 and HLM1 which might constitute upstream components of the signalling pathway leading to HR. The resulting increase in cytosolic Ca^{2+} concentration could initiate signalling events leading to defence-related gene expression including genes encoding PR proteins, phytoalexin and SA biosynthetic enzymes, antioxidant proteins and proteins putatively involved in HR cell death. CaM, calmodulin; GC: guanylate cyclase, P: phosphatase.

involves the NO-p21Ras-MAPK pathway (Zheng *et al.*, 1999). It has become increasingly clear, therefore, that a potential feedback loop exists between NO, MAPK and NOS.

In plants, diverse stimuli are transduced through MAP kinases cascades (MAPK group, 2002). In particular, MAPK cascades are involved in the defence response to various avirulent pathogens and elicitors, and may make the difference between cell death or survival (Asai *et al.*, 2002; Frye *et al.*, 2000; Yang *et al.*, 2001). In this regard, the possibility that NO may deliver signals into MAPK cascades was highlighted by the demonstration that artificially-generated NO can activate an unidentified 47 kDa MAPK in *Arabidopsis* cell suspensions, and the MAPK salicylic acid-induced protein kinase (SIPK) in tobacco leaves and cell suspensions (Clarke *et al.*, 2000; Kumar and Klessig, 2000). Because the *Arabidopsis* genome does not seem to have a Ras-like gene, the NO-dependent MAPK activation in plants might not resemble that used in animals. In this context, SA appears to be a potential plant-specific mediator of NO-mediated induction of MAPK cascades. This conclusion is based on the finding that NO's ability to activate SIPK was suppressed in plants expressing the salicylate hydroxylase-encoding *nahG* gene from *Pseudomonas putida* (Kumar and Klessig, 2000). These plants rapidly degrade SA to catechol, thereby preventing significant accumulation of SA (Friedrich *et al.*, 1995). The precise mechanism by which SA activates SIPK, as well as the physiological significance of this NO-dependent pathway, are far from understood.

The demonstration that NO produced physiologically in plants in response to biotic stress may modulate MAPK cascades has not been reported to date. For example, NO appears not to be required for cryptogein-induced MAPK activation in tobacco cell suspensions (Lamotte *et al.*, 2003). Similarly, NO seems not to act upstream of the MAPK cascade activated by a purified glycoprotein elicitor from *Botrytis cinerea* in grapevine (Vandelle E., Bentéjac M. and Pugin A., personal communication). Interestingly, Kumar and Klessig (2000) reported that NO-donor treatment of tobacco cell suspensions strongly delayed the activation of both SIPK and the MAPK wound-induced protein kinase (WIPK) by cryptogein. This observation raises the hypothesis that NO may also activate MAPK-counteracting components in plants. In mammalian studies, NO-dependent protein phosphatases appear to represent key negative regulators of plant

MAPK cascades. Indeed, it has been reported in various mammalian cell types that NO decreases MAPK activity through the induction of protein phosphatases including MAPK phosphatase-1 (Beck *et al.*, 1999). The occurrence of a similar regulation in plants clearly warrants further investigation.

2.4. NO-mediated Plant Defence Responses

2.4.1. Modulation of Defence Gene Expression by Nitric Oxide

Over the past few years, increasing numbers of plant genes have been shown to be under regulatory control by NO. Evidence is accumulating that NO preferentially alters those genes associated with defence and oxidative stress. Indeed, in addition to *PAL* and *PR* genes (see above), the application of exogenous NO to tobacco, soybean and *Arabidopsis thaliana* suspensions cells resulted in the induction of classical defence genes such as *CHS*, *ACC oxidase* and of an array of genes encoding antioxidant proteins including various peroxidases, catalase, superoxide dismutase, metallothionein and GST as determined by cDNA microarray strategy and Northern blotting analysis (Delledonne *et al.*, 1998; Durner *et al.*, 1998; Huang *et al.*, 2002a, b). In comparison, only the genes encoding PR-1, PAL, the heat-shock protein TLHS-1 and the ethylene-forming enzyme cEFE-26 have been shown to be up-regulated through NO-dependent pathways in response to avirulent pathogens or elicitors (Delledonne *et al.*, 1998; Klessig *et al.*, 2000; Lamotte *et al.*, 2003). The observation that the accumulation of the corresponding transcripts was not completely inhibited by NOS inhibitors or NO scavengers suggests that the activation of these genes in response to avirulent pathogen and/or elicitors may occur through both NO-dependent and NO-independent pathways.

2.4.2. Nitric Oxide Involvement in Phytoalexin Synthesis

The phenylpropanoid pathway leading to phytoalexins production is regulated at the level of the enzymes PAL and CHS (Dixon, 2001). The finding that NO induces genes for both of these enzymes suggests that NO may confer to plants a selective advantage against microbial attack by contributing to phytoalexin synthesis. In support of this hypothesis, artificially-generated NO was shown to activate phytoalexin accumulation in potato tuber tissue and soybean cotyledons (Modolo *et al.*, 2002; Noritake *et al.*, 1996). Furthermore, NO produced via NOS-like enzyme activity appeared to be a key mediator of the signalling pathway(s) leading to flavones (apigenin and luteolin), isoflavones (daidzein and genistein) and pterocarpans (glyceollins) synthesis in soybean cotyledons challenged by a fungal elicitor purified from *Diaporthe phaseolorum* f.sp. *meridionalis* (Modolo *et al.*, 2002). Interestingly, although the role of flavones such as apigenin and luteolin in plant defence remains unclear, these compounds inhibit NOS enzyme expression in lipopolysaccharide-stimulated macrophages (Kim *et al.*, 1999; Schroeter *et al.*, 2002). According to Modolo *et al.* (2002), a similar role for apigenin and luteolin in countering NOS-like activity could be envisioned in plants. Such potential feedback loop between NO and flavones may help in reducing NO-mediated nitrosative stress during the defence response to pathogens.

2.4.3. Nitric Oxide, Salicylic Acid and Systemic Acquired Resistance

Several lines of evidence point to the importance of the relationship between NO and SA in the expression of plant defences. The most convincing demonstration arose from the study of Durner *et al.* (1998) who showed that treatment of tobacco leaves with NO induced a significant increase in the endogenous SA and that this increase was required for *PR-1* induction. Because SA biosynthesis proceeds via PAL enzyme activity in various plant species (Dixon *et al.*, 2001), this process may first occur through NO-dependent *PAL* gene induction as detailed above. Recently, Song and Goodman (2001) exploited this functional link to study the role of NO in the signalling pathway leading to SAR in tobacco. Surprisingly, they found that NOS inhibitors and the NO scavenger cPTIO attenuated the SAR induced by SA in tobacco. These data imply that not only is NO involved in SA biosynthesis, but also the NOS-like enzyme is activated by SA. More importantly, this study supports the contention that NO exerts an important function in SAR, being required for the full function of SA as an SAR inducer.

How NO and SA interact in the induction of defence responses in plants challenged by avirulent pathogens is currently unresolved. Of interest, it was recently demonstrated that NO inhibits the redox-related enzyme aconitase and the two H_2O_2 scavenging enzyme catalase and ascorbate peroxidase, and thus enhancing ROS levels (Clark *et al.*, 2000; Navarre *et al.*, 2000). Because these enzymes are also inhibited by SA, it was proposed that SA may potentiate the

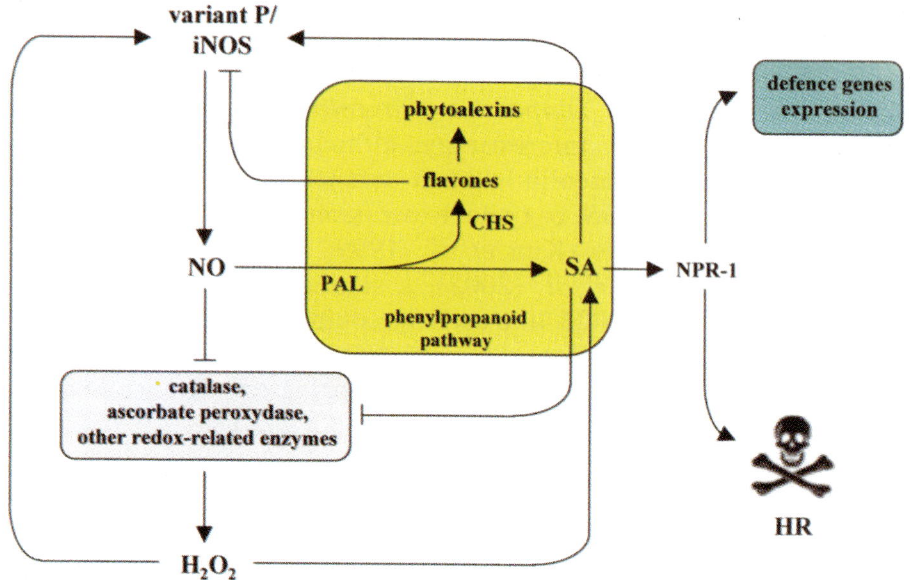

Figure 5 : Cross-talk between nitric oxide, salicylic acid and reactive oxygen species in plant defence
The phenylpropanoid pathway leading to phytoalexins and SA production is regulated at the level of the enzymes PAL and CHS, for which the corresponding genes are activated through NO-dependent pathways. Both SA and NO inhibit the two H_2O_2 scavenging enzyme catalase and ascorbate peroxidase, and thus enhance ROS levels. In turn, H_2O_2 amplifies NO production through a salicylic-dependent or –independent potentiation loop. The resulting signals activate NPR1, which may positively regulate defence gene expression and a cell death pathway. Flavones such as apigenin and luteolin may negatively regulate this process by attenuating NOS activity or synthesis. For further details see Dixon (2001); Durner and Klessig (1999); Loake (2001).

effects of NO, suggesting that both molecules may act synergistically to transduce the defence signal together with ROS (figure 5; Durner and Klessig, 1999; Wendehenne *et al.*, 2001). According to Durner and Klessig (1999), this mechanism should play a significant role in the activation of defence genes and/or induction of hypersensitive cell death.

3. NITRIC OXIDE AS A RESPONSE TO ABIOTIC STRESSES

Nitric oxide is not only produced by pathogen-infected plants. Healthy, well-fertilised shoots also release significant amounts of NO into the atmosphere. The NO emissions reported for a variety of agricultural crops, weeds, and a gymnosperm species are in the order of 1×10^{-15} to 5×10^{-14} mol Cm^{-1} s^{-1}, and represent an estimated global loss of 0.23 Tg nitrogen per year (Stutte and Weiland, 1978; Weiland and Stutte, 1979; Wildt *et al.* 1997). NO emissions, however, sharply escalate when plants are subjected to one or more abiotic stresses. Several different types of chemical, mechanical, and environmental stressor have been reported to induce rapid, substantial surges in NO evolution in a variety of plant species (Table 2).

Table 2. Reports of nitric oxide induction by abiotic stresses

Stressor	Species	Reference
Abscisic acid	*Pisum sativum*	Neill *et al.* (2002)
	Arabidopsis thaliana	Desikan *et al.* (2002)
	Vicia faba	García-Mata and Lamattina (2002)
Nitrate application	Various	Wildt *et al.* (1997)
Heavy metal toxicity	*Scenedesmus obliquus*	Mallick *et al.* (2000)
Herbicides	*Glycine max*	Klepper (1979)
	Scenedesmus obliquus	Mallick *et al.* (2000)
	Chlamydomonas reinhardtii	Sakihama *et al.* (2002)
High temperature	*Medicago sativa*	Leshem *et al.* (1998)
	Nicotiana tabacum	Gould *et al.* (2003)
Low temperature	*Scenedesmus obliquus*	Mallick *et al.* (2000)
Mechanical injury	*Arabidopsis thaliana*	Garcês *et al.* (2001)
	Taxus brevifolia	Pedroso *et al.* (2000b)
Nutrient deficiency	*Scenedesmus obliquus*	Mallick *et al.* (2000)
Drought/osmotic stress	*Nicotiana tabacum*	Gould *et al.* (2003)
	Pisum sativum	Leshem and Haramaty (1996)
Salinity	*Nicotiana tabacum*	Gould *et al.* (2003)

Unlike that of the reactive oxygen species, the induction of NO cannot be regarded as a generalised stress response. The responses to some of the stressors appear to be specific to certain plant species. For example, mechanical injury imposed by centrifugation or wounding induces NO in *Taxus* and *Arabidopsis* (Garcês *et al.* 2001; Pedroso *et al.* 2000b), but not in *Lycopersicon* or *Nicotiana* (Gould *et al.* 2003; Orozco-Cárdenas and Ryan 2002). Similarly, high temperatures promote NO formation in *Medicago* and *Nicotiana* (Gould *et al.* 2003; Leshem *et al.* 1998), but not in *Scenedesmus* (Mallick *et al.* 2000). Moreover, and in sharp contrast to its effect on ROS levels, strong light does not appear to stimulate NO production, though the evolution of NO is possible under light when plants are challenged by an additional stressor such as heat or herbicide (Sakihama *et al.* 2002). Indeed, algal cell cultures, which generate only low levels of NO in the light, yield significant pulses of NO when they are transferred to darkness (Mallick *et al.* 2000; Sakihama *et al.* 2002).

3.1. Nitric Oxide Production in Response to Abiotic Stresses

Relatively little is known of the mechanism(s) by which abiotic stressors generate NO in plants. The involvement of an NOS-like enzyme has been implicated in some studies (Garcês *et al.* 2001; Leshem and Haramaty 1996; Pedroso *et al.* 2000b). However, an increasing body of evidence suggests that much of the stress-induced NO results from perturbed nitrate metabolism mediated by NR. Under favourable conditions, cytosolic nitrate is reduced to nitrite by NR, using the reducing power of NAD(P)H. The nitrite is then transported across a proton gradient into chloroplasts, where it is reduced to ammonium ions by nitrite reductase (NiR), and assimilated into amino acids. NiR-mediated conversion of nitrite to NH_4^+ requires reducing equivalents, which are normally supplied by reduced ferredoxin (Fd_{red}) located at the terminus of the electron transport chain of photosystem I (Crawford, 1995). Under various environmental stresses, however, photosynthetic electron transport ceases to function efficiently (Mano, 2002), supplies of Fd_{red} become limited, and in the absence of a proton gradient to drive nitrite transport, nitrite accumulates in the cytoplasm (Klepper 1975, 1976). Under those conditions, NR can reduce nitrite to NO (Sakihama *et al.* 2002; Yamasaki *et al.* 1999; Yamasaki 2000; Yamasaki and Sakihama, 2000).

Yamasaki and Sakihama (2000) considered that the NO produced by NR-reduction of nitrite be a by-product of an "uncontrolled pathway", potentially leading to a dysfunction of cellular metabolism, in contrast to the stoicheometrically-synthesized and highly-regulated NO produced by the NOS route. The NR hypothesis is attractive because it potentially explains why NO can be induced by disparate abiotic stressors (since all can impair photosynthetic electron transport), and why surges in NO have been reported when plants are transferred from the light to the dark (when supplies of Fd_{red} are no longer replenished). It does not, however, adequately explain why photoinhibitory fluxes of light have no apparent effect on rates of NO production (Gould *et al.* 2003). Nonetheless, there is substantive evidence for the NR route of NO synthesis from *in vitro* experiments, from pharmacological approaches, and from studies involving NR-deficient mutants (reviewed by Yamasaki, 2000). In the NR double mutant *nia1, nia2* of *Arabidopsis*, for example, guard cells do not synthesise NO in response to abscisic acid, in contrast to those of the wild-type plants (Desikan *et al.* 2002). Similarly, the NR-deficient *nr1* mutant of *Glycine max* failed to generate NO during *in vivo* assays (Nelson *et al.* 1983), and the cc-2929 mutant of *Chlamydomonas reinhardtii* did not show the normal NO surges observed in response to applications of nitrite or of DCMU, a photosynthetic electron transport inhibitor (Sakihama *et al.* 2002). Moreover, because inhibitors of NOS did not abolish NO production in response to various abiotic stressors (Mallick *et al.* 2000; Sakihama *et al.* 2002), NR would instead appear to be the primary enzymatic candidate for the "uncontrolled", stress-induced synthesis of NO.

3.2. Physiological Consequences of Nitric Oxide Production in Plants

Loss of nitrogen through the emission of NO is unlikely to impact significantly on the nitrogen economy of a plant. The maximum emissions, recorded in the dark for various species by Wildt *et al.* (1997), constituted only 3% of the amount of nitrogen taken up through the roots. Nonetheless, it is worth considering whether NO emissions represent a gratuitous loss of nitrogen resulting from dysfunctional metabolism, or if, instead, they might assist to mitigate the effects of abiotic stressors.

As a free radical, NO is a potent oxidant with a proven capacity to injure membranes, proteins, and nucleic acids in the plant cell (Yamasaki, 2000). In haploid cultures of *Taxus*, high NO levels were associated with irreversible DNA fragmentation and cell death (Pedroso *et al.*, 2000b). Reacting with thiol groups in proteins or glutathione (figure 3), NO can cause the formation of thiyl radicals, which can then react with reduced thiols to produce disulphide bonds. Mitochondria and chloroplasts contain catalytically functional thiols that are inactivated by the formation of disulphides (Asada, 1999). It is not surprising, therefore, that exposures to NO have been reported to lead to reductions in net photosynthesis (Hill and Bennett, 1970) and respiration (Zottini *et al.*, 2002). NO concentrations above 10^{-6} M inhibited the expansion of leaf laminae, increased the viscosity of simulated thylakoid lipid monolayers, and potentially impaired photosynthetic electron transport in peas (Leshem *et al.* 1997; Leshem *et al.*, 1998).

In contrast to the reports of NO cytotoxicity in plants, there are also good indications that exposures to low levels of NO improve the performance of plants under environmental stress. Tolerance to drought stress, for example, was enhanced in wheat seedlings that were loaded first with sodium nitroprusside (SNP), an NO-donor (García-Mata and Lamattina, 2001; 2002). Similarly, SNP-treated potato plants were more resistant to methylviologen herbicides than were untreated plants (Beligni and Lamattina 1999). Tolerance to UV-B radiation has also been implicated; treatment of *Arabidopsis thaliana* with the NO scavenger carboxy-PTIO prevented the up-regulation of transcription of the gene encoding CHS that is believed to be important in conferring UV-B protection (Mackerness *et al.*, 2001).

At least three hypotheses have been propounded to explain the possible beneficial effects of NO on stress tolerance. First, NO may confer antioxidant protection to plants under stress, supplementing other detoxification mechanisms. Almost all abiotic stressors have been shown to generate free radicals and other oxidants in plant cells, particularly from the chloroplasts, mitochondria and peroxisomes (Mano, 2002). NO is more diffusible and less cytotoxic than many oxygen radicals; the preferential production of NO, and its emission from leaves via stomata and/or trichomes (Foissner *et al.*, 2000; Gould *et al.*, 2003), would directly moderate the overall oxidative load. NO also reacts rapidly with lipid alcoxyl and peroxyl radicals, and thus may serve to break chain reactions of lipid peroxidation (Rubbo and Freeman, 1996). Superoxide radicals, which are generated via electron leakage to molecular oxygen from transport chains in the chloroplasts and mitochondria (reviewed by Alscher *et al.*, 1997; Elstner, 1991; Foyer *et al.*, 1994), are scavenged by NO at near diffusion-controlled rates (Koppenol, 1998; Squadrito and Pryor, 1998). Effective removal of superoxide would prevent the formation of the highly toxic hydroxyl radicals via Haber-Weiss reactions.

There is some empirical support for an antioxidant role of NO *in vivo*. NO has been shown significantly to decrease superoxide formation in microsome suspensions of *Glycine max* (Caro and Puntarulo, 1998). Applications of NO donors to *Solanum tuberosum* prevented chlorophyll bleaching by a methyl viologen herbicide, and diminished levels of DNA fragmentation in leaves infected with *Phytopthora infestans* (Beligni and Lamattina, 1999). In contrast, however, an excess of NO has been suggested to inhibit electron flow through cytochrome c oxidase, leading to enhanced production of superoxide in mitochondria (Millar and Day, 1996; 1997). Moreover, the reaction between NO and superoxide is not necessarily beneficial because it leads to the formation of $ONOO^-$, itself considered to be a major cytotoxic agent of active nitrogen species (Arteel *et al.* 1999; Koppenol, 1998; Squadrito and Pryor, 1998; Stamler *et al.*, 1992; Wink and Mitchell, 1998), and which can react with H_2O_2 to produce singlet oxygen, another toxic ROS (Di Mascio *et al.*, 1996).

A second hypothesis for NO effects on stress tolerance concerns the elimination of excess nitrite from cells. Nitrite, which can be present in the cytoplasm at significant concentrations when photosynthetic electron transport is not operational (Shingles *et al.* 1996), is toxic to plant cells (Wellburn, 1990). The conversion of nitrite to NO by NR, and its rapid diffusion into the atmosphere or conversion to $ONOO^-$ could therefore avoid the effects of nitrite toxicity. Direct experimental support for this hypothesis is lacking, however.

Finally, NO may ameliorate plant defence responses to abiotic stress via its properties as a signalling molecule. There is evidence both for positive and negative regulation of signalling cascades by NO. In wounded leaves of *Lycopersicon esculentum*, NO apparently interacts with the signalling pathway downstream from jasmonic acid synthesis, but upstream of H_2O_2 synthesis, leading to the down-regulation of antiherbivory proteinase-inhibitor genes (Orozco-Cárdenas and Ryan, 2002). In *Pisum sativum*, NO appears to be an effective signal leading to the closure of stomatal pores in response to abscisic acid (Neill *et al.*, 2002), and may, therefore, be an essential element in the mechanism of drought resistance.

4. CONCLUSIONS AND FUTURE PROSPECTS

In this review, we have considered the mechanisms for the production of NO and its functions in plant pathways related to defence and abiotic stress. The current state of knowledge indicates that NO is likely to play a crucial role in plant defence signalling. Indeed, NO is emerging as an important component of signal transduction in the defence network, leading to the activation of effective arsenal of inducible responses of phytoalexins and *PR* gene expression. There are evidently complex interactions between NO and the cyclic nucleotides cGMP and cADPR, and it is tempting to speculate that this cross-talk critically influences Ca^{2+}-permeable channels activities. Biochemical and electrophysiological approaches are expected to yield further insights into the nature and control of channels that underlie the generation of Ca^{2+} signals in response to NO. Moreover, because of its pre-eminence as a mechanism for signal transduction, particular consideration needs to be given to protein S-nitrosylation and tyrosine nitrosation and their role in plant cellular signalling. Surprisingly, despite the considerable advances made in understanding NO functions in plant defence, this aspect of NO research remains poorly explored. This may reflect the technical difficulties in identifying nitros(yl)ated proteins in a biological context. Therefore, improved methods for detection and identification of NO proteinaceous targets are essential.

The identification of the pathogen-inducible NOS in plants is indisputably a great step in our understanding of NO functions. Generation of NOS knockout mutants, or mutants over-expressing this enzyme should lead to the identification of many more NO signalling components and a clearer picture of biotic stress signalling networks. Moreover, these mutants may prove useful for the study of NOS involvement in other plant physiological processes. Finally, *in situ* localization of plant NOS and identification of its partners by two-hybrid screens or affinity methods are expected to greatly increase our knowledge of the spatio-temporal aspects of NO in plant cells.

Observations of enhanced drought tolerance in laboratory plants treated with NO donors, and the potential for improved detoxification of free radicals by NO produced under abiotic stress, present an exciting prospect for the future manipulation of crop species. It is easily foreseeable that plants which evolve NO at the greater rates, either having been selected through conventional breeding techniques or else generated by targeted genetic transformation, could provide a significant advantage to agricultural productivity in an era of global climate change. It should be noted, however, that as yet there is no evidence to indicate that NO confers a competitive edge to plants grown in the field. Future research should address the ecophysiological importance of NO under natural conditions.

In the 1970s a famous British punk rock group proclaimed in the chorus of one of its songs, “NO FUTURE”. In contrast, we believe that the NO future in plants is certain!

ACKNOWLEDGEMENTS

We would like to thank Daniel F. Klessig and his colleagues who kindly provided preprint of their study regarding the identification of the plant NOS (Chandok *et al.*, 2003). We thank E. Vandelle and M. Bentéjac for providing unpublished results. Our research discussed in this review is supported by the Institut National de la Recherche Agronomique (Grant SPE 1008a), Ministère de l’Education Nationale, de la Recherche et de la Technologie, and the Conseil Régional de Bourgogne.

REFERENCES

Alamillo, J.M. and Garcia-Olmedo, F. (2001). Effects of urate, a natural inhibitor of peroxynitrite-mediated toxicity, in the response of *Arabidopsis thaliana* to the bacterial pathogen Pseudomonas syringae. *Plant J.*, **25** : 529-540.

Allen, G.J., Muir, S.R. and Sanders, D. (1995). Release of Ca^{2+} from individual plant vacuoles by both InsP3 and cyclic ADP-ribose. *Science*, **268** : 735-737.

Allen, G.J. and Schroeder, J.I. (1998). Cyclic ADP-ribose and ABA signal transduction. *Trends Plant Sci.*, **3** : 123-124.

Alscher, R.G., Donahue, J.L. and Cramer, C.L. (1997). Reactive oxygen species and antioxidants: relationships in green cells. *Physiol. Plant.*, **100** : 224-233.

Arteel, G.E., Briviba, K. and Sies, H. (1999). Protection against peroxynitrite. *FEBS Lett.*, **445** : 226-230.

Asai, T., Tena, G., Plotnikova, J., Willmann, M.R., Chiu, W-L., Gomez-Gomez, L., Boller, T., Ausubel, F.M. and Sheen, J. (2002). MAP kinase signalling cascade in *Arabidopsis* innate immunity. *Nature*, **415** : 977-983.

Asada, K. (1999). The water-water cycle in chloroplasts: scavenging of active oxygens and dissipation of excess photons. *Ann. Rev. Plant Physiol. Plant Mol. Biol.*, **50** : 601-639.

Balague, C., Lin, B., Alcon, C., Flottes, G., Malmstrom, S., Kohler, C., Neuhaus, G., Pelletier, G., Gaymard, F. and Roby, D. (2003). HLM1, an essential signaling component in the hypersensitive response, is a member of the cyclic nucleotide-gated ion channel family. *Plant Cell*, **15** : 365-379.

Barroso, J.B., Corpas, F.J., Carreras, A., Sandalio, L.M., Valderrama, R., Palam, J.M., Lupianez, J.A. and del Rio, L.A. (1999). Localization of nitric oxide synthase in plant peroxisomes. *J. Biol. Chem.*, **274** : 36729-36733.

Beck, K-F., Eberhardt, W., Frank, S., Huwiler, A., Messmer, U.K., Mühl, H. and Pfeilschifter, J. (1999). Inducible NO synthase: role in cellular signalling. *J. Exp. Biol.*, **202** : 645-653.

Beligni, M.V. and Lamattina, L. (1999). Nitric oxide counteracts cytotoxic processes mediated by reactive oxygen species in plant tissues. *Planta*, **208** : 337-334.

Binet, M-N., Humbert, C., Lecourieux, D., Vantard, M. and Pugin, A. (2001). Disruption of microtubular cytoskeleton induced by cryptogein, an elicitor of hypersensitive response in tobacco cells. *Plant Physiol.*, **125** : 564-572.

Blume, B., Nurnberger, T., Nass, N. and Scheel, D. (2000). Receptor-mediated increase in cytoplasmic free calcium required for activation of pathogen defense in parsley. *Plant Cell*, **12** : 1425-1440.

Blumwald, E., Aharon, G.S. and Lam, B.C-H. (1998). Early signal transduction pathways in plant-pathogen interactions. *Trends Plant Sci.*, **3** : 342-346.

Bourque, S., Binet, M-N., Ponchet, M., Pugin, A. and Lebrun Garcia, A. (1999). Characterization of the cryptogein binding sites on plant plasma membranes. *J. Biol. Chem.*, **274** : 34699-34705.

Bourque, S., Lemoine, R., Sequeira-Legrand A., Fayolle, L., Delrot, S. and Pugin, A. (2002). The elicitor cryptogein blocks glucose transport in tobacco cells. *Plant Physiol.*, **130** : 2177-2187.

Bowler, C., Neuhaus, G., Yamagata, H. and Chua, N.H. (1994). Cyclic GMP and calcium mediate phytochrome phototransduction. *Cell*, **77**: 73–81.

Bowler, C. and Chua, N-H. (1994). Emerging themes of plant signal transduction. *Plant Cell*, **6** : 1529-1541.

Broillet, M.C. (2000). A single intracellular cysteine residue in responsible for the activation of the olfactory cyclic nucleotide-gated channel by NO. *J. Biol. Chem.*, **275** : 15135-15141.

Butt, Y.K.C., Lum, J.H.K. and Lo, S.C.L. (2003). Proteomic identification of plant proteins probed by mammalian nitric oxide synthase antibodies. *Planta*, **216** : 762-771.

Campbell, D.L., Stamler, J.S. and Strauss, H.C. (1996). Redox modulation of L-type calcium channels in ferret ventricular myocytes. Dual mechanisms regulation by nitric oxide and S-nitrosothiols. *J. Gen. Physiol.*, **108** : 277-293.

Caro, A. and Puntarlo, S. (1999). Nitric oxide generation by soybean embryonic axes. Possible effect on mitochondrial function. *Free Rad. Res.*, **31** : 205-212.

Chandok, M.R., Ytterberg, A.J., van Wijk, K.J. and Klessig, D.F. (2003). The pathogen-inducible nitric oxide synthase (iNOS) in plants is a variant of the P protein of the glycine decarboxylase complex. Cell, *in press.*

Chen, J., Dagget, H., De Waard, M., Heinemann, S.H. and Hoshi, T. (2002). Nitric oxide augments voltage-gated P/Q-type Ca^{2+} channels constituting a putative positive feedback loop. *Free Rad. Biol. Med.*, **32** : 638-649.

Choi, Y.B., Tenneti, L., Le, D.A., Ortiz, J., Bai, G., Chen, H.S. and Lipton, S.A. (2000). Molecular basis of NMDA receptor-coupled ion channel modulation by S-nitrosylation. *Nat. Neurosci.*, **3** : 15-21.

Clarke, A., Desikan, R., Hurst, R.D., Hancock, J.T. and Neill, S.J. (2000). NO way back: nitric oxide and programmed cell death in *Arabidopsis thaliana* suspension cultures. *Plant J.*, **24** : 667-677.

Clark, D., Durner, J., Navarre, D.A. and Klessig, D.F. (2000). Nitric oxide inhibition of tobacco catalase and ascorbate peroxidase. *Mol. Plant Microbe Interact.*, **13** : 1380-1384.

Clough, S.J., Fengler, K.A., Yu, I.C., Lippok, B., Smith, R.K. and Bent, A.F. (2000). The *Arabidopsis* dnd1 "defense, no death" gene encodes a mutated cyclic nucleotide-gated ion channel. *Proc. Natl. Acad. Sci. USA*, **97** : 9323-9328.

Corpas, F.J., Barroso, J.B. and del Rio, L.A. (2001). Peroxisomes as a source of reactive oxygen species and nitric oxide signal molecules in plant cells. *Trends Plant Sci.*, **6** : 145-150.

Crawford, N.M. (1995). Nitrate: nutrient and signal for plant growth. *Plant Cell*, **7** : 859-868.

Dangl, J.L. and Jones, J.D. (2001). Plant pathogens and integrated defence responses to infection. *Nature*, **411** : 826-833.

Delledonne, M., Xia, Y., Dixon, R.A. and Lamb, C. (1998). Nitric oxide functions as signal in plant disease. *Nature*, **394** : 585-588.

Desikan, R., Griffiths, R., Hancock, J. and Neill, S. (2002). A new role for an old enzyme: Nitrate reductase-mediated nitric oxide generation is required for abscisic acid-induced stomatal closure in *Arabidopsis thaliana. Proc. Natl. Aca. Sci. USA*, **99** : 16314-16318.

Di Mascio, P., Briviba, K., Bechara, E.J.H., Medeiros, M.H.G. and Sies, H. (1996). Reaction of peroxynitrite and hydrogen peroxide to produce singlet molecular oxygen (^{1}?g). *Meth. Enzymol.*, **269** : 395-400.

Dixon, R.A. (2001). Natural products and plant disease resistance. *Nature*, **411** : 843-847.

Durner, J. and Klessig, D.F. (1999). Nitric oxide as a signal in plants. *Curr. Opin. Plant Biol.*, **2** : 369-374.

Durner, J., Wendehenne, D. and Klessig, D.F. (1998). Defense genes induction in tobacco by Nitric Oxide, Cyclic GMP and cyclic ADP ribose. *Proc. Natl. Acad. Sci. USA*, **95** : 10328-10333.

Elstner, E.F. (1991). Mechanisms of oxygen activation in different compartments of plant cells. In *Active Oxygen/Oxidative Stress and Plant Metabolism* (Eds Pell, E.J., Steffen, K.L.) American Society of Plant Physiologists, Rockville pp 13-25.

Fan, W. and Dong, X. (2002). *In vivo* interaction between NPR1 and transcription factor TGA2 leads to salicylic acid-mediated gene activation in *Arabidopsis*. *Plant Cell*, **14** : 1377-1389.

Foissner, I., Wendehenne, D., Langebartels, C. and Durner, J. (2000). In vivo imaging of an elicitor-induced nitric oxide burst in tobacco. *Plant J.*, **23** : 817-824.

Foyer, C.H., Lelandais, M. and Kunert, K. H. (1994). Photooxidative stress in plants. *Physiol. Plant.*, **92** : 696-717.

Friedrich, L., Vernooij, B., Gaffney, T., Morse, A. and Ryals, J. (1995). Characterization of tobacco plants expressing a bacterial salicylate hydroxylase gene. *Plant Mol. Biol.*, **29** : 959-968.

Fritig, B., Heitz, T. and Legrand, M. (1998). Antimicrobial proteins in induced plant defense. *Curr. Opin. Immuno.*, **10** : 16-22.

Frye, C.A., Tang, D. and Innes, R.W. (2000). Negative regulation of defense responses in plants by a conserved MAPKK kinase. *Proc. Natl. Acad. Sci. USA*, **98** : 373-378.

Garcês, H., Durzan, D. and Pedroso, M.C. (2001). Mechanical stress elicits nitric oxide formation and DNA fragmentation in *Arabidopsis thaliana*. *Ann. Bot.*, **87** : 567-574.

García-Mata, C. and Lamattina, L. (2001) Nitric oxide induces stomatal closure and enhances the adaptive plant responses against drought stress. *Plant Physiol.*, **126** : 1196-1204.

García-Mata, C. and Lamattina, L. (2002). Nitric oxide and abscisic acid cross talk in guard cells. *Plant Physiol.*, **128** : 790-792.

Gómez-Gómez, L. and Boller, T. (2002). Flagellin perception: a paradigm for innate immunity. *Trends Plant Sci.*, **7** : 251-256.

Gould, K., Lamotte, O., Klinguer, A., Pugin, A. and Wendehenne, D. (2003). Nitric oxide production by tobacco leaves: a general stress response? *Plant Cell Environ.*, **26** : 1851-1862.

Hanafy, K.A., Krumenacker, J.S. and Murad, F. (2001). NO, nitrotyrosine, and cyclic GMP in signal transduction. *Med. Sci. Monit.*, **7** : 801-819.

Hill, A.C. and Bennett, J.H. (1970). Inhibition of apparent photosynthesis by nitrogen oxides. *Atmos. Environ.*, **4** : 341-348.

Hoeberichts, F.A. and Woltering, E.J. (2002). Multiple mediators of plant programmed cell death: interplay of conserved cell death mechanisms and plant-specific regulators. *BioEssays*, **25** : 47-57.

Hooper, D.C., Spitsin, S., Kean, R.B., Champion, J.M., Dickson, G.M., Chaudhry, I. and Koprowski, H. (1998). Uric acid, a natural scavenger of peroxynitrite, in experimental allergic encephalomyelitis and multiple sclerosis. *Proc. Natl. Acad. Sci. USA*, **95** : 675-680.

Huang, J-S. and Knopp, J.A. (1998). Involvement of nitric oxide in *Ralstonia solanacearum*-induced hypersensitive reaction in tobacco. In: *bacterial Wilt disease: Molecular and Ecological Aspects* (Eds Prior, P., Elphinstone, J., Allen, C.) INRA and Springer Editions, Berlin pp 218-224.

Huang, X., von Rad, U. and Durner, J. (2002a). Nitric oxide induces transcriptional activation of the nitric oxide-tolerant alternative oxidase in *Arabidopsis* suspension cells. *Planta*, **215** : 914-923.

Huang, X., Kiefer, E., von Rad, U., Ernst, D., Foissner, I. and Durner, J. (2002b). Nitric oxide burst and nitric oxide-dependent gene induction in plants. *Plant Physiol. Biochem.*, **40** : 625-631.

Jaffrey, S.R., Erdjument-Bromage, H., Ferris, C.D., Tempst, P. and Snyder, S.H. (2001). Protein S-nitrosylation: a physiological signal for neuronal nitric oxide. *Nat. Cell. Biol.*, **3** : 193-197.

Kaiser, W.M. and Huber, S. (2001). Post-translational regulation of nitrate reductase: mechanism, physiological relevance and environmental triggers. *J. Exp. Bot.*, **52** : 1981-1989.

Kim, H.K., Cheon, B.S., Kim, Y.H., Kim, S.Y. and Kim, H.P. (1999). Effects of naturally occurring flavonoids on nitric oxide production in the macrophage cell line RAW 264.7 and their structure-activity relationships. *Biochem. Pharmacol.*, **58** : 759-765.

Klepper, L.A. (1975). Inhibition of nitrite reduction by photosynthetic inhibitors. *Weed Sci.*, **23** : 188-190.

Klepper, L.A. (1976). Nitrite accumulation within herbicide-treated leaves. *Weed Sci.*, **24** : 533-535.

Klepper, L.A. (1979). Nitric oxide (NO) and nitrogen dioxide (NO_2) emissions from herbicide-treated soybean plants. *Atmos. Environ.*, **13** : 537-542.

Klessig, D.F., Durner, J., Noad, R., Navarre, D.A., Wendehenne, D., Kumar, D., Zhou, J.M., Shah, J., Zhang, S., Kachroo, P., Trifa, Y., Pontier, D., Lam, E. and Silva, H. (2000). Nitric oxide and salicylic acid signaling in plant defense. *Proc. Natl. Acad. Sci. USA*, **97 :** 8849-8855.

Knight, M.R., Campbell, A.K., Smith, S.M. and Trewavas, A.J. (1991). Transgenic plant aequorin reports the effects of touch and cold-shock and elicitors on cytoplasmic calcium. *Nature*, **352** : 524-526.

Koppenol, W.H. (1998). The basic chemistry of nitrogen monoxide and peroxynitrite. *Free Rad. Biol. Med.*, **25** : 385-391.

Kumar, D. and Klessig, D.F. (2000). Differential induction of tobacco MAP kinases by the defense signals nitric oxide, salicylic acid, ethylene, and jasmonic acid. *Mol. Plant Microb. Interact.*, **13** : 347-351.

Lam, E., Kato, N. and Lawton, M. (2001). Programmed cell death, mitochondria and the plant hypersensitive response. *Nature*, **411** : 848-853.

Lamotte, O., Gould, K., Lecourieux, D., Sequeira-Le Grand, A., Lebrun-Garcia, A., Durner, J., Pugin, A. and Wendehenne, D. (2003). Analysis of nitric oxide signalling functions in tobacco cells challenged by the elicitor cryptogein. *Plant Physiol.*, submitted.

Lander, H.M., Jacovina, A.T., Davis, R.J. and Tauras, J.M. (1996). Differential activation of mitogen-activated protein kinases by nitric oxide-related species. *J. Biol. Chem.*, **271** : 19705-19709.

Lebrun-Garcia, A., Ouaked, F., Chiltz, A. and Pugin, A. (1998). Activation of MAPK homologues by elicitors in tobacco cells. *Plant J.*, **15** : 773-781.

Lebrun-Garcia, A., Bourque, S., Binet, M-N., Ouaked, F., Wendehenne, D., Chiltz, A., Shäffner, A. and Pugin, A. (1999). Involvement of plasma membrane proteins in plant defense responses. *Biochimie* **81** : 1-6.

Lecourieux, D., Mazars, C., Pauly, N., Ranjeva, R. and Pugin, A. (2002). Analysis and effects of cytosolic free calcium increases in response to elicitors in *Nicotiana plumbaginifolia* cells. *Plant Cell*, **14** : 2627-2641.

Loake, G. (2001). Plant cell death: unmasking the gatekeepers. *Curr. Biol.*, **11** : 1028-1031.

Lucas, K.A., Pitari, G.M., Kazerounian, S., Ruiz-Stewart, I., Park, J., Schulz, S., Chepenik, K.P. and Waldman, S. (2000). Guanylyl cyclases and signalling by cyclic GMP. *Pharmacol. Rev.*, **52** : 375-413.

Ludidi, N. and Gehring, C. (2003). Identification of a novel protein with guanylyl cyclase activity in *Arabidopsis thaliana. J. Biol. Chem.*, **278** : 6490-6494.

Leng, Q., Mercier, R.W., Yao, W. and Berkowitz, G.A. (1999). Cloning and first functional characterization of a plant cyclic nucleotide-gated cation channel. *Plant Physiol.*, **121** : 753-761.

Leshem, Y.Y., Wills, R.B.H. and Ku, V.V-V. (1998). Evidence for the function of the free radical gas - nitric oxide ($NO^{\cdot}$) - as an endogenous maturation and senescence regulating factor in higher plants. *Plant Physiol. Biochem.*, **36** : 825-833.

Lum, H.K., Butt, Y.K.C. and Lo, S.C.L. (2001). Hydrogen peroxide induces a rapid production of nitric oxide in Mung Bean (*Phaseolus aureus*). *Nitric Oxide*, **6** : 205-213.

Mackerness, S.A.H., John, C.F., Jordan, B. and Thomas, B. (2001). Early signaling components in ultraviolet-B responses: distinct roles for different reactive oxygen species and nitric oxide. *FEBS Lett.*, **489** : 237-242.

Maleck, K., Levine, A., Eulgem, T., Morgan, A., Schmid, J., Lawton, K.A., Dangl, J.L. and Dietrich, RA. (2000). The transcriptome of *Arabidopsis thaliana* during systemic acquired resistance. *Nat Genet.*, **26** : 403-410.

Mallick, N., Mohn, F.H., Rai, L. and Soeder, C.J. (2000). Impact of physiological stresses on nitric oxide formation by green alga, *Scenedesmus obliquus. J. Microbiol. Biotechnol.*, **10** : 300-306.

Mano, J. (2002). Early events in environmental stresses in plants. Induction mechanisms of oxidative stress. In: *Oxidative Stress in Plants* (Eds. Inzé, D. and Van Montagu, M.) Taylor & Francis, London pp 217-246.

MAPK group. (2002). Mitogen-activated protein kinase cascades in plants: a new nomenclature. *Trends Plant Sci.*, **7** : 301-308.

Millar, A.H. and Day, D.A. (1996). Nitric oxide inhibits the cytochrome oxidase but not the alternative oxidase of plant mitochondria. *FEBS Lett.*, **398** : 155-158.

Millar A.H. and Day D.A. (1997). Alternative solutions to radical problems. *Trends Plant Sci.*, **2** : 289-290.

Mithöfer, A. and Mazars, C. (2002). Aequorin-based measurements of intracellular Ca^{2+}-signatures in plant cells. *Biol. Proced. Online*, **4** : 105-118.

Modolo, L.V., Cunha, F.Q., Brage, M.R. and Salgado, I. (2002). Nitric oxide synthase-mediated phytoalexin accumulation in soybean cotyledons in response to the *Diaporthe phaseolorum* f.sp. *meridionalis* elicitor. *Plant Physiol.*, **130** : 1288-1297.

Morot-Gaudry-Talarmain, Y., Rockel, P., Moureaux, T., Quilleré, I., Leydecker, M.T., Kaiser, W.M. and Morot-Gaudry, J.F. (2002). Nitrite accumulation and NO emission in relation to cellular signalling in NiR antisense tobacco. *Planta*, **215** : 708-15.

Nathan, C. and Xie, Q.W. (1994). Nitric oxide synthases: roles, tolls, and controls. *Cell*, **78** : 915-918.

Navarre, D.A., Wendehenne, D., Durner, J., Noad, R. and Klessig, D.F. (2000). Nitric oxide modulates the activity of tobacco aconitase. *Plant Physiol.*, **122** : 573-582.

Neill, S.J., Desikan, R., Clarke, A. and Hancock, J.T. (2002). Nitric oxide is a novel component of abscisic acid signalling in stomatal guard cells. *Plant Physiol.*, **128** : 13-16.

Nelson, R.S., Ryan, S.A. and Harper, J.E. (1983). Soybean mutants lacking constitutive nitrate reductase activity. I. Selection and initial plant characterization. *Plant Physiol.*, **72** : 503-509.

Neuhaus, G., Bowler, C., Hiratsuka, K., Yamagata, H. and Chua, N.H. (1997). Phytochrome-regulated repression of gene expression requires calcium and cGMP. *EMBO J.*, **6** : 2554-2564.

Noritake, T., Kaxakita, K. and Doke, N. (1996). Nitric oxide induces phytoalexin accumulation in potato tuber tissues. *Plant Cell Physiol.*, **37** : 113-116.

Ohkuma, S., Katsura, M., Hibino, Y., Xu, J., Shirotani, K. and Kuriyama, K. (1998). Multiple actions of nitric oxide on voltage-dependent Ca^{2+} channels in mouse cerebral cortical neurons. *Mol. Brain. Res.*, **54** : 133-140.

Orozco-Cárdenas, M.L and Ryan, C.A. (2002). Nitric oxide negatively modulates wound signaling in tomato plants. *Plant Physiol.*, **130** : 487-493.

Pauly, N., Knight, M.R., Thuleau, P., van der Luit, A.H., Moreau, M., Trewavas, A.J., Ranjeva, R. and Mazars, C. (2002). Control of free calcium in plant cell nuclei. *Nature*, **405** : 754-755.

Pedroso, M.C., Magalhaes, J.R. and Durzan, D. (2000a). A nitric oxide burst precedes apoptosis in angiosperm and gymnosperm callus cells and foliar tissues. *J. Exp. Bot.*, **51** : 1027-1036.

Pedroso, M.C., Magalhaes, J.R. and Durzan, D. (2000b). Nitric oxide induces cell death in *Taxus* cells. *Plant Sci.*, **157** : 173-180.

Peunova, N. and Enikolopov, G. (1993). Amplification of calcium-induced gene transcription by nitric oxide in neuronal cells. *Nature*, **364** : 450-453.

Pfeiffer, S., Janistyn, B., Jessner, G., Pichorner, H. and Ebermann, R. (1994). Gaseous nitric oxide stimulates guanosine-3',5'-cyclic monophosphate (cGMP) formation in spruce needles. *Phytochem.*, **36** : 259-262.

Pieterse, C.M., van Wees, S.C., van Pelt, J.A., Knoester, M., Laan, R., Gerrits, H., Weisbeek P.J. and van Loon, L.C. (1998). A novel signaling pathway controlling induced systemic resistance in Arabidopsis. *Plant Cell*, **10** : 1571-1580.

Poteser, M., Romanin, C., Schreibmayer, W., Mayer, B. and Groschner, K. (2001). S-nitrosation controls gating and conductance of the a1 submit of class C L-type Ca^{2+} channels. *J. Biol. Chem.*, **18** : 14797-14803.

Pugin, A. and Guern, J. (1996). Mode of action of elicitors: involvement of plasma membrane functions. *C. R. Acad. Sci. Paris*, **319** : 1055-1061.

Pugin, A., Frachisse, J-M., Tavernier, E., Bligny, R., Gout, E., Douce, R. and Guern, J. (1997). Early events induced by the elicitor cryptogein in tobacco cells: involvement of a plasma membrane NADPH oxidase and activation of glycolysis and pentose phosphate pathway. *Plant Cell,* **9** : 2077-2091.

Reyes-Harde, M., Potter, B.V.L., Galione, A. and Stanton, P.K. (1999). Induction of hippocampal LTD requires nitric oxide-stimulated PKG activity and Ca^{2+} release from cyclic ADP-ribose-sensitive stores. *J. Neurophysiol.*, **82** : 1569-1576.

Ribeiro, E.A., Cunha, F.Q., Tamashiro, W.M.S.C. and Martins, I.S. (1999). Growth phase-dependent subcellular localization of nitric oxide synthase in maize cells. *FEBS Lett.*, **445** : 283-286.

Ricci, P. (1997). Induction of the hypersensitive response and systemic acquired resistance by fungal proteins: the case of elicitins. In: *Plant-Microbe Interactions, Vol. 3* (Eds. Stacey, G. and Keen, N.T.) Chapman and Hall, New York pp 53-75.

Rubbo, H. and Freeman, B. (1996). Nitric oxide regulation of lipid oxidation reactions: formation and analysis of nitrogen-containing oxidized lipid derivatives. *Meth. Enzymol.*, **269** : 385-394.

Ryals, J.A., Neuenschwander, U.H., Willits, M.G., Molina, A., Steiner, H-Y. and Hunt, M.D. (1996). Systemic acquired resistance. *Plant Cell*, **8** : 1809-1819.

Sakihama, Y., Nakamura, S. and Yamasaki, H. (2002). Nitric oxide production mediated by nitrate reductase in the green alga *Chlamydomonas reinhardtii*: an alternative NO production pathway in photosynthetic organisms. *Plant Cell Physiol.*, **43** : 290-297.

Scheel, D. (1998). Resistance response physiology and signal transduction. *Curr. Opin. Plant Biol.*, **1** : 305-310.

Schenk, P.M., Kazan, K., Wilson, I., Anderson, J.P., Richmond, T., Somerville, S.C. and Manners, J.M. (2000). Coordinated plant defense responses in *Arabidopsis* revealed by microarray analysis. *Proc. Natl. Acad. Sci. U SA*, **97** : 11655-11660.

Schroeter, H., Boyd, C., Spencer, J.P.E., Williams, R.J., Cadenas, E. and Rice-Evans, C. (2002). MAPK signalling in neurodegeneration: influences of flavonoids and of nitric oxide. *Neur. Aging.*, **23** : 861-880.

Sentenac, H., Bonneaud, N., Minet, M., Lacroute, F., Salmon, J.M., Gaymard, F. and Grignon, C. (1992). Cloning and expression in yeast of a plant potassium ion transport system. *Science*, **256** : 663-665.

Shingles, R., Roh, M.H. and McCarty, R.E. (1996). Nitrate transport in chloroplast inner envelope vesicles. *Plant Physiol.*, **112** : 1375-1381.

Simon-Plas, F., Elmayan, T. and Blein, J-P. (2002). The plasma membrane oxidase NtrbohD is responsible for AOS production in elicited tobacco cells. *Plant J.*, **31** : 137-147.

Song, F. and Goodman, R.M. (2001). Activity of nitric oxide is dependent on, but is partially required for function of, salicylic acid in the signalling pathway in tobacco systemic acquired resistance. *Mol. Plant Microb. Interact.*, **14** : 1458-1462.

Squadrito, G.L. and Pryor, W.A. (1998). Oxidative chemistry of nitric oxide: the roles of superoxide, peroxynitrite, and carbon dioxide. *Free Rad. Biol. Med.*, **25** : 392-403.

Stamler, J.S., Singel, D.J. and Loscalzo, J. (1992). Biochemistry of nitric oxide and its redox-activated forms. *Science*, **258** : 1898-1902.

Stamler, J.S., Lamas, S. and Fang, F.C. (2001). Nitrosylation: the prototypic redox-based signalling mechanism. *Cell*, **106** : 675-683.

Stutte, C.A. and Weiland, R.T. (1978). Gaseous nitrogen loss and transpiration of several crop and weed species. *Crop Sci.*, **18** : 887-889.

Sun, J., Xu, L., Eu, J.P., Stamler, J.S. and Meissner, G. (2001). Classes of thiols that influence the activity of the skeletal muscle calcium release channel. *J. Biol. Chem.*, **276** : 15625-15630.

The *Arabidopsis* Genome Initiative. (2000). Analysis of the genome sequence of the flowering plant *Arabidopsis thaliana*. *Nature*, **408** : 796-815.

Tritsaris, K., Looms, D.K., Nauntoffe, B. and Dissing, S. (2000) Nitric oxide synthesis causes inositol phosphate production and Ca^{2+} release in rat parotid acinar cells. *Pflügers Arch.*, **440** : 223-228.

Volotovski, I.D., Sokolovsky, S.G., Molchan, O.V. and Knight, M.R. (1998). Second messengers mediate increases in cytosolic calcium in tobacco protoplasts. *Plant Physiol.*, **117** : 1023-1030.

Wang, Y-G., Wagner, M.B., Joyner, R.W. and Kumar, R. (2000). cGMP-dependent protein kinase mediates stimulation of L-type calcium current by cGMP in rabbit atrial cells. *Cardio. Res.*, **48** : 310-322.

Weiland, R.T. and Stutte, C.A. (1979). Pyro-chemiluminescent differentiation of oxidised and reduced N forms evolved from plant foliage. *Crop Sci.*, **19** : 545-547.

Wellburn, A.R. (1990). Why are atmospheric oxides of nitrogen usually phytotoxic and not alternative fertilizers? *New Phytol.*, **115** : 395-429.

Wendehenne, D., Pugin, A., Klessig, D.F. and Durner, J. (2001). Nitric oxide: comparative synthesis and signaling in animal and plant cells. *Trends Plant Sci.*, **6** : 177-183.

Wendehenne, D., Dussably, L., Jeannin, J.F. and Pugin, A. (2002). Nitric oxide: chemistry and bioactivity in animal and plant cells. In: *Studies in Natural Products Chemistry, Vol 26* (Ed Atta-ur-Rahman) Elsevier Science, Amsterdam pp 909-963.

Wendehenne, D., Lamotte, O., Frachisse, J-M., Barbier-Brygoo, H. and Pugin, A. (2002). Nitrate efflux is an essential component of the cryptogein signaling pathway leading to defense responses and hypersensitive cell death in tobacco. *Plant Cell*, **14** : 1937-1951.

Wendehenne, D., Lamotte, O. and Pugin, A. (2003). Plant iNOS : Conquest of the Holy Grail. *Trends Plant Sci.*, **8** : 465-468.

Wildt, J., Kley, D., Rockel, A., Rockel, P. and Segschneider, H.J. (1997). Emission of NO from several higher plant species. *J. Geophys. Res.*, **102** : 5919-5927.

Willmott, N., Sethi, J.K., Walseth, T.F., Lee, H., White, A.M. and Galione, A. (1996). Nitric oxide-induced mobilization of intracellular calcium via the cyclic ADP-ribose signalling pathway. *J. Biol. Chem.*, **271** : 3699-3705.

Willmott, N.J., Wong, K. and Strong, A.J. (2000). Intercellular Ca^{2+} waves in rat hippocampal slice and dissociated glial-neuron cultures mediated by nitric oxide. *FEBS Lett.*, **487** : 239-247.

Wink, D.A. and Mitchell, J.B. (1998). Chemical biology of nitric oxide: insights into regulatory, cytotoxic, and cytoprotective mechanisms of nitric oxide. *Free Rad. Biol. Med.*, **25** : 434-456.

Xu, L., Eu, J.P., Meissner, G. and Stamler, J.S. (1998). Activation of the cardiac calcium release channel (ryanodine receptor) by poly-S-nitrosylation. *Science*, **279** : 234-237.

Yamasaki, H. (2000). Nitrite-dependent nitric oxide production pathway: implications for involvement of active nitrogen species in photoinhibition in *vivo. Phil. Trans. R. Soc. Lond.*, **355** : 1477-1488.

Yamasaki, H. and Sakihama, Y. (2000). Simultaneous production of nitric oxide and peroxynitrite by plant nitrate reductase: in vitro evidence for the NR-dependent formation of active nitrogen species. *FEBS Lett.*, **468** : 89-92.

Yamasaki, H., Sakihama, Y. and Takahashi, S. (1999). An alternative pathway for nitric oxide production in plants: new features of an old enzyme. *Trends Plant Sci.*, **4** : 128-129.

Yang, K-Y., Liu, Y. and Zhang, S. (2001). Activation of a mitogen-activated protein kinase pathway is involved in disease resistance in tobacco. *Proc. Natl. Acad. Sci. USA*, **98** : 741-746.

Yoshimura, N., Seki, S. and De Groat, W.C. (2001). Nitric oxide modulates Ca^{2+} channels in dorsal root ganglion neurones innervating rat urinary bladder. *J. Neurophysiol.*, **86** : 304-311.

Zheng, J., Bird, I.M., Melsaether, A.N. and Magness, R.R. (1999). Activation of the mitogen-activated protein kinase cascade is necessary but not sufficient for basic fibroblast growth-factor- and epidermal growth factor-stimulated expression of the endothelial nitric oxide synthase in ovine fetoplacental artery endothelial cells. *Endocryn.*, **140** : 1399-1407.

Zottini, M., Formentin, E., Scattolin M., Carimi, F., Lo Schiavo, F. and Terzi, M. (2002). Nitric oxide affects plant mitochondrial functionality *in vivo*. *FEBS Lett.*, **515** : 75-78.

Chapter 3

NITRIC OXIDE : BRIDGING THE GAP BETWEEN ENVIRONMENTAL STIMULI AND ENDOGENOUS SIGNALS IN PLANTS

Carlos García-Mata•, Magdalena Graziano•, Gabriela C. Pagnussat• and Lorenzo Lamattina★

Instituto de Investigaciones Biológicas, Facultad de Ciencias Exactas y Naturales, Universidad Nacional de Mar del Plata, CC1245, 7600 Mar del Plata, Argentina
★Corresponding author : E-mail : lolama@mdp.edu.ar; Fax : (+54)-223-4753150
•Authors that have equally contributed to this work

Summary

Nitric oxide (NO) is emerging as a new chemical messenger in plants that functions in a broad spectrum of pathophysiological and developmental processes. An intriguing cross talk between environmental stimuli, hormone actions and NO functions seems to be the key for understanding how plant responses are synchronized for adaptation to a continuous changing environment. In this chapter, we have focused our attention on three unrelated plant physiological processes in which NO is unambiguously implicated: (i) adventitious root formation, (ii) drought stress responses and (iii) iron nutrition and homeostasis. The data we are presenting as well as the available data described in other chapters of this book strongly support the interaction of NO, plant hormones and regulatory proteins as major players for determining the success of specific plant responses to improve the tolerance to environmental cues.

Key words : ABA, Auxin, drought stress, hormone, iron deficiency, iron overload, nitric oxide, rooting, stomata closure.

In : Nitric Oxide Signaling in Higher Plants, 2004
(Eds Jose R. Magalhaes, Rana P. Singh and Leonidas P. Passos)
Studium Press, LLC, Houston, USA, pp 65-79

1. INTRODUCTION

Nitric oxide (NO) is a wide spread intracellular and intercellular signaling molecule involved in the regulation of a notable spectrum of diverse cellular functions. Nitric oxide is a free radical lipophilic diatomic gas under atmospheric conditions. Its small Stokes' radius and neutral charge allows rapid membrane diffusion (Stamler, 1994). However, molecular mechanisms for precise and controlled NO delivery from stored cellular compartments have nevertheless been reported in animals (Pawlosky *et al.*, 2001). Cellular responses to NO are variable, ranging from cell death to controlled physiological responses. Its specificity depends on many complex regulatory mechanisms given by the rate of NO production and diffusion, the cellular type or the environment surrounding the cell. Here, we present an update of recent reported and unpublished results in which NO plays a central role as a linkage between exogenous or endogenous stimuli and plant cell responses involved in adaptive physiological changes.

2. NITRIC OXIDE : AN EMERGING SIGNAL MOLECULE IN PLANTS

Although research on NO biology in plants is not as advanced as in animals, during the past years compelling evidence on the role of NO in a variety of plant physiological processes such as growth, organogenesis, and response to biotic and abiotic stresses came from experiments carried out in different systems (Gouvea *et al.*, 1997; Huang and Knop, 1997; Laxalt *et al.*, 1997; Durner *et al.*, 1998; Leshem *et al.*, 1998; Beligni and Lamattina, 2000; Beligni and Lamattina, 2001a; Beligni and Lamattina, 2001b; Pagnussat *et al.*, 2002).

The signaling effect of NO is determined by its chemical reactivity. In animal cells, NO signaling typically involves both cGMP-dependent and independent pathways (McDonald and Murad, 1995). In the cGMP-dependent pathway, NO induces the activation of soluble guanylate cyclase (GC) by binding to the heme group of the enzyme, which results in a transient increase of the cellular concentration of cGMP (Stamler, 1994). The binding of NO to the ferrous heme group is reversible and GC turn off immediately after the NO gradient has been dissipated (Beckman, 1996). In plants, the presence of cGMP was demonstrated by mass spectrometry and radio-immune assays

(Brown and Newton, 1992). Probably, the first evidence for the occurrence of cGMP in plants is related to a change in the concentration of this cyclic mono-phosphate during cell division and enlargement in tobacco (Lundeen *et al.*, 1973). A cyclic nucleotide phosphodiesterase (PDE) activity has been also reported in plants (Brown *et al.*, 1975; Reggiani, 1997), although neither homologue sequence to GC nor the PDE have been cloned in plants yet. A range of stimuli, such as giberellic acid in barley aleurone (Penson *et al.*, 1996), light stimulation of bean cells (Brown and Newton, 1992) or NO treatment in spruce needles (Pfeiffer *et al.*, 1994) cause transient increases in cGMP levels. NO treatment was also reported to induce a transient increase of cGMP levels in tobacco (Durner *et al.*, 1998). On the other hand, cGMP was reported to activate anthocyanins production and in combination with calcium to induce the development of mature chloroplasts (Bowler *et al.*, 1994). Furthermore, specific GC inhibitors suppress the NO-induced expression of phenylalanine ammonia-lyase (PAL) genes in tobacco (Durner *et al.*, 1998). In addition, the specific inhibition of GC activity blocked the NO-induced cell death in *Arabidopsis* cells, and this inhibition is reversed by the cell-permeable cGMP analogue, 8-Br-cGMP (Clarke *et al.*, 2000).

The cGMP-independent pathways of NO signaling usually involve protein nitrosation. NO signaling targets possess transition metals and/or thiol and tyrosine residues at regulatory or active sites (Stamler, 1994). Thus, NO attack leads to either a reduction or an increase in protein activity. In animal cells, NO induces different mitogen-activated protein kinases (MAPK) cascades by both the activation of an upstream factor such as the small GTP-binding protein $p21^{ras}$ or by directly nitrosation of MAPK causing either the activation or the inhibition of the enzymes (Lander *et al.*, 1996). The MAPK signal transduction cascades are important mediators in signal transmission, connecting the perception of external stimuli to cellular responses. MAPKs are involved in signaling various biotic and abiotic stresses, and have been implicated in the regulation of cell cycle and developmental processes. Recent studies have revealed that NO is involved in the activation of SIPK, a tobacco MAPK, in the NO signaling pathway for plant defense responses (Kumar and Klessig, 2000).

All together, these data suggest that the components of animal NO signaling are also functional in plants and reveal the high responses complexity of the cellular responses to NO, which may be mediated by phosphorylation-, nitrosylation-, or cGMP-controlled mechanisms. Figure 1 illustrates the involvement of NO in growth, developmental and adaptive processes of plants, probably as a mediator between environmental or endogenous stimuli and cell messengers.

3. NITRIC OXIDE AND HORMONE SIGNAL NETWORKS

Plant responses to hormones involve complex mechanisms of signal cell communication. The elucidation of the molecular basis of these signaling systems has been a long-standing objective of the plant hormone research. The known steps along hormones´ signal transduction pathways usually involve protein-protein interactions, activation of gated channels, regulation of catalytic activities and a combination of these. Since NO can regulate processes related to plant growth and development, it has been postulated as a non-traditional regulator of plant growth (Beligni and Lamattina, 2001a). Therefore, the cross talk between NO and classical plant hormones and the common cellular messengers involved in their signaling pathways is becoming a field of intense research.

4. NITRIC OXIDE IS INVOLVED IN THE AUXIN SIGNALING DURING ADVENTITIOUS ROOT DEVELOPMENT

The plant hormone auxin regulates most aspects of plant growth and development, including cell division, elongation and differentiation. Although genetic approaches have revealed some discrete aspects of auxin transport, signaling and response, the understanding of the molecular mechanisms of auxin action in the establishment of new root meristems are not completely elucidated yet (Doerner, 2000; Berleth and Sachs, 2001).

Some years ago, Gouvêa *et al.* (1997) demonstrated the capability of root cells to respond to NO, which induced the elongation of maize root segments in a dose-dependent manner. Since IAA and NO elicited the same plant response, the investigators proposed that IAA and NO might share common steps in the signal transduction pathways. Recent results in cucumber showed that NO mediates auxin response during adventitious rooting process (Pagnussat *et al.*, 2002). A transient increase in endogenous NO level after IAA treatment was reported in the basal region of the hypocotyl, where the new

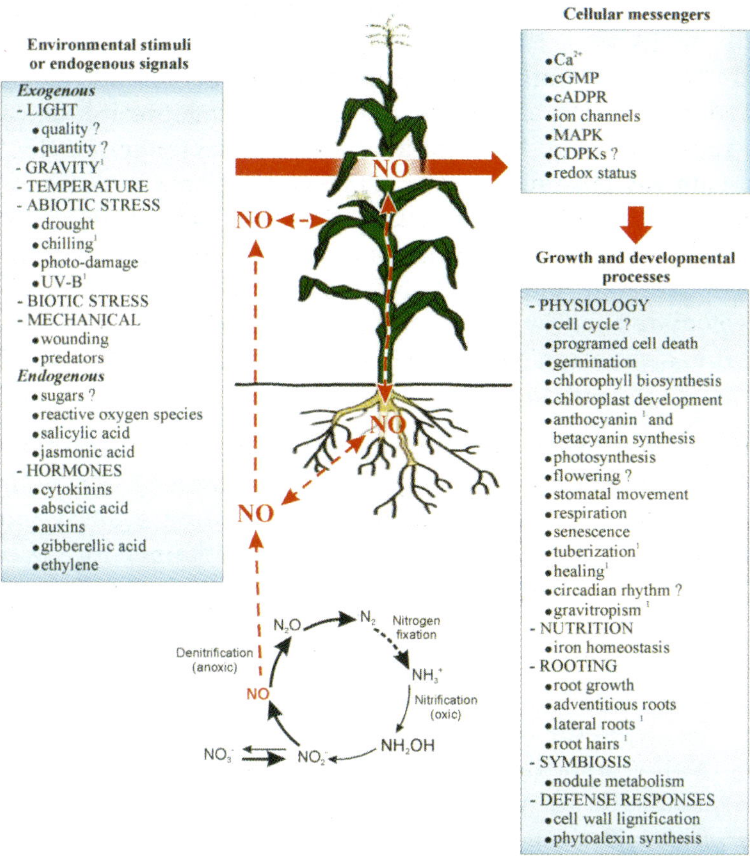

Figure 1 : NO regulates growth, developmental and adaptive processes in plants. Plants produce NO by the enzyme nitrate reductase (NR) and a nitric oxide synthase activity (for a review see Lamattina *et al.*, 2003) in both aerial organs and roots. NO is also produced during nitrification process by microorganisms in the rhizosphere (Colliver and Stephenson, 2000). Exchange of NO between the atmosphere and aerial plant organs as well as between rhizosphere and roots occurs. Plants are sensing external and internal signals and integrate them into a signal-transduction system operated by hormones and cellular messengers, in which NO is involved. Then, plants act through gene expression or metabolic changes in order to assess a balanced response during growth and developmental processes. The circuitry of the complex network driving the chemical signals possesses control and safeguard points and is thought to be permanently working to successfully adapt plants to a changing environment. Where no references are given, examples are reviewed in Refs. Beligni and Lamattina (2001b) and Wendehenne *et al* (2001) or within the text. (1) Lamattina *et al.* (unpublished data); (?) No direct experimental evidence on the NO involvement are available yet. With permission, from the Annual Review of Plant Biology, Volume 54 © 2003 by Annual Reviews www.annualreviews.org.

root meristems develop. Given the presence of endogenous auxins, a serial linkage IAA → NO → rooting, cannot be distinguished from a situation in which both IAA and NO are needed to act in parallel on a third intermediate. However, an auxin depletion approach strongly supported that NO acts as a second messenger in the IAA-mediated pathway (Pagnussat *et al.*, 2003a). Pharmacological evidence suggested that a localized NO bulk might stimulate the GC-catalyzed synthesis of cGMP. Using inhibitors of both GC and PDE, it was demonstrated that in cucumber, NO acts as a second messenger in the IAA-mediated pathway that induces adventitious root formation through the activation of the GC-catalyzed synthesis of cGMP (Pagnussat *et al.*, 2003a).

On the other hand, a cGMP-independent pathway is also present in stimulating adventitious rooting. Evidence that a MAPK signaling cascade is activated during the adventitious rooting process induced by IAA in a NO-mediated pathway was also obtained in cucumber explants (Pagnussat *et al.*, 2003b). The activation of a MAPK activity in response to auxins had been described before by Mizoguchi *et al.* (1994) and Mockaitis and Howell (2000). In addition, studies in tobacco have revealed that NO is involved in the activation of SIPK, a tobacco MAPK, in the NO signaling pathway as part of plant defense responses (Kumar and Klessig, 2000). The MAPK signal transduction in plants seems to be really complex. Different stimuli activate different but overlapping MAPK pathways. A useful tool in the study of MAPK cascades is the use of specific inhibitors of precise steps in the pathway. PD098059 is a MAPK kinase (MAPKK) inhibitor that acts by binding to its inactive de-phosphorylated form (Alessi *et al.*, 1995). As demonstrated in *Arabidopsis*, a PD098059-sensitive MAPKK activity appears to be necessary for the expression of some auxin-induced genes (Mockaitis and Howell, 2000). However, a PD098059-insensitive MAPKK pathway that appears to be regulating an auxin-induced MAPK cascade was also described. Moreover, the NPK-1 class of MAPKKKs, which is specifically induced by H_2O_2 and activates stress-induced MAPKs, not only activates stress response genes, but also represses auxin-inducible promoters (Kovtun *et al.*, 1998). Given this scenario, the specificity of the signaling could be supplied by the cellular type or the environment surrounding the cell. All together, these data indicate that convergent signal pathways are probably orchestrating the formation of a new root system and expose the unexpected complexity of the cellular responses regulated by NO.

5. NITRIC OXIDE AND ABA CROSS-TALK DURING DROUGHT STRESS RESPONSES

Water stress is considered as one of the most serious environmental factors constraining crop production. Under hydric stress conditions, ABA is accumulated in leaves triggering different physiological responses (Giraudat *et al.*, 1994). Among plant physiological adaptations, the control of stomatal movement is of great importance for controlling the water loss through the transpiration stream while balancing the requirement of gas exchange for photosynthesis (Blatt and Grabov, 1997; Xiong and Zhu, 2001). As discussed by Bray (1997), resistance to water deficit occurs when a plant withstand the imposed stress, and may arise from either tolerance or a mechanism that permits avoidance of the situation. In that sense, it was recently reported that NO enhances plant tolerance to drought stress conditions by decreasing the rate of transpirational water loss through the induction of stomatal closure (García-Mata and Lamattina, 2001). This first report has become a milestone for the involvement of NO within the guard cell signaling network.

The transition from "pathway" to "network" in plant signaling has deviated the attention towards the description of cross-talks between different signaling components. Guard cell ABA-signaling is not the exception. New and diverse components that are filling the gaps in the guard cell signaling network have been appearing very often during the last few years. Such is the case of NO, which was very recently reported to be generated during ABA-induction of stomatal closure (García-Mata and Lamattina, 2002; Neill *et al.*, 2002). Even more, the presence of endogenous NO was demonstrated to be required for the ABA-dependent induction of stomatal closure (García-Mata and Lamattina, 2002; Neill *et al.*, 2002).

Once in the guard cell, ABA first induces a depolarization of the plasma membrane potential that leads to the generation of a driving force for K^+ outflow. To achieve this K^+ loss, ABA induce the following changes: (i) inactivates K^+ influx through inward rectifying K^+ channels (K^+_{in}), (ii) generates a K^+ efflux through the activation of outward-rectifying K^+ channels (K^+_{out}), and (iii) produces and anion efflux (mainly Cl^- and malate) via activation of anion channels.

Overall, the long-term efflux of both anions and K^+ from guard cells results in a net loss of guard cell turgor, leading to the stomatal closure response (Blatt, 2000a; MacRobbie, 1998). In addition, both cytosolic free Ca^{2+} concentrations ($[Ca^{2+}]_{cyt}$) and pH have been reported to participate as second messengers of this response (Schroeder *et al.*, 2001). ABA-induced elevation of guard cell $[Ca^{2+}]_{cyt}$ can be generated either by influx from the extracellular space and by release from internal stores (Sanders *et al.*, 1999). At least four different Ca^{2+} channel types have been identified at vacuolar membranes. Two of these channels are ligand gated, one by inositol 1,4,5-triphosphate (IP3) and the other by cyclic ADP-ribose (cADPR). The pharmacological properties of plant cADPR-gated channels resemble those of animal cells ryanodine receptors which, along with IP3 receptors, are responsible of the mobilization of ER-based Ca^{2+} pools during signaling (Sanders *et al.*, 1999). In plants, IP3 is rapidly produced within guard cells in response to ABA and interacts with IP3-sensitive Ca^{2+} channels to potentiate a rise in $[Ca^{2+}]_{cyt}$ (Blatt, 2000b) and to inhibit K^+_{in} channels at the same time (Schroeder *et al.*, 2001). In animal cells, the role of NO in the control of Ca^{2+} release from IP3-sensitive and ryanodine-sensitive stores has been clearly ascertained (for review see Clementi, 1998).

In animals, NO is recognized to be a regulator of both plasma membrane and endomembrane ion channels (Clementi and Meldolesi, 1997; Broillet, 2000). NO regulation of K^+ and Ca^{2+} channels activity can be achieved either by triggering the cGMP signaling cascade through activation of GC (Clementi, 1998; Fagni and Bockaert, 1996), or by cGMP-independent pathway that generally involves a direct interaction between NO (or one of its redox activated species) with the channel protein (Bolotina *et al.*, 1994; Shin *et al.*, 1997).

All these evidence point toward a putative role of ion channels as link bridging external stimuli, ABA, NO and stomatal closure in guard cell signaling network. Even though conclusive evidence connecting the serial linkage in this process is lacking, it has been demonstrated that NO-mediated stomatal closure is Ca^{2+} dependent in different plant species (Garcia-Mata and Lamattina, 2001) as occurs for ABA-induced response. In a recent work it has been suggested that NO may be acting in ABA-induced stomatal closure via the modulation of cyclic nucleotide-gated ion channels (Neill *et al.*,

2002), although, no direct measurement of NO effect on ion channel activity was reported in plants until now.

Two recent reports have recently introduced the enzyme nitrate reductase (NR) as a key player responsible of the NO production required for ABA-induced stomatal closure (Desikan *et al.*, 2002; Garcia-Mata and Lamattina, 2003). Further details about the involvement of NO and its regulatory role in ABA-induced stomatal closure, are extensively discussed by Prof. S. Neill in another chapter of this book.

6. ROLE OF NITRIC OXIDE IN IRON DEFICIENCY AND OVERLOAD

Iron is essential for fundamental cellular processes such as electron transfer in photosynthesis, respiration, nitrogen fixation and DNA synthesis (Welch, 1995). Iron deficiency leads to important physiological disorders in plant cells, especially within the photosynthetic apparatus. On the contrary, excessive iron accumulation causes severe damage to cellular components due to the formation of the highly reactive hydroxyl radical by the Fenton reaction (Briat *et al.*, 1995). Thus, the precise control of iron homeostasis is a basic prerequisite for cellular function.

Recently, some evidences that indicate a participation of NO in plant iron metabolism were shown (Graziano *et al.*, 2002; Murgia *et al.*, 2002). Under limited iron concentration in the nutrient solution, maize plants develop the iron-deficiency phenotype characterized by low chlorophyll content and interveinal chlorosis in leaves (Thoiron *et al.*, 1997). NO was able to restore the chlorophyll content and to avoid the interveinal chlorosis in maize plants growing in iron concentrations as low as 10 μM Fe-EDTA (Graziano *et al.*, 2002). This effect was achieved either when NO was applied as a NO-donor in aqueous solution or in gaseous form. While iron-deficient plants have disorganized chloroplasts and very few grana (Thoiron *et al.*, 1997), chloroplast ultrastructure in NO-treated plants resembles that of iron-sufficient plants (Graziano *et al.*, 2002). Moreover, the use of specific NO scavengers in iron sufficient plants resulted in interveinal chlorosis of leaves, suggesting that endogenous NO plays a physiological role in iron nutrition. Hence, NO appears to improve iron availability within the cell and/or the chloroplast, although the mechanisms involved are still unknown.

In other report, NO was shown to mediate the induction of the iron storage protein ferritin (Murgia *et al.*, 2002). Exogenous applied NO induces the accumulation of ferritin transcript and protein in *Arabidopsis* leaves (Murgia *et al.*, 2002). Moreover, endogenous NO is necessary for the iron-mediated induction of ferritin, since NO scavenging abolishes the response even under high iron levels. Ferritins play a central role in the regulation of iron homeostasis through the control of free iron levels in cells. Plant ferritins are regulated mainly at the transcriptional level. It is known that free iron induces ferritin gene transcription (Lescure *et al.*, 1991) probably acting through promoter derepression (Wei and Theil, 2000). There are two pathways proposed to explain the iron-induced effect on ferritin gene expression: (a) mediated by ROS (Lobreaux *et al.*, 1995; Gaymard *et al.*, 1996) and (b) mediated by ABA and drought stress (Lobreaux *et al.*, 1993; Forbis-Loisy *et al.*, 1995). Different ferritin genes in *Arabidopsis* (Petit *et al.*, 2001) and maize (Forbis-Loisy *et al.*, 1995) respond to one or the other pathway. It was also shown that activities of Ser/Thr phosphatases are involved in ferritin gene induction (Savino *et al.*, 1997). Since activity of Ser/Thr phosphatases is also required for the NO-induced ferritin accumulation (Murgia *et al.*, 2002), NO might be one of the regulatory molecules acting downstream of iron to induce ferritin gene expression.

As stated above, NO was shown to be involved in the ABA signal transduction pathway that directs stomatal closure during drought stress conditions (Garcia-Mata and Lamattina 2002; Neill *et al.*, 2002). Additionally, NO was also reported to increase as a response to hydrogen peroxide treatment (Lum *et al.*, 2002). Therefore, it would be interesting to check whether NO is required for ABA- or ROS-mediated ferritin gene induction.

Up to date, the available evidence for NO participation in plant iron metabolism indicates that the role of NO could be dual. Under low iron level NO increases the availability of metabolically active iron required for cellular processes. On the other hand, when iron is at a high level, NO mediates the induction of ferritin protein and therefore, contributes to the maintenance of low and constant levels of free and catalytic cellular iron. Overall, the involvement of NO in iron nutrition and homeostasis is a breakthrough for the state of the knowledge in this area and opens a wide field of research with fundamental and applied implications in plant biology.

7. CONCLUSIONS AND FUTURE PROSPECTS

Although our understanding of the signaling function of NO in plants remains incomplete and very limited, the available data strongly support the presence of NO and its biological functions in a wide spectrum of growth and developmental processes. Thus, a deep analysis of how NO affects the cellular metabolism in both normal growth conditions and stress situations will permit us to envisage the obtention of transgenic plants with a regulated production of NO. This endogenous NO might serve as a general synchronizing molecule useful for adaptive responses directed to improve the fitness of plant growth in a permanent changing environment (Lamattina *et al.*, 2001).

As future directions of research, it remains to be investigated among others: (a) the physiological significance of endogenous NO production, (b) the intermediate molecules leading the NO-mediated regulation of gene expression, (c) the target molecules of the NO-dependent post translational modifications and (d) the NO involvement in cellular redox homeostasis. Molecular, chemical and genetic approaches will surely contribute to raise those goals in this new and fascinating field of the plant biology.

AKNOWLEDGEMENTS

This work was supported by grants to L.L. from ANPCyT (Agencia Nacional de Promoción Científica y Tecnológica), Fundación Antorchas, CONICET (Consejo Nacional de Investigaciones Científicas y Técnicas), UNMdP (Universidad Nacional de Mar del Plata), Argentina.

REFERENCES

Alessi, D.R., Cuenda, A., Cohen, P., Dudley, D.T. and Saltiel, A.R. (1995). PD098059 is a specific inhibitor of the activation of mitogen-activated protein kinase kinase in vitro and in vivo. *J. Biol. Chem.,* **270** : 27489-27494.

Allen, G.J., Muir, S.R. and Sanders, D. (1995). Release of Ca^{2+} from individual plant vacuoles by both InsP3 and cyclic ADP-ribose. *Science,* **268** : 735-737.

Beckman, J.S. (1996). The physiological and pathological chemistry of nitric oxide. In: Lancaster, J. ed. Nitric oxide: principles and actions. New York: Academic Press; pp 1-83.

Beligni, M.V. and Lamattina, L. (2000). Nitric Oxide induces seed germination and de-etiolation, and inhibits hypocotyls elongation, three light-inducible responses in plants. *Planta,* **210** : 215-221.

Beligni, M.V. and Lamattina, L. (2001a) Nitric Oxide: a non traditional regulator of plant growth. *Trends Plant Sci.,* **6** : 508-509.

Beligni, M.V. and Lamattina, L. (2001b). Nitric Oxide in Plants: The History is Just Beginning. *Plant Cell Environ.,* **24** : 267-278.

Berleth, T. and Sachs, T. (2001). Plant morphogenesis: long-distance coordination and local patterning. *Curr. Opin. Plant Biol.,* **4** : 57-62.

Blatt, M. (2000a). Ca^{2+} signalling and control of guard-cell volume in stomatal movements. *Curr. Opin. Plant Biol.,* **3** : 196-204.

Blatt, M. (2000b). Cellular signaling an volume control in stomatal movements in plants. *Annu. Rev. Cell Dev. Biol.,* **16** : 221-241.

Blatt, M. and Grabov, A. (1997). Signalling gates in abscisic acid-mediated control of guard cell ion channels. *Physiol. Plant.,* **100** : 481-490.

Bolotina, V.M., Najibi, S., Palacino, J.J., Pagano, P.J. and Cohen, R.A. (1994). Nitric oxide directly activates calcium-dependent potassium channels in vascular smooth muscle. *Nature,* **368** : 850-853.

Bowler, C., Neuhaus, G., Yamagata, H. and Chua, N.H. (1994). Cyclic GMP and calcium mediate phytochrome phototransduction. *Cell,* **77** : 73-81.

Bray, E.A. (1997). Plant responses to water deficit. *Trends Plant Sci.,* **2** : 48-54.

Briat, J.F., Fobis-loisy, I., Grignon, N., Lobréaux, S., Pascal, N., Savino, G., Thoiron, S., v. Wirén N. and v. Wuytswinkel O. (1995). Cellular and molecular aspects of iron metabolism in plants. In: Biol. Cell, Band 84, Heft, S. 69-81.

Broillet, M.C. (2000). A Single intracellular Cysteine residue is responsible for the activation of the olfactory cyclic nucleotide-gated channel by NO. *J. Biol. Chem.,* **275** : 15135–15141.

Brown, E.G. and Newton, R.P. (1992). Analytical procedures for cyclic nucleotides and their associated enzymes in plant tissue. *Phytochem. Anal.,* **3** : 1-3.

Brown, E.G., Al-Najafi, T. and Newton, R.P. (1975). Partial purification of adenosine 3':5'-cyclic monophosphate phosphodiesterase from *Phaseolus vulgaris* l.: associated activator and inhibitors. *Biochem. Soc. Trans.,* **3** : 393-395.

Clarke, A., Desikan, R., Hurst, R.D., Hancock, J.T. and Neill, S.J. (2000). NO way back: nitric oxide and programmed cell death in *Arabidopsis thaliana* suspension cultures. *Plant J.,* **24** : 667-677.

Clementi, E. (1998). Role of nitric oxide and its intracellular signaling pathway in the control of Ca^{2+} homeostasis. *Biochem. Pharmacol.,* **55** : 713-718.

Clementi, E. and Meldolesi, J. (1997). The cross-talk between nitric oxide and Ca^{2+}: a story with a complex past and a promising future. *Trends Biochem. Sci.,* **18** : 266-269.

Colliver, B.B. and Stephenson, T. (2000). Production of nitrogen oxide and dinitrogen oxide by autotrophic nitrifiers. *Biotechnol. Adv.,* **18** : 219-32.

Desikan, R., Griffiths, R., Hancock, J. and Neill, S. (2002). A new role for an old enzyme: nitrate reductase-mediated nitric oxide generation is required for abscisic acid-induced stomatal closure in *Arabidopsis thaliana. Proc. Natl. Acad. Sci. USA,* **99** : 16314-16318.

Doerner, P. (2000). Plant stem cells: the only constant thing is change. *Curr. Biol.,* **10** : 201-203.

Durner, J., Wendehenne, D. and Klessig, D. (1998). Defense gene induction in tobacco by nitric oxide, cyclic GMP and cyclic ADP-ribose. *Proc. Natl. Acad. Sci. USA,* **95** : 10328-10333.

Fagni, L. and Bockaert, J. (1996). Effects of nitric oxide on glutamate-gated channels and other ionic channels. *J. Chem. Neuroanat.*, **10** : 231-240.

Forbis-Loisy, I., Loridon, K., Lobreaux, S., Lebrun, M. and Briat, J.F. (1995). Structure and differential expression of two maize ferritin genes in response to iron and absisic acid. *Eur. J. Biochem.*, **231** : 609-619.

García Mata, C. and Lamattina, L. (2001). Nitric oxide induces stomatal closure and enhances the adaptive plant responses against drought stress. *Plant Physiol.*, **126** : 1196-1204.

García-Mata, C and Lamattina, L. (2002). Nitric oxide and abscisic acid cross talk in guard cells. *Plant Physiol.*, **128** : 790-792.

Garcia-Mata, C. and Lamattina, L. (2003). Abscisic acid, nitric oxide and stomatal closure – is nitrate reductase one of the missing links? *Trends Plant. Sci.*, **8** : 20-26.

Gaymard, F., Boucherez, J. and Briat, J.F. (1996). Characterization of a ferritin mRNA from Arabidopsis thaliana accumulated in response to iron through an oxidative pathway independent of abscisic acid. *Biochem. J.*, **318** : 67-73.

Giraudat, J., Parcy, F., Bertauche, N., Gosti, F., Leung, J., *et al.* (1994). Current advances in abscisic acid action and signaling. *Plant. Mol. Biol.*, **26** : 1557-1577.

Gouvea, C.M., Souza, J.F., Magalhaes, A.C.N. and Martins, I.S. (1997). NO-releasing substances that induce growth elongation in maize root segments. *Plant Growth Reg.*, **21** : 183-187.

Graziano, M., Beligni, M.V. and Lamattina, L. (2002). Nitric oxide improves internal iron availability in plants. *Plant Physiol.*, **130** : 1852-1859.

Huang, J.S. and Knopp, J.A. (1997). Involvement of nitric oxide in *Ralstonia solanacearum* induced hypersensitive reaction in tobacco. In: Proceedings of the Second International Wilt Symposium (eds P.Prior, J. Elphinstone and C. Allen). INRA, Versailles, France

Kovtun, Y., Chiu, W.L., Zeng, W. and Sheen, J. (1998). Suppression of auxin signal transduction by a MAPK cascade in higher plants. *Nature*, **395** : 716-720.

Kumar, D. and Klessig, D.F. (2000). Differential induction of tobacco MAP kinases by the defense signals nitric oxide, salicylic acid, ethylene and jasmonic acid. *Mol. Plant-Microbe. Interact.*, **13** : 347-351.

Lamattina, L., Beligni, M.V., García-Mata, C., Laxalt, A.M. (2001). Method of enhancing the metabolic function and the growing conditions of plants and seeds. US Patent No. 6 242 384 B1.

Lamattina, L., Garcia-Mata, C., Graziano, M. and Pagnussat, G. (2003). Nitric oxide: the versatility of an extensive signal molecule. *Annu. Rev. Plant Biol.*, **54** : 109-136.

Lander, H.M., Jacovina. A.T., Davis, R.J. and Tauras, J.M. (1996). Differential activation of mitogen-activated protein kinases by nitric oxide-related species. *J. Biol. Chem.*, **271** : 19705-19709.

Laxalt, A.M., Beligni, M.V. and Lamattina, L. (1997). Nitric oxide preserves the level of chlorophyll in potato leaves infected by *Phytophthora infestans. Eur. J. Plant Pathol.*, **73** : 643-651.

Lescure, A.M., Proudhon, D., Pesey, H., Ragland, M., Theil, E.C. and Briat, J.F. (1991). Ferritin gene transcription is regulated by iron in soybean cell cultures. *Proc. Natl. Acad. Sci. USA*, **88** : 8222-8226.

Leshem, Y.Y., Wills, R.B.H. and Ku, V.V.V. (1998). Evidence for the function of the free radical gas -nitric oxide (NO)- as an endogenous maturation and senescence regulating factor in higher plants. *Plant Physiol. Biochem.*, **36** : 825-833.

Lobreaux, S., Hardy, T. and Briat, J.F. (1993). Abscisic acid is involved in the iron-induced synthesis of maize ferritin. *EMBO J.*, **12** : 651-657.

Lobreaux, S., Thoiron, S. and Briat, J.F. (1995). Induction of ferritin synthesis in maize leaves by an iron-mediated oxidative stress. *Plant J.*, **8** : 443-449.

Lum, H.K., Butt, Y.K. and Lo, S.C. (2002). Hydrogen peroxide induces a rapid production of nitric oxide in mung bean (*Phaseolus aureeus*). *Nitric Oxide*, **6** : 205-213.

Lundeen, C.V., Wood, N.H. and Braun, A.C. (1973). Intracellular levels of cyclic nucleotides during cell enlargement and cell division in excised tobacco piths. *Differentiation,* **1** : 225-260.

MacRobbie, E.A. (1998). Signal transduction and ion channels in guard cells. *Philos. Trans. R. Soc. Lond. B. Biol. Sci.*, **353** : 1475-1488.

McDonald, L.J. and Murad, F. (1995). Nitric oxide and cGMP signaling. *Adv. Pharmacol.*, **34** : 263-276.

Mizoguchi, T., Gotoh, Y., Nishida, E., Yamaguchi-Shinozaki, K., Hayashida, N., Iwasaki, T., Kamada, H. and Shinozaki, K. (1994). Characterization of two cDNAs that encode MAP kinase homologues in *Arabidopsis thaliana* and analysis of the possible role of auxin in activating such kinase activities in cultured cells *Plant J.*, **5** : 111-22.

Mockaitis, K. and Howell S.H. (2000). Auxin induces mitogenic activated protein kinase (MAPK) activation in roots of *Arabidopsis* seedlings. *Plant J.*, **24** : 785-796.

Murgia, I., Delledone, M. and Soave, C. (2002). Nitric oxide mediates iron-induced ferritin accumulation in *Arabidopsis*. *Plant J.*, **30** : 521-528.

Neill, S.J., Desikan, R., Clarke, A. and Hancock, J.T. (2002). Nitric oxide is a novel component of abscisic acid signaling in stomatal guard cells. *Plant Physiol.*, **128** : 13-16.

Pagnussat, G. C., Lanteri, M.L. and Lamattina, L. (2003a). Nitric oxide and cyclic GMP are messengers in the IAA-induced adventitious rooting process. *Plant Physiol.* In press.

Pagnussat, G. C., Lombardo M: C., Lanteri, M.L. and Lamattina, L. (2003b). Nitric oxide is required for a MAPK activation involved in the IAA-induced adventitious root development. (submitted).

Pagnussat, G.C., Simontacchi, M., Puntarulo, S. and Lamattina, L. (2002). Nitric oxide is required for root organogenesis. *Plant Physiol.*, **129** : 954-956.

Pawlosky J.R., Hess D. and Stamler J.S. (2001) Export by red blood cells of NO bioactivity. *Nature,* **409** : 622-626.

Penson, S.P., Schuurink, R.C., Fath, A., Gubler, F., Jacobsen, J.V. and Jones, R.L. (1996). cGMP is required for gibberellic acid-induced gene expression in barley aleurone. *Plant Cell,* **8** : 2325-2333.

Petit, J.M., Briat, J.F., Lobreaux, S. (2001). Structure and differential expression of the four members of the *Arabidopsis thaliana* ferritin gene family. *Biochem J.*, **359** : 575-582.

Pfeiffer, S., Janistyn, B., Jessner, G., Pichorner, H. and Ebermann, R.E. (1994). Gaseous NO stimulates guanosine 3-5-cyclic monophosphate formation in spruce needles. *Photochemistry,* **36** : 259-262.

Reggiani, R. (1997). Alteration of levels of cyclic nucleotides in response to anaerobiosis in rice seedlings. *Plant Cell Physiol.,* **38** : 740-742.

Sanders, D., Brownlee, C. and Harper, J.F. (1999). Communicating with calcium. *Plant Cell,* 11 : 691-706.

Savino, G., Lobreaux, J.S. and Briat, J.F. (1997). Inhibition of the iron-induced ZmFer1 maize ferritin gene expression by antioxidants and serine/threonine phosphatase inhibitors. *J. Biol. Chem.,* **272** : 33319-33326.

Schroeder, J.I., Allen, G.J., Hugouvieux, V., Kwak, J.M. and Waner, D. (2001). Guard cell signal transduction. *Annu. Rev. Plant Physiol. Plant Mol. Biol.,* **52** : 627-658.

Sethi, J.K., Willmott, N., Walseth, T.F., Lee, H.C., White, A.M. and Galione, A. (1996). Nitric oxide-induced mobilization of intracellular calcium via the cyclic ADP-ribose signaling pathway. *J. Biol. Chem.,* **271** : 3699-3705.

Shin, J.H., Chung, S., Park, E.J., Uhm, D.Y. and Suh, C.K. (1997). Nitric oxide directly activates calcium-activated potassium channels from rat brain reconstituted lipid bilayer. *FEBS Lett.,* **415** : 299-302.

Stamler, J.S. (1994). Redox signaling: nitrosylation and related targets interactions of nitric oxide. *Cell,* **78** : 931-936.

Thoiron, S., Pascal, N. and Briat, J.F. (1997). Impact of iron deficiency and iron re-supply during the early stages of vegetative development in maize (*Zea mays* L.). *Plant Cell Environ.,* **20** : 1051-1060.

Wei, J. and Theil, E. (2000). Identification and characterization of the iron regulatory element in the ferritin gene of a plant (soybean). *J. Biol Chem.,* **275** : 17488-17493.

Welch, R.M. (1995). Micronutrient nutrition in plants. *Crit Rev Plant Sci.,* **14** : 49-82.

Wendehenne, D., Pugin, A., Klessig, D.F. and Durner, J. (2001). Nitric oxide: comparative synthesis and signaling in animal and plant cells. *Trends Plant Sci.,* **6** : 177-183.

Xiong, L. and Zhu, J.K. (2001). Abiotic stress signal transduction in plants: molecular and genetic perspectives. *Physiol. Plant.,* **112** : 152-166.

Chapter 4

NITRIC OXIDE SIGNAL TRANSDUCTION IN PLANT PROGRAMMED CELL DEATH

Ione Salgado*[1], Jose R. Magalhaes[2], Elzira E. Saviani[1] and Luzia V. Modolo[1]

[1]*Departamento de Bioquímica, Instituto de Biologia, Universidade Estadual de Campinas (UNICAMP), Campinas, SP, 13083-970, Brazil;* [2]*EMBRAPA-CNPGL, 36038-330, Juiz de Fora, MG, Brazil.*

*Corresponding author : E-mail : ionesm@unicamp.br; Fax : 55-19-3788-6129

Summary

Recent evidences have shown that nitric oxide (NO) plays key roles in various developmental processes and also in defense mechanisms in plants. In addition to activating the pathway for the synthesis of antimicrobial compounds in plant defense responses against invading pathogens, NO is also a necessary signal for triggering cell death during the hypersensitive response (HR). Cell death induced in HR is a form of programmed cell death (PCD) that shares features with apoptosis observed in animals. Elucidation of the biochemical signaling pathway of NO-induced PCD in plants is central to understanding the mechanism of resistance to pathogens. Some of the biochemical events in NO-induced PCD in plants have been identified and indicate a signaling cascade by which NO induces apoptotic-like cell death that begins with the mitochondria. By inhibiting mitochondrial electron transport, NO causes collapse of the inner membrane electrical potential, which leads to opening of the mitochondrial permeability transition pore and the consequent release of cytochrome c. Caspase-like activity is increased following

In : Nitric Oxide Signaling in Higher Plants, 2004
(Eds. Jose R. Magalhaes, Rana P. Singh and Leonidas P. Passos)
Studium Press, LLC, Houston (USA), pp 81-109

exposure to NO, although it is unclear whether this activity precedes or follows cytochrome c release. Subsequent to mitochondrial dysfunction and the activation of caspase-like proteases, NO causes chromatin condensation, nuclear DNA fragmentation and cell death. Cytochrome c oxidase and the mitochondrial permeability transition pore are primary targets of NO during plant PCD. However, based on data from animals, various other NO-reactive sites exist in mitochondrial proteins, although the sensitivity of these sites to NO has not yet been examined. In contrast to cytochrome c oxidase, alternative oxidase (AOX), the secondary terminal oxidase of plant mitochondria, is barely affected by NO. AOX appears to counteract the deleterious effects of NO in plant mitochondria by preventing the inhibition of mitochondrial respiration, the trigger for NO-induced PCD. Thus, although NO-induced PCD in plants apparently involves a unique mechanism, the principal site of action appears to be the mitochondrial membrane, as is also the case with the Bcl-2 protein family in animal cells.

Keywords : Alternative oxidase, apoptosis, caspase-like activity, cytochrome *c* release, DNA fragmentation, mitochondrial permeability transition, nitric oxide, plant programmed cell death.

1. INTRODUCTION

The important role of nitric oxide (NO) in neurotransmission, smooth muscle relaxation, blood pressure regulation, and platelet aggregation is well established (Denninger and Marletta, 1999). These physiological actions of NO are mediated through the NO/cyclic GMP (cGMP) signaling pathway (Warner *et al.*, 1994; Buechler *et al.*, 1994; Jaffrey and Snyder, 1995). NO binds to the heme group of soluble guanylate cyclase (sGC) and increases the formation of the second messenger cGMP, which in turn can activate various downstream events in a cGMP-dependent cascade (Denninger and Marletta, 1999). However, there is growing evidence that the effects of NO on cell signaling pathways are not restricted to cGMP-dependent processes. Furthermore, excessive NO production can contribute to cell death in neurodegenerative disorders, in ischaemia-reperfusion injury and in inflammatory diseases (Murphy, 1999; Li and Förstermann, 2000).

NO is a relatively stable, charge-neutral radical with a strong tendency to react with species possessing unpaired electrons. Apart from its reactivity, the neutral charge of NO facilitates its diffusion across cell membranes (Kelm, 1999). NO may be converted to a number of more reactive derivatives, known collectively as reactive nitrogen species (RNS). Thus, inappropriate or excessive production of NO can cause cell death following the interaction of NO and RNS with key cellular components. Such a disturbance can lead to either apoptotic or necrotic cell death, depending on the severity of the damage (Murphy, 1999). Although NO has several potential mechanisms of toxicity, this toxicity is highly dependent on the rate of NO formation and decomposition, and on the particular microenvironment in which NO and its derivatives are produced and/or metabolized. In this context, an important role is played by anti-oxidant defense systems, involving enzymes and reductants, which prevent oxidative damage following NO generation. The diversity of factors affecting the fate of NO partly explains the double behavior of this radical as a physiological messenger and as a cytotoxic agent.

Recently, a broad spectrum of NO-mediated effects has also been reported in the plant kingdom. NO has been shown to influence various plant developmental processes and also to act as a cytotoxic agent during plant-microbe interactions. In this chapter, we review recent findings which show how NO can induce cell death in plants, and consider the similarities and differences between NO-induced PCD in plants and apoptosis in animals.

2. SYNTHESIS AND ACTIONS OF NITRIC OXIDE IN PLANTS

In mammalian cells, the main mechanism of NO generation is through enzymes known as nitric oxide synthases (NOS), which metabolizes L-arginine to L-citrulline with the concomitant formation of NO with reasonable similarity to the mammalian counterparts (Pollock *et al.*, 1991). Although no NOS gene with reasonable similarity to the mammalian counter parts has been found yet, various studies have suggested the occurrence of an NOS-like enzyme in plant cells. NOS-like activity, measured by L-citrulline formation from L-arginine and/or by its sensitivity to mammalian NOS inhibitors, has been detected in several plant species (Cueto *et al.*, 1996; Ninnemann and Maier, 1996; Delledonne *et al.*, 1998; Durner *et al.*, 1998; Barroso *et al.*, 1999; Ribeiro *et al.*, 1999; Modolo *et al.*, 2002). Additionally, using antibodies produced against mammalian NOS isoforms, NOS-like proteins have been localized in the cytosol and nucleus of maize root tips (Ribeiro *et al.*, 1999), and in peroxisomes and chloroplasts of pea leaves (Barroso *et al.*, 1999). Western blot analysis has shown that the NOS immunoreactive protein in maize and pea extracts has a molecular mass of approximately 166 kDa and 130 kDa, respectively (Ribeiro *et al.*, 1999; Barroso *et al.*, 1999), which is in the same range reported for mammalian NOS (Pollock *et al.*, 1991). Recently, NOS enzymes with no sequence similarities to mammalian isoforms were identified in tobacco (Chandok *et al.*, 2003) and *Arabidopsis* (Guo *et al.*, 2003) suggesting that those enzymes may use different chemistry to produce NO. Additional sources of NO in plants include endogenous nitrate reductase activity (Yamasaki *et al.* 1999; Magalhaes *et al.*, 2002) and exogenous nitrification and denitrification processes of soil microorganisms (Bollmann and Conrad, 1998; Bollmann *et al.*, 1999). Regardless of the source of NO, this radical has recently emerged as a pleiotropic effector in plants (reviewed in Beligni and Lamattina, 2001; Salgado *et al.*, 2002; Delledonne *et al.*, 2003).

There is increasing evidence that NO influences various plant developmental processes. Thus, exogenous NO induces root growth (Gouvêa *et al.*, 1997), leaf expansion (Leshem and Haramaty, 1996) and seed germination (Keeley and Fotheringham, 1997; Giba *et al.*, 1998), stimulates de-etiolation and inhibits hypocotyl elongation (Beligni and Lamattina, 2000), delays fruit maturation and flower senescence (Leshem and Haramaty, 1996), and affects plant peroxidases involved in the lignification of xylem vessels (Ferrer and Ros Barceló, 1999).

NO is also an important signaling molecule in plant defense responses against attack by pathogens. This response appears to be mediated by an NOS-like enzyme since increased NOS-like activity occurs in *Arabidopsis thaliana-Pseudomonas syringae* (Delledonne *et al.*, 1998), tobacco-tobacco mosaic virus (Durner *et al.*, 1998), tobacco-*Ralstonia solanacearum* (Huang and Knopp, 1998) and soybean-*Diaporthe phaseolorum* (Modolo *et al.*, 2002) interactions.

The response of plants to attack by pathogens involves the activation of a wide variety of defenses in order to prevent pathogen replication and dissemination. These responses include the increased expression of various defense genes, such as that for phenylalanine ammonia lyase (PAL). PAL is the first enzyme in the phenylpropanoid pathway that leads to the synthesis of phytoalexins which are flavonoids with antimicrobial activity (Kuc, 1995). Increased levels of transcripts encoding for PAL and other defense-related proteins and the accumulation of phytoalexins occur during plant-pathogen interactions, or following the treatment of plant cells with NO donors, mammalian NOS infiltration, or pathogen elicitors (Noritake *et al.*, 1996; Beligni *et al.*, 1997; Delledonne *at al.*, 1998; Durner *et al.*, 1998; Foissner *et al.*, 2000; Klessig *et al.*, 2000; Modolo *et al.*, 2002). These results have revealed that NO is an important activator of the phenylpropanoid pathway related to disease resistance.

In addition to the role of NO in the accumulation of antimicrobial compounds during plant-pathogen interactions, NO production by the host has been proposed to be a key signal for the development of the hypersensitive response (HR). The HR is characterized by rapid, localized cell death that often develops in host tissues in order to prevent the spread of the pathogen from the site of infection (Dangl *et al.*, 1996; Lamb and Dixon, 1997). An early event observed in the HR is the generation of reactive oxygen species (ROS), including superoxide anions (O_2^-) and hydrogen peroxide (H_2O_2), in an oxidative burst that is thought to be a central component in establishing plant immunity. During the HR in soybean cells, NO acts synergistically with ROS to potentiate hypersensitive cell death (Delledonne *et al.*, 1998). In contrast, NO alone is a sufficient signal to induce cell death in suspension cultures of *Arabidopsis thaliana* when present at concentrations similar to those generated following challenge by avirulent bacteria; there is no apparent synergism with ROS (Clarke *et al.*, 2000).

There is evidence that the HR observed in several plant-pathogen interactions is a form of programmed cell death (PCD) that shares a subset of features with apoptotic cell death seen in animals (Ryerson and Heath, 1996; Wang *et al.*, 1996a; Gilchrist, 1998; Lam *et al.*, 1999). There are also many examples of cell death occurring in plants during developmental processes or in response to abiotic stress that resembles apoptosis (Orzáez and Granell, 1997; Wang *et al.*, 1996a,b; Ryerson and Heath, 1996; Katsuhara, 1997; Stein and Hansen, 1999). However, despite its importance, the genetic and biochemical basis underlying PCD in plants is much less understood than in animals. The recent findings that NO mediates the HR during plant-microbe interactions suggest that this radical could be a signaling molecule to trigger apoptosis-like cell death in plants. Thus, elucidating the signaling events during NO-induced PCD is important for understanding the molecular pathways activated during plant PCD.

3. NITRIC OXIDE AND APOPTOSIS IN ANIMAL CELLS

Programmed cell death is a genetically-controlled process by which cells direct their own death during organ development, or when they reach the end of their life span, or when they are damaged. Apoptosis is a distinct form of PCD frequently encountered in animal cells and was initially described based on its morphological features, which are distinct from those observed in cells undergoing necrosis (Kerr *et al.*, 1972; Tomei and Cope, 1994; Hale *et al.*, 1996). The morphological hallmarks of apoptosis include cell shrinkage, rearrangement of the chromatin structure with DNA fragmentation, and the formation of apoptotic bodies that are ingested by phagocytes (Kerr *et al.*, 1972). Following this early morphological characterization, various biochemical pathways leading to apoptosis were subsequently identified. Most of the morphological changes that characterize apoptotic cell death are caused by caspases, a family of cysteine proteases that are specifically activated in apoptotic cells (Thornberry and Lazebnik, 1998; Earnshaw *et al.*, 1999). Caspases cleave proteins important for cell integrity; including the protein inhibitor of the DNase that causes internucleosomal cleavage of nuclear DNA to produce the DNA fragmentation typical of apoptosis (Enari *et al.*, 1998).

The discovery that mitochondria store a variety of important proapoptotic factors in their intermembranous space, including cytochrome *c* and pro-caspases, and that the release of these factors triggers the activation of caspase family proteases, revealed an important role for these organelles in controlling of apoptosis (Liu *et al.*, 1996; Li *et al.*, 1997; Green and Reed, 1998). Another proapoptotic factor present in the mitochondrial intermembranous space is a flavoprotein known as apoptosis-inducing factor (AIF), which promotes PCD through a caspase-independent pathway. Once released from mitochondria, AIF rapidly translocates to the cytoplasm and nucleus, where it induces chromatin condensation and the movement of phosphatidylserine into the outer leaflet of the plasma membrane (Susin *et al.*, 1999; Candé *et al.*, 2002). The key role of mitochondria in apoptosis was reinforced by the discovery that Bcl-2 family proteins, which have pro- and anti-apoptotic activities, act at the level of the mitochondrial membrane to regulate the mitochondrial permeability transition, cytochrome *c* release and caspase activation (see Hengartner, 2000).

Mitochondria play a central role in the regulation of apoptosis, particularly for pro-apoptotic stimuli that act independently of death receptors on the cell membrane. NO is an intrinsic pro-apoptotic stimulus that induces mitochondria-dependent apoptosis in a variety of animal cell types (Green and Reed, 1998; Adrain and Martin, 2001). Various mechanisms have been suggested to explain NO-induced apoptosis in animal cells. Although these mechanisms differ in some aspects, most of them consider that NO (or its derivatives) acts directly on mitochondria to disrupt electron transport through the respiratory chain and alter its normal functioning. As a result, the permeability of the mitochondrial membrane is increased, and this leads to the release of intramitochondrial pro-apoptogenic factors, including cytochrome *c*, a key component of the electron transport chain (Liu *et al.*, 1996). The cells subsequently enter a proteolytic degradation phase in which they acquire the morphological hallmarks of late apoptosis.

4. NITRIC OXIDE-INDUCED PCD IN PLANTS RESEMBLES APOPTOSIS

Although PCD in plants is not as well understood as in animals, a number of morphological and biochemical events associated with apoptosis in animal cells also occur in plants (Greenberg *et al.*, 1994; Mittler and Lam, 1996; Wang *et al.*, 1996a,b; Lamb and Dixon, 1997). That PCD in plants is a genetically-controlled process is seen in lesion-mimic mutants of *Arabidopsis* and maize that spontaneously simulate a disease-resistance response (Dietrich *et al.*, 1994; Greenberg *et al.*, 1994). In addition, the effect of over-expression in plants of animal genes implicated in apoptosis has been tested, and some plant homologues of these genes have been found. Such studies have suggested that at least some common components of a core mechanism involved in PCD in plants are similar to those of their animal counterparts (reviewed in Danon *et al.*, 2000). Apoptosis in animal cells has been defined using cytological criteria that include the appearance of apoptotic bodies and phagocytosis. However, this phenotype has not been frequently observed during plant PCD and would be functionally irrelevant for walled-cells. Thus, the term apoptosis appears to be inappropriate for qualifying plant PCD at the morphological level. For this reason, the term "apoptotic-like" PCD will be used here.

Recent studies, mostly done in cell culture systems, have identified several of the components in the signaling pathway by which NO induces PCD in plants. As noted below, these studies have revealed important biochemical mediators of NO-induced PCD in plants that are hallmarks of NO-induced apoptosis in animals.

4.1. Nitric Oxide-Induced DNA Fragmentation

Although NO is a pro-apoptotic agent in a number of animal cell types, this radical also frequently induces cell death by necrosis. The NO concentration and the redox state of the cells are among the factors that influence the type of cell death induced by NO. High nitrosative stress can deplete cellular ATP and, since apoptosis is an energy-requiring process, cells fail to undergo apoptosis and eventually dye by necrosis (Brown and Borutaite, 2001). Thus, in studying the effect of NO on cell viability it is important to distinguish whether the treatment causes necrosis or apoptosis. The major morphological criterion that can distinguish between these possibilities is the analysis of nuclear morphology following exposure to NO. In cells undergoing a defined program of cellular events, such as apoptosis, shrinkage of the nuclei is observed. Condensation and fragmentation of the heterochromatin is followed by the fragmentation of whole nuclei. In contrast, the nuclei of cells dying by necrosis appear swollen and the heterochromatin is uniformly dispersed through the nuclei. Using the criteria of changes in chromatin structure and DNA fragmentation, apoptosis-like cell death has been demonstrated in various developmental processes in plants and in the HR triggered by abiotic stress or attempted infection by pathogens (Greenberg *et al.*, 1994; Greenberg, 1996; Mittler and Lam, 1995; Tenhaken *et al.*, 1995; Lamb and Dixon, 1997).

Alterations in nuclear morphology can be identified *in situ* using specific fluorescent-labeled nuclear probes, such as Hoechst-33342 and DAPI (4,6-diamino-2-phenylindole). Recently, the NO donors

SNP (sodium nitroprusside) and SNAP (*S*-nitroso-N-acetylpenicillamine) were shown to inhibit the proliferation and reduce the survival of suspension-cultured cells of *Citrus sinensis* (Saviani *et al.*, 2002). To determine the type of cell death induced by NO donors, *Citrus* cells exposed to SNP were labeled with the fluorescent DNA probe Hoechst-33342. The condensation and fragmentation of chromatin, as well as, the shrinkage and fragmentation of whole cell nuclei, were evident in *Citrus* cells exposed to NO. SNP also caused chromatin condensation in carrot cells in suspension culture, as seen using the nuclear fluorescent probe DAPI (Zottini *et al.*, 2002). DAPI was also used to visualize chromatin condensation in *Arabidopsis* cells following treatment with the NO donor, Roussin's black salt (Clarke *et al.*, 2000).

The hallmark of the end stage of apoptosis is the internucleosomal fragmentation of nuclear DNA following increased nuclease activity that yields multiples of 180 base-pair fragments (Wyllie *et al.*, 1984). Separation of these fragments by agarose gel electrophoresis reveals the characteristic DNA ladder pattern of apoptosis. In contrast, the nuclear DNA of necrotic cells is randomly degraded and results in DNA fragments of heterogeneous length that produce a smudge pattern on gel electrophoresis. DNA laddering has been reported during various developmental processes, and in stress and death induced by pathogens or a pathogen toxin (reviewed in Danon *et al.*, 2000). However analysis by electrophoresis does not allow the identification of individual cells undergoing apoptosis and, in some cases, the ladder pattern can be difficult to detect, especially if the proportion of cells undergoing apoptosis is very small. Alternatively, internucleosomal DNA fragmentation can be detected at the level of individual cells using an *in situ* method, the terminal deoxynucleotidyl transferase-mediated dUTP nick-end labeling reaction (TUNEL) that detects free 3'-OH DNA breaks (Gavrieli *et al.*, 1992; Ben-Sasson *et al.*, 1995). The TUNEL method is a powerful method for identifying apoptosis in single cells. Using this method, nuclear DNA fragmentation was identified in the leaves and callus of *Kalanchoë daigremontiana* and *Taxus brevifolia* following treatment with SNP (Pedroso *et al.*, 2000a,b). TUNEL-positive nuclei were also observed in *Kalanchoë* and *Taxus* cells submitted to mechanical stress (centrifugation). To determine whether NO mediated the nuclear fragmentation in stressed cells, a specific NO fluorescent probe, 4,5-

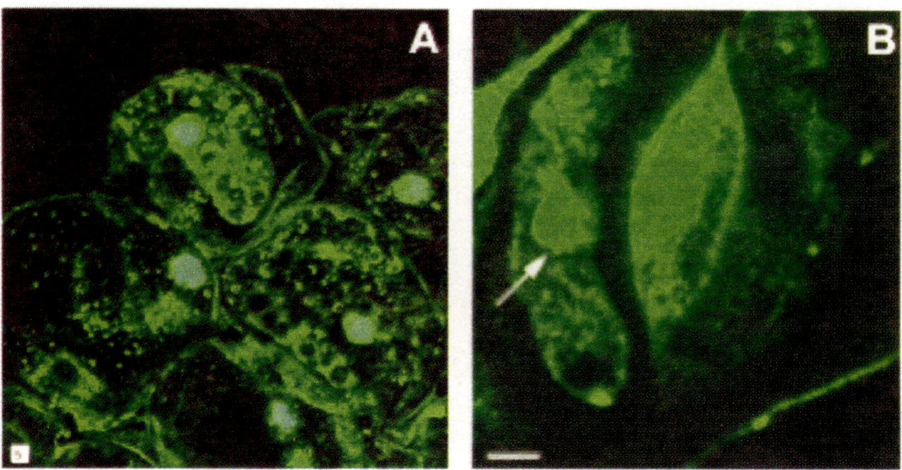

Figure 1 : Visualization of nitric oxide (NO) distribution and nuclear DNA fragmentation in *Taxus brevifolia* cells (A) and *Kalanchoë daigremontiana* leaf sections (B) folowing mechanical stress (centrifugation). NO was visualized using the DAF-2DA probe and DNA fragmentation was detected using the TUNEL assay. (A) NO formation (green cytoplasm) and TUNEL-positive nuclei (bright green), indicative of DNA fragmentation. Note the marked plasmolysis, a feature of cell death by apoptosis. Bar=5 mm (B) Stoma in a leaf section double stained for NO and DNA fragmentation. Note the TUNEL-positive nucleus (arrow) and the NO-positive cytosol in the guard cell on the left. Bar=6 mm. (A) Reproduced from Plant Science, 157, Pedroso MC, Magalhaes JR and Durzan D, Nitric oxide induces cell death in *Taxus* cells, 173-180, 2000, with permission from Elsevier. (B) Reproduced from Pedroso MC, Magalhaes JR and Durzan D, A nitric oxide burst precedes apoptosis in angiosperm and gymnosperm callus cells, Journal of Experimental Botany, 2000, 51, 347, 1027-1036, with permission from Oxford University Press.

diaminofluorescein diacetate (DAF-2 DA), was used. Double labeling of the tissues with DNA and NO probes showed that a burst of NO production precede DNA fragmentation in cells under stress (Fig. 1). The involvement of a putative NOS-like enzyme in the apoptotic-like cell death under stressful conditions in *Kalanchoë* and *Taxus* cells was further suggested by the use of an NOS inhibitor, L-NMMA, which significantly decreased the NO burst, nuclear DNA fragmentation and subsequent cell death induced by mechanical stress. D-NMMA, the inactive enantiomer of L-NMMA, had no significant effect on the NO burst and DNA fragmentation (Pedroso *et al.*, 2000a,b). Similar findings were obtained in leaves and leaf-plantlets of *K. daigremontiana* under gravity stimulation in which variations in the number of cells showing DNA fragmentation was directly correlated with NO formation (Pedroso and Durzan, 2000).

Table 1 summarizes the plant systems in which NO has been shown to induce apoptotic-like cell death based on changes in chromatin structure and DNA fragmentation.

NO has also been reported to counteract oxidative stress in plants (Caro and Puntarulo, 1998) and to reduce DNA fragmentation and cell death spots (HR) in *Phytophthora infestans*-infected potato leaves (Beligni and Lamatina, 1999a,b). These protective effects were attributed to the ability of NO to scavenge excess ROS produced during the HR. In animals, NO regulates apoptosis in a number of cell types and may be considered a unique among the regulators of apoptosis since it can act as either a pro- or an anti-apoptotic agent (Dimmerler and Zeiher, 1997; Nicotera *et al.*, 1997; Shen *et al.*, 1998). This dual behavior of NO in apoptosis reflects the variety of interactions between NO and its molecular targets. The extent of

Table 1. Chromatin condensation and DNA fragmentation in NO-induced plant cell death.

Species	NO donor[a]	Method[b]	Cell, tissue	Reference
Arabdopsis thaliana	RBS	DAPI	Cells in culture	Clarke *et al.* (2000)
Citrus sinensis	SNP, SNAP	Hoechst-33342	Cells in culture	Saviani *et al.* (2002)
Daucus carota	SNP	DAPI	Cells in culture	Zottini *et al.* (2002)
Kalanchoë daigremontiana	SNP	TUNEL	Leaves and callus	Pedroso *et al.* (2000b); Pedroso & Durzan (2000)
Taxus brevifolia	SNP	TUNEL	Leaves and callus	Pedroso *et al.* (2000a,b)

[a]RBS, Roussin's black salt; SNAP, *S*-nitroso-*N*-acetylpenicillamine; SNP, sodium nitroprusside.
[b]DAPI, 4,6-diamino-2-phenylindole; Hoechst-33342, 2'-(4-ethoxyphenyl)-5-(4-methyl-1-piperazinyl)-2,5'-bi-1H-benzimidazole; TUNEL, terminal deoxynucleotidyl transferase-mediated dUTP nick-end labeling.

such interactions depends on the redox-state of each particular system and can influence the actions of NO in the control of plant cell death.

4.2. Nitric Oxide-Induced Caspase-Like Activity

NO-induced apoptosis is always a caspase-mediated process and involves cytochrome *c* release from mitochondria. Cytochrome *c* is fundamental in orchestrating the assembly of proteins with initiator caspases that then activate and trigger a downstream cascade of effector caspases. The activation of downstream caspases during NO-induced apoptosis in animal cells can be blocked by caspase inhibitors (Borutaite *et al.*, 2000; Bal-Price and Brown, 2000; Moriya *et al.*, 2000).

Caspases are a family of cysteine proteases which selectively cleave a restricted set of target proteins, usually after an Asp residue, and have a recognition sequence near to this cleavage site (Sleath *et al.*, 1990). These proteases normally exist in the cytosol as inactive zymogens and must be activated either by autoprocessing or by other caspases. In general, apoptotic cell death involves a sequence of caspase activation events in which initiator caspases activate downstream executioner caspases. As an example, initiator caspases, such as caspase-8 and –9, directly or indirectly activate executioner caspases, such as caspase-3, -6 and –7 (Cohen, 1997). Once activated, these proteolytic enzymes attack various cytosolic and nuclear substrates, resulting in protein degradation and DNA fragmentation and, consequently, in the apoptotic phenotype. Identification of the key regulatory role of these proteases in animal apoptosis came from results showing that mutation or over-expression of specific cysteine protease genes protected or induced apoptosis in a variety of cell types (Dorstyn *et al.*, 1998; Earnshaw *et al.*, 1999). In addition, viral proteins capable of inhibiting apoptosis in host cells were identified as cysteine proteases (Xue and Horvitz, 1995). Although no functional homologs with the substrate specificity (absolute requirement for cleavage after an Asp residue and a recognition sequence) of animal caspases have been identified in plants, caspase-like activities have been measured during HR cell death. In addition, inhibitors of various mammalian caspases markedly suppress plant PCD (reviewed in Woltering *et al.*, 2002).

Caspase activity can be measured using specific fluorogenic or colorimetric tetrapeptide substrates that mimic the recognition sites of the different classes of the caspase family. The corresponding caspase activity can be blocked with the same peptide substrates coupled with an aldehyde (CHO: reversible inhibitor) or a methylketone radical (CMK, FMK: irreversible inhibitor). Using this approach, caspase-like protease activity was reported in *Nicotiana tabacum* leaves undergoing HR induced by the bean pathogen *Pseudomonas syringae* pv. *phaseolicola*, and specific inhibitors of human caspase-1 (Ac-YVAD-CMK) and caspase-3 (Ac-DEVD-CHO) effectively blocked HR in tobacco leaves induced by bacteria (Woltering *et al.*, 2002). Similarly, menadione-induced PCD in tobacco protoplasts was prevented by Ac-DEVD-CHO, a caspase-3 inhibitor (Sun *et al.*, 1999). Using synthetic fluorogenic substrates for caspase-1, caspase-like activity was demonstrated in extracts from tobacco mosaic virus-infected tobacco leaves, and this activity was inhibited by caspase-1 and caspase-3 inhibitors, but not by caspase-unrelated protease inhibitors (Lam and Del Pozo, 2000). These findings indicate the existence of functional caspase activity in plant extracts and suggest a key role for caspase-like proteases in PCD in plants.

The involvement of caspase-like proteases in NO-induced PCD in *Arabidopsis* cells in culture has been described (Clarke *et al.*, 2000). NO donors such as SNP and Roussin's black salt (RBS) induced *Arabidopsis* cell death when applied at concentrations that released NO in amounts similar to those generated by cells in response to challenge by avirulent bacteria (Clarke *et al.*, 2000). NO-induced cell death in *Arabidopsis* cultures under such conditions is an active process that requires gene expression, since cell death is reduced by cordycepin and cycloheximide, inhibitors of RNA processing and translation, respectively. In addition, treating *Arabidopsis* cells with RBS results in alterations in nuclear morphology and the condensation of chromatin. The involvement of caspase-like proteases in this form of PCD in *Arabidopsis* cells was demonstrated by pretreating the cells with Ac-YVAD-CMK, an irreversible inhibitor of mammalian caspase-1 (Villa *et al.*, 1997), which significantly reduced RBS-induced *Arabidospis* cell death. Furthermore, increased caspase-like activity in RBS-treated cells was detected with Ac-YVAD-AMC, a synthetic fluorogenic substrate for caspase-1 (Del Pozo and Lam,

1998). Although the exact nature of the protease activity modulated by NO was not elucidated, these findings suggested that NO-induced cell death in *Arabidopsis* suspension cultures involved a caspase-like protease activity when NO was applied in amounts similar to those causing the hypersensitive response.

4.3. Nitric Oxide-Induced Alterations in Mitochondrial Bioenergetics

During oxidative phosphorylation, the transport of electrons across the mitochondrial respiratory chain generates a proton electrochemical gradient across the mitochondrial inner membrane ($\Delta\mu_H^+$) which is used to drive ATP synthesis and transport of metabolites. Thus, maintenance of the $\Delta\mu_H^+$ is critical for the normal functioning of mitochondria. An alteration in the redox state of the respiratory chain can damage mitochondrial membranes and lead to the release of apoptogenic factors (Green and Reed, 1998; Cai and Jones, 1998). This release represents the point of no return and triggers the execution phase of cell death because it directly activates the latent apoptogenic machinery involved in the caspase signaling cascade.

The mitochondrial inner membrane electrical potential ($\Delta\Psi_m$) is one of the components of $\Delta\mu_H^+$, along with ΔpH, generated by the mitochondrial respiratory chain during electron transport. The collapse of $\Delta\Psi_m$ always precedes cytochrome *c* release and is a primary indicator of the deleterious effect of NO on the mitochondrial respiratory chain. Thus, alterations in $\Delta\Psi_m$ can be followed in intact cells in order to determine the participation of mitochondria in the apoptotic process. Cell permeant potential-sensitive fluorescent probes, such as chloromethyl X-rosamine (CMXRos) and tetramethylrhodamine (TMRM), can be used to follow changes in $\Delta\Psi_m$ in cells undergoing apoptosis. CMXRos has been used to examine the involvement of mitochondria in NO-induced PCD in *Citrus sinensis* cells in culture. Treatment with SNP led to dissipation of the $\Delta\Psi_m$ that preceded nuclear DNA fragmentation in *Citrus* cells (Fig. 2; Saviani *et al.*, 2002). Using TMRM, NO was also shown to affect mitochondrial functions in carrot cells in suspension cultures. Following collapse of the mitochondrial $\Delta\Psi_m$, NO induced the condensation of chromatin and decreased the viability of carrot cells in suspension cultures. In addition, western blot analysis of the cytosolic extracts of SNP-treated carrot cells showed that this

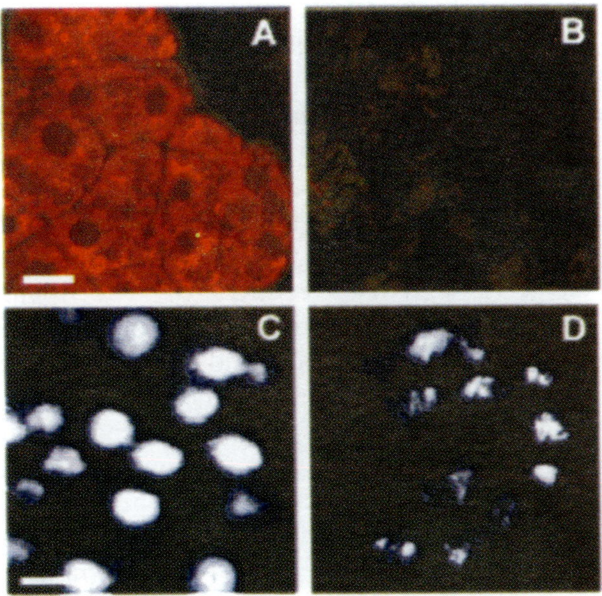

Figure 2 : Mitochondrial $\Delta\Psi_m$ loss and nuclear apoptosis induced by nitric oxide (NO) in suspension-cultured cells of *Citrus sinensis*. Cells were cultured in the absence (control) or presence of 10 mM SNP. After 2 or 3 days in culture, the cells were stained with the potential-sensitive dye CMXRos (A, B) or with the nuclear probe Hoechst-33342 (C, D), respectively. A and C, control. B and D, 10 mM SNP. Bar = 10 mm. Reproduced with permission from Saviani *et al.* (2002).

treatment also lead to cytochrome *c* release from mitochondria (Zottini *et al.*, 2002). The translocation of cytochrome *c* from mitochondria to the cytosol has been observed during PCD induced by heat in cucumber plants (Balk *et al.*, 1999), by D-mannose in maize cells in culture (Stein and Hansen, 1999), and by menadione in tobacco protoplasts (Sun *et al.*, 1999).

Various stress signals cause a sudden increase in the mitochondrial permeability to small solutes, leading to rapid mitochondrial swelling. The changes in mitochondrial permeability are mediated by the permeability transition pore (PTP) located at the site of contact of the inner and outer mitochondrial membranes (Petit *et al.*, 1996). Although massive increase in the mitochondrial permeability can disrupt the outer membrane, reversible PTP opening allows for the release of matrix and intermembrane proteins. The mechanism by which cytochrome *c* and other apoptogenic intermembrane proteins are released from mitochondria has not yet been established unequivocally, but may result from the breakdown of the outer mitochondrial membrane or opening of the PTP (Petit *et al.*, 1996; Green and Reed, 1998; Loeffler and Kroemer, 2000). Although the full structure of the PTP has not yet been elucidated, this pore is a polyprotein complex that includes inner membrane proteins such as the adenine nucleotide translocator (ANT), outer membrane proteins such as the voltage-dependent anion channel (VDAC), and cyclophilin D from the matrix (Green and Reed, 1998). The opening of this nonselective channel in animal mitochondria is regulated by the mitochondrial redox potential (Zoratti and Szabò, 1995), as well as by endogenous effectors, including members of the *bcl-2* gene family (Petit *et al.*, 1996). Cyclosporin A (CsA) is a specific inhibitor of PTP formation through its binding to cyclophilin D, a proposed component of PTP. The ability of CsA to prevent NO-induced apoptosis in some animal cell types has been attributed to its inhibitory activity on mitochondrial PTP formation (Hortelano *et al.*, 1997). The mitochondrial PTP also appears to be involved in NO-induced plant cell death since CsA prevented the SNP-induced death of *Citrus* cells in the culture (Saviani *et al.*, 2002). CsA also prevented loss of the mitochondrial $\Delta\Psi_m$ and the subsequent nuclear degradation observed after exposure to NO. In agreement with these protective effects on mitochondrial and nuclear integrity, CsA substantially delayed the antiproliferative action and cell death induced by NO in *Citrus* suspension cultures (Saviani *et al.*, 2002). These findings suggested that PTP formation and the release of cytochrome *c* are involved in the signaling pathway by which NO induces PCD in plants.

5. TARGETS FOR NITRIC OXIDE IN MITOCHONDRIA

NO and reactive nitrogen species (RNS) inhibit electron flux through the mitochondrial respiratory chain in a number of ways that are relevant to their role in cell death (Brown and Borutaite, 2001; Boyd and Cadenas, 2002; Brookes *et al.*, 2002). Identification of the targets for NO in mitochondria is based mainly on studies done in animals. In view of the highly conserved structure of the main components of the respiratory chain and other mitochondrial proteins in different kingdoms, analysis of the interactions of NO with such complexes may be valuable for understanding the mechanism (s) by which NO induces PCD in plants.

Mitochondria are the main site for the generation of cellular ROS. During normal mitochondrial respiration, there is a small but continuous leakage of electrons, mostly at the level of ubiquinone and also of complex I (NADH:ubiquinone oxidoreductase). Electrons leaking from the mitochondrial respiratory chain can cause a monoelectronic reduction of O_2 resulting in superoxide radical (O_2^-) formation. Superoxide can be dismutated to hydrogen peroxide (H_2O_2) by intramitochondrial superoxide dismutase, and subsequently to the hydroxyl radical ($OH^\bullet$) in the presence of Fe^{2+}. Under normal physiological conditions, ROS generated by mitochondria are efficiently degraded by the mitochondrial antioxidant systems, without causing any permanent damage to mitochondria. However, inhibition of the electron flux through the mitochondrial respiratory chain has, as a primary effect, an overreduction of the electron carriers, thereby favoring electron leakage and, consequently, ROS generation. Thus, NO favors the generation of ROS by inhibiting electron transport through the mitochondrial respiratory chain. In addition, the reaction of NO and O_2^- to form $OH^\bullet$ and peroxynitrite ($ONOO^-$) and other RNS is also enhanced. The extent of this formation will depend on the steady-state NO concentration and the cellular redox-state of the cell. These more reactive oxygen- and nitrogen-centered radicals can then react with various complexes of the mitochondrial respiratory chain and other mitochondrial proteins to alter their normal functioning.

NO is a potent inhibitor of cytochrome *c* oxidase (complex IV), the final electron acceptor of the mitochondrial respiratory chain. At nanomolar concentrations, NO specifically and reversibly inhibits complex IV by reacting with the enzyme's reduced or oxidized binuclear (oxygen-binding) site to give either cytochrome a_3^{2+}-NO or cytochrome a_3^{3+}-NO_2^-. This inhibition is competitive with O_2 and, under normal conditions, NO is considered to be a physiological regulator of mitochondrial respiration (Cleeter *et al.*, 1994; Brown 1995; Torres *et al.*, 1995; Sarti *et al.*, 1999). The core of the respiratory components in plant mitochondria is very similar to that in mitochondria from other eukaryotes, and studies with plant mitochondria have confirmed that the corresponding cytochrome *c* oxidase is very sensitive to NO (Millar and Day, 1996). However, in addition to the standard complexes of the respiratory chain, plant mitochondria also contain a secondary terminal oxidase, known as the alternative oxidase (AOX) which, in contrast to cytochrome *c* oxidase, is not inhibited by NO (Millar and Day, 1996). The importance of this alternative mitochondrial electron transport pathway in counteracting the effect of NO in inducing PCD in plants will be discussed later.

In animal mitochondria, NO also inhibits the bc_1 segment of ubiquinone:cytochrome *c* oxidoreductase (complex III), thereby increasing O_2^- generation, the fate of which depends on the NO concentration. At low physiological levels of NO, superoxide dismutation to H_2O_2 is favored, but when the concentration of NO reaches a certain threshold, conversion to $ONOO^-$ is favored because of the diffusion controlled reaction of NO and O_2^- (Poderoso *et al.*, 1999a,b; Cadenas *et al.*, 2000). Peroxynitrite can oxidize ubiquinol, thereby enhancing O_2^- generation, and possibly its own formation. This situation may lead to the irreversible inhibition of mitochondrial respiration (Schöpfer *et al.*, 2000).

Prolonged exposure to NO can also inhibit complexes I and IV through the conversion of NO to RNS (Clementi *et al.*, 1998). Although a precise inhibitory site for NO within complex I was not yet identified, the inactivation of complex I was originally proposed to involve *S*-nitrosylation of a critical thiol in an iron-sulfur center of the complex (Clementi *et al.*, 1998). More recently, it was suggested that NO-mediated inhibition of complex I proceeds via $ONOO^-$ formation (Riobó *et al.*, 2001), although the mechanism of inhibition is still unknown. Succinate:ubiquinone oxidoreductase (complex II) and aconitase are also inhibited by NO, probably by the NO-induced removal of iron from iron-sulfur centers (Grangner and Lenhinger, 1982; Drapier and Hibbs, 1988; Stuehr and Nathan, 1989).

Peroxynitrite can inhibit complex II, cytochrome oxidase, ATP synthase, aconitase, Mn-SOD, and other proteins (Cassina and Radi, 1996; Brown, 1999). Peroxynitrite is also thought to alter the membrane permeability transition through its direct action on PTP, thereby causing lipid peroxidation or thiol cross-linking (Gadelha *et al.*, 1997). RNS may directly oxidize thiols of the adenine nucleotide translocator, a key component of the mitochondrial permeability transition pore (Viveira *et al.*, 2001).

The NO-induced, irreversible inhibition of key mitochondrial proteins at multiple sites affects mitochondrial bioenergetics by promoting the mitochondrial permeability transition, cytochrome *c* release, and the initiation of downstream signaling cascades for apoptosis. Thus, the oxido-reduction and additive reactions of NO with dioxygen in its various redox forms and with transition metals are the most relevant pathways for inhibiting mitochondrial respiration and initiating apoptotic cell death. However, if the damage is very intense and leads to ATP depletion, the cells will die by necrosis.

6. ALTERNATIVE OXIDASE PROTECTS PLANTS FROM NO

In plant mitochondria, electrons from ubiquinol can be diverted to the alternative oxidase (AOX) which reduces dioxygen to water, thus bypassing complexes III and IV (McIntosh, 1994; Vanlerberghe and McIntosh, 1997). Unlike the cytochrome pathway, electron transport from ubiquinol to AOX is not coupled to proton translocation, and the energy of electron flow through the alternative pathway is lost as heat instead of being conserved to drive ATP synthesis. AOX is thought to have a role in preventing oxidative stress because it prevents over reduction of the ubiquinol pool, thereby limiting electron leakage from ubiquinol and the subsequent monoelectron reduction of oxygen.

Unlike cytochrome *c* oxidase, AOX is barely affected by NO so that AOX may have a role in counteracting the deleterious inhibitory effects of NO in the cytochrome pathway (Millar and Day, 1996; Yamasaki *et al.*, 2001). Indeed, recent studies with plant cells in culture have shown that AOX counteracts the effect of NO in inducing PCD. NO reduces the total respiration of carrot cell suspensions through its strong inhibition of the cytochrome pathway. Using salicylhydroxamic acid (SHAM) and antimycin-A to inhibit the AOX and cytochrome pathway, respectively, treatment with SNP decreased the activity of the cytochrome pathway and enhanced the contribution of the alternative pathway to total respiration. In addition, western blot analysis using anti-AOX antibodies has shown that induction of the alternative pathway is associated with an increase in AOX protein expression which cannot be accounted for solely by the diversion of electrons to AOX (Zottini *et al.*, 2002).

NO also affects cytochrome-dependent respiration and increases respiration through the alternative pathway in *Arabidopsis* cells in culture. The addition of an AOX inhibitor to *Arabidopsis* cells resulted in a dramatic increase in sensitivity to NO and cell death. Using cDNA microarrays and northern blot analysis, the treatment of *Arabidopsis* cells with NO was shown to strongly induce AOX1a transcription (Huang *et al.*, 2002).

Interestingly, increased alternative respiratory activity has been observed during various conditions of stress. Cyanide-resistant plant respiration is increased by chilling, wounding, and during microbial attack (Uritani and Asahi, 1980; Vanlerberghe and McIntosh, 1992; Purvis and Shewfelt, 1993). AOX protein levels (Lennon *et al.*, 1997) and AOX transcript levels (Simons *et al.*, 1999) are enhanced during plant defense responses to pathogens. The alternative pathway may be the major contributor to the increased respiration frequently seen in plant pathogen-interactions (Lennon *et al.*, 1997). These observations, as well the recent findings that NO is generated under abiotic stress and during HR, and that NO induces AOX expression in cells undergoing PCD, suggest that NO could be the signal for the induction of AOX during plant abiotic stress and microbe interactions. The extent to which AOX is induced would be important in overcoming the nitrosative stress that leads to PCD in such conditions.

In mammalian cells, members of the Bcl-2 family of proteins exert their anti-apoptotic actions by controlling the opening of the mitochondrial PTP and the subsequent release of cytochrome *c* (Green and Reed, 1998; Hengartner, 2000). Plants apparently do not have Bcl-2 homologues but instead use AOX to counteract the cytotoxic effects of NO on mitochondria. Although AOX is different from the Bcl-2 protein family and prevents apoptosis-like cell death by a different mechanism, it nevertheless acts at the level of the mitochondrial membrane. These observations indicate that mitochondria have played a central role in the evolution of the mechanisms controlling cell survival in eukaryotes.

7. CONCLUSIONS AND FUTURE PROSPECTS

As discussed above, there is now substantial evidence for the involvement of NO in PCD in plants, and for a central role of mitochondria in regulating this process. However, many issues remain to be addressed. One of these is the need to identify other mitochondrial sites that are susceptible to oxidative and nitrosative damage by NO. The plant mitochondrial complex III core proteins differ from those of other eukaryotes (Braun *et al.*, 1993), and plants also express unique electron transport components, such as NAD(P)H-dehydrogenases (Siedow and Umbach, 1995), that are not found in the mitochondria of most other organisms. It will therefore be necessary to determine whether NO interacts with these complexes and the relevance of such an interaction in inhibiting cellular respiration.

Most of the NO-donors that have been used to induce cell death in plants are not true NO donors, but rather are donors of the nitrosonium cation (NO^+) which has different properties from those of NO. NO^+ is a very effective nitrosating agent, particularly at sulfur centers (Hughes, 1999), and its effects on cell death need to be compared with those of NO. Furthermore, the ability of a particular cell type to alter the redox state of the NO group may have a role in determining the fate of NO and its effects on cell death. Since the exposure of *Arabidopsis* cells to NO induces the expression of an array of anti-oxidant genes (Huang *et al.*, 2002), it will be important to identify the strategies used by plant cells, other than AOX expression, to overcome the cytotoxic effects of NO. It would also be useful to know the level of NO necessary to trigger AOX expression. In this context, differences in the AOX content prior to stimulation by NO could partly explain the different sensitivities of plant cells to NO.

In animal cells, physiological concentrations of NO can inhibit caspase activity through *S*-nitrosation of critical residues (Kim *et al.*, 1997; Liu and Stamler, 1999) indicating that NO can act as an anti-apoptotic agent. On the other hand, caspase inactivation by severe oxidative stress resulting from excessive NO production may lead to cell death by necrosis, instead of apoptosis (Murphy, 1999). The extent to which caspase-like proteases in plant cells can regulate NO-induced PCD and how these enzymes interact with mitochondria during apoptosis remains to be determined.

Finally, although the participation of NO in plant PCD has been studied mainly following abiotic stress and during HR, this form of cell death in plants is not restricted to these responses. It will be of interest to determine whether this radical is also involved in PCD observed during certain developmental processes. Thus, a proper understanding of the role of NO in plant PCD will require the identification of how NO is synthesized under different physiological conditions.

ACKNOWLEDGEMENTS

This work was supported by the Fundação de Amparo à Pesquisa do Estado de São Paulo (FAPESP)

REFERENCES

Adrain, C. and Martin, S.J. (2001). The mitochondrial apoptosome: a killer unleashed by the cytochrome seas. *Trends Biochem. Sci.*, **26** : 390-397.

Bal-Price, A. and Brown, G.C. (2000). Nitric oxide induced necrosis and apoptosis in PC12 cells mediated by mitochondria. *J. Neurochem.*, **75** : 1455-1464.

Balk, J., Leaver, C.J. and McCabe, P.F. (1999). Translocation of cytochrome *c* from the mitochondria to the cytosol occurs, during heat-induced programmed cell death in cucumber plants. *FEBS Lett.*, **463** : 151-154.

Barroso, J.B., Corpas, F.J., Carreras, A., Sandalio, L.M., Valderrama, R., Palma, J.M., Lupianez, J.A. and Del Rio, L.A. (1999). Localization of nitric-oxide synthase in plant peroxisomes. *J. Biol. Chem.*, **274** : 36729-36733.

Beligni, M.V., Laxalt, A.M. and Lamattina, L. (1997). Putative role of nitric oxide in plant-pathogen interactions. In: *The Biology of Nitric Oxide* (Eds. Moncada, S., Toda, N., Maeda, H. and Higgs, E.A.) Portland Press, London, U.K. pp 250.

Beligni, M.V. and Lamattina, L. (1999a). Is nitric oxide toxic or protective? *Trends Plant Sci.*, **4** : 299–300.

Beligni, M.V. and Lamattina, L. (1999b). Nitric oxide counteracts cytotoxic processes mediated by reactive oxygen species in plant tissues. *Planta*, **208** : 337-344.

Beligni, M.V. and Lamattina, L. (2000). Nitric oxide stimulates seed germination and de-etiolation, and inhibits hypocotyl elongation, three light-inducible responses in plants. *Planta*, **210** : 215-221.

Beligni, M.V. and Lamattina, L. (2001). Nitric oxide in plants: the history is just beginning. *Plant Cell Environ.*, **24** : 267-278.

Ben-Sasson, S., Sherman, Y. and Gavrieli, Y. (1995). Identification of dying cells– *in situ* staining. In: *Methods in Cell Biology* (Eds.Schwartz L. and Osborne, B.) Academic, San Diego, pp 29-39.

Bollmann, A. and Conrad, R. (1998). Influence of O_2^- availability on NO and N_2O release by nitrification and denitrification in soil. *Global Change Biology*, **4** : 387-396.

Bollmann, A., Koschorreck, M. and Meuser, K. (1999). Comparison of two different methods to measure nitric oxide turnover in soils. *Biol. Fertil. Soils*, **29** : 104-110.

Borutaite, V., Morkuniene, R. and Brown, G.C. (2000). Nitric oxide donors, nitrosothiols and mitochondrial respiration inhibitors induce caspase activation by different mechanisms. *FEBS Lett.*, **467** : 155–159.

Boyd, C.S. and Cadenas, E. (2002). Nitric oxide and cell signaling pathways in mitochondrial-dependent apoptosis. *Biol. Chem.*, **383** : 411-423.

Braun, H.P., Emmermann, M. and Schmitz, U.K. (1993). Cytochrome *c* reductase from potato mitochondria: a protein complex involved in respiration and protein import. In: *Plant Mitochondria* (Eds. Brennicke, A. and Kück, U.) Weinheim, Germany: VCH, pp 307-313.

Brookes, P.S., Levonen, A.L., Shiva, S., Sarti, P. and Darley-Usmar, V.M. (2002). Mitochondria: regulators of signal transduction by reactive oxygen and nitrogen species. *Free Radical Biol. & Med.*, **33** : 755–764.

Brown, G.C. (1995). NO regulates mitochondrial respiration and cell functions by inhibiting cytochrome oxidase. *FEBS Lett.*, **369** : 136-139.

Brown, G.C. (1999). Nitric oxide and mitochondrial respiration. *Biochim. Biophys. Acta*, **1411** : 351-369.

Brown, G.C. and Borutaite, V. (2001). Nitric oxide, mitochondria, and cell death. *IUBMB Life*, **52** : 189-195.

Buechler, W.A., Ivanova, K., Wolfram, G., Drummer, C., Heim, J.M. and Gerzer, R. (1994). Soluble guanylate cyclase and platelet function. *Ann. NY Acad. Sci.*, **714** : 151-157.

Cadenas, E., Poderoso, J.J., Antunes, F. and Boveris, A. (2000). Analysis of the pathways of nitric oxide utilization in mitochondria. *Free Radical Res.*, **33** : 747–756.

Cai, J. and Jones, D.P. (1998). Superoxide in apoptosis: mitochondrial generation triggered by cytochrome *c* loss. *J. Biol. Chem.*, **273** : 11401-11404.

Candé C., Cohen, I., Daugas, E., Ravagnan, L., Larochette, N., Zamzami, N., Kroemer, G. (2002). Apoptosis-inducing factor (AIF): a novel caspases-independent death effector released from mitochondria. *Biochimie*, **84** : 215-222.

Caro, A. and Puntarulo, S. (1998). Nitric oxide decreases superoxide anion generation by microsomes from soybean embryonic axes. *Physiol. Plant.*, **104** : 357–364.

Cassina, A. and Radi, R. (1996). Different inhibitory actions of NO and peroxynitrite on mitochondrial electron transport. *Arch. Biophys. Biochem.*, **328** : 309-316.

Chandok, M.R, Ytterberg, A.J., van Wijk, K.J. and Klessig, D.F. (2003). The pathogen-inducible nitric oxide synthase (iNOS) in plants is a variant of the P protein of the glycine decarboxylase complex. *Cell*, **113** : 469-482.

Clarke, A., Desikan, R., Hurst, R.D., Hancock, J.T. and Neill, S.J. (2000). NO way back: nitric oxide and programmed cell death in *Arabidopsis thaliana* suspension cultures. *Plant J.*, **24** : 667–77.

Cleeter, M.W., Cooper, J.M., Darley-Usmar, V.M., Moncada, S. and Schapira, A.H. (1994). Reversible inhibition of cytochrome *c* oxidase, the terminal enzyme of the mitochondrial repiratory chain, by nitric oxide. Implications for neurodegenerative diseases. *FEBS Lett.*, **345** : 50-54.

Clementi, E., Brown, G.C., Feelisch, M. and Moncada, S. (1998). Persistent inhibition of cell respiration by nitric oxide: crucial role of *S*-nitrosylation of mitochondrial complex I and protective role of glutathione. *Proc. Natl. Acad. Sci. USA*, **95** : 7631–7636.

Cohen, G.M. (1997). Caspases: The executioners of apoptosis. *Biochem. J.*, **326** : 1-16.

Cueto, M., Hernández-Perera, O., Martin, R., Bentura, M.L., Rodrigo, J., Lamas, S. and Golvano, M.P. (1996). Presence of nitric oxide synthase activity in roots and nodules of *Lupinus albus*. *FEBS Lett.*, **398** : 159-164.

Dangl, J.L., Dietrich, R.A. and Richberg, M.H. (1996). Death don't have no mercy: cell death programs in plant-microbe interations. *Plant Cell*, **8** : 1793-1807.

Danon, A., Delorme, V., Mailhac, N. and Gallois P. (2000). Plant programmed cell death: a commom way to die. *Plant Physiol. Biochem.*, **38** : 647-655.

Delledonne, M., Xia, Y.J., Dixon, R.A. and Lamb, C. (1998). Nitric oxide functions as a signal in plant disease resistance. *Nature*, **394** : 585-588.

Delledonne, M., Polverari, A. and Murgia, I. (2003). The functions of nitric oxide-mediated signaling and changes in gene expression during the hypersensitive response. *Antioxid. Redox Signal.*, **5** : 33-41.

Del Pozo, O. and Lam, E. (1998). Caspases and programmed cell death in the hypersensitive response of plants to pathogens. *Curr. Biol.,* **8** : 1129-1132.

Denninger, J.W. and Marletta, M.A. (1999). Guanylate ciclase NO/cGMP signaling pathway. *Biochim. Biophys. Acta*, **1411** : 334-350.

Dimmerler, S. and Zeiher, A.M. (1997). Nitric oxide and programmed cell death: another paradigm for the doubled edged role of nitric oxide. *Nitric Oxide: Biol. Chem.,* **1** *:* 275–281.

Dietrich, R.A., Delaney, T.P., Uknes, S.J., Ward, E.R., Ryals, J.A. and Dangl, J.L. (1994). Arabidopsis mutants simulating disease resistance response. *Cell*, **77** : 565-577.

Dorstyn, L., Kinoshita, M. and Kumar, S. (1998). Caspases in cell death. In: *Apoptosis: Mechanisms and Role in Disease* (Ed. Kumar, S.) Springer-Verlag, Berlin pp 1-24.

Drapier, J.C. and Hibbs, J.B. Jr. (1988). Differentiation of murine macrophages to express non-specific cytotoxicity for tumour cells results in L-arginine-dependent inhibition of mitochondrial iron-sulphur in the macrophage effector cells. *J. Immunol.*, **140** : 2829-2838.

Durner, J., Wendehenne, D. and Klessig, D.F. (1998). Defense gene induction in tobacco by nitric oxide, cyclic GMP, and cyclic ADP-ribose. *Proc. Natl. Acad. Sci. USA*, **95** : 10328-10333.

Earnshaw, W.C., Martins, L.M. and Kaufmann, S.H. (1999). Mammalian caspases: structure, activation, substrates, and functions during apoptosis. *Annu. Rev. Biochem.*, **68** : 383-424.

Enari, M., Sakahira, H., Yokoyama, H., Okawa, K., Iwamatsu, A. and Nagata, S. (1998). A caspase-activated DNase that degrades DNA during apoptosis, and its inhibitor ICAD. *Nature*, **391** : 43-50.

Ferrer, M.A. and Ros Barceló, A.R. (1999). Differential effects of nitric oxide on peroxidase and H_2O_2 production by the xylem of *Zinnia elegans. Plant Cell Environ.*, **22** : 891-897.

Foissner I., Wendehenne, D., Langebartels, C. and Durner, J. (2000). *In vivo* imaging of an elicitor-induced nitric oxide burst in tobacco. *Plant J.*, **23** : 817-824.

Gadelha, F.R., Thomson, L., Fagian, M.M., Costa, A.D.T., Radi, R. and Vercesi, A.E. (1997). Calcium-independent permeabilization of the inner mitochondrial membrane by peroxynitrite is mediated by membrane protein thiol cross-linking and lipid peroxidation. *Arch. Biochem. Biophys.*, **345** : 243–250.

Gavrieli, Y., Sherman, Y. and Ben-Sasson, S.A. (1992). Identification of programmed cell death *in situ* via specific labeling of nuclear DNA fragmentation. *J. Cell Biol.*, **119** : 493-501.

Giba, Z., Grubisic, D., Todorovic, S., Sajc, L., Stojakovic, D. and Konjevic, R. (1998). Effect of nitric oxide-releasing compounds on phytochrome-controlled germination of empress tree seeds. *Plant Growth Regul.*, **26** : 175-181.

Gilchrist, D.G. (1998). Programmed cell death in plant disease: the purpose and promise of cellular suicide. *Annu. Rev. Phytopathol.*, **36** : 393-414.

Gouvêa, C.M.C.P., Souza, J.F., Magalhães, A.C.N. and Martins, I.S. (1997). NO-releasing substances that induce growth elongation in maize root segments. *Plant Growth Regul.*, **21** : 183-187.

Grangner, D.L. and Lehninger, A.L. (1982). Sites of inhibition mitochondrial electron transport in macrophage-injured neoplastic cells. *J. Cell Biol.*, **95** : 527-535.

Green, D.R. and Reed, J.C. (1998). Mitochondria and apoptosis. *Science*, **281** : 1309-1312.

Greenberg, J.T., Guo, A., Klessig, D.F. and Ausubel, F.M. (1994). Programmed cell death in plants: a pathogen-triggered response activated coordinately with multiple defense functions. *Cell*, **77** : 551-563.

Greenberg, J.T. (1996). Programmed cell death: a way of life for plants. *Proc. Natl. Acad. Sci. USA*, **93** : 12094-12097.

Guo, F.Q., Okamoto, M. and Crawford, N.M. (2003). Identification of a plant nitric oxide synthase gene involved in hormonal signaling. *Science,* **302** : 100-103.

Hale, A.J., Smith, C.A., Sutherland, L.C., Stoneman, V.E.A., Longthorne, V.L., Culhane, A.C. and Williams, G.T. (1996). Apoptosis: molecular regulation of cell death. *Eur. J. Biochem.*, **236** : 1-26.

Hengartner, M.O. (2000). The biochemistry of apoptosis. *Nature*, **407** : 770-776.

Hortelano, S., Dallaporta, B., Zamzami, N., Hirsch, T., Susin, S.A., Marzo, I., Boscá, L. and Kroemer, G. (1997). Nitric oxide induce apoptosis via triggering mitochondrial permeability transition. *FEBS Lett.*, **410** : 373-377.

Huang, J.S. and Knopp, J.A. (1998). Involvement of nitric oxide in *Ralstonia solanacearum*-induced hypersensitive reaction in tobacco. In: *Bacterial Wilt Disease: Molecular and Ecological Aspects* (Eds. Prior, P., Elphinstone, J. and Allen, C.) INRA and Springer Editions, Berlin pp 218-224.

Huang, X., von Rad, U. and Durner, J. (2002). Nitric oxide induces transcriptional activation of the nitric oxide-tolerant alternative in *Arabidopsis* suspension cells. *Planta*, **215** : 997-1005.

Hughes, M.N. (1999). Relationships between nitric oxide, nitroxyl ion, nitrosonium cation and peroxynitrite. *Biochim. Biophys. Acta*, **1411** : 263-272.

Jaffrey, S.R. and Snyder, S.H. (1995). Nitric oxide: a neural messenger. *Ann. Rev. Cell Dev. Biol.*, **11** : 417-440.

Katsuhara, M. (1997). Apoptosis-like cell death in barley roots under salt stress. *Plant Cell Physiol.*, **38** : 1091-1093.

Keeley, J.E. and Fotheringham, C.J. (1997). Trace gas emissions and smoke-induced seed germination. *Science*, **276** : 1248-1250.

Kelm, M. (1999). Nitric oxide metabolism and breakdown. *Biochim. Biophys. Acta*, **1411** : 273-289.

Kerr, J.F.R., Wyllie, A.H. and Currie, A.R. (1972). Apoptosis: a basic biological phenomenon with wide-ranging implications in tissue kinetics. *Br. J. Cancer*, **26** : 239-257.

Kim, Y.M., Talanian, R.V. and Billiar, T.R. (1997). Nitric oxide inhibits apoptosis by preventing increases in caspase-3-like activity via two distinct mechanisms. *Proc. Natl. Acad. Sci. USA*, **272** : 31138–31148.

Klessig, D.F., Durner, J., Noad, R., Navarre, D.A., Wendehenne, D., Kumar, D. Zhou, J.M., Shah, J., Zhang, S.Q., Kachroo, P., Trifa, Y., Pontier, D., Lam,

E. and Silva, H. (2000). Nitric oxide and salicylic acid signaling in plant defense. *Proc. Natl. Acad. Sci. USA*, **97** : 8849-8855.

Kuc, J. (1995). Phytoalexins, stress metabolism, and disease resistance in plants. *Annu. Rev. Phytopathol.*, **33** : 275-297.

Lam, E., Pontier, D. and Del Pozo, O. (1999). Die and let live: programmed cell death in plants. *Curr. Opin. Plant Biol.*, **2** : 502-507.

Lam, E. and Del Pozo, O. (2000). Caspase-like protease involvement in the control of plant cell death. *Plant Mol. Biol.*, **44** : 417-428.

Lamb, C. and Dixon, R.A. (1997). The oxidative burst in plant disease resistance. *Annu. Rev. Plant Physiol. Plant Mol. Biol.*, **48** : 251-275.

Lennon, A.M., Neuenschwander, U., Ribas-Carbo, M., Giles, L., Ryals, J.A. and Siedow, J.N. (1997). The effects of salicylic acid and tobacco mosaic virus infection on the alternative oxidase of tobacco. *Plant Physiol.*, **115** : 783–791.

Leshem, Y.Y. and Haramaty, E.J. (1996). The characterization and contrasting effects of the nitric oxide free radical in vegetative stress and senescence of *Pisum sativum* Linn. foliage. *J. Plant Physiol.*, **148** : 258-263.

Li, F., Nijhawan, D., Budihardjo, I., Srinivasula, S.M., Ahmad, M., Alnemri, E.S. and Wang, X. (1997). Cytochrome *c* and dATP-dependent formation of Apaf-1/caspase-9 complex initiates an apoptotic protease cascade. *Cell*, **91** : 479-489.

Li, H.G. and Förstermann, U. (2000). Nitric oxide in the pathogenesis of vascular disease. *J. Pathol.*, **190** : 244-254.

Liu, X., Kim, C.N., Yang, J., Jemmerson, R. and Wang, X. (1996). Induction of apoptotic program in cell-free extracts: requirement for dATP and cytochrome *c. Cell*, **86** : 147-157.

Liu, L. and Stamler, J.S. (1999). NO: an inhibitor of cell death. *Cell Death Differ.*, **6** : 937–942.

Loeffler, M. and Kroemer, G. (2000). The mitochondrion in cell death control: certainties and incognita. *Exp. Cell Res.*, **256** : 19-26.

Magalhães, J.R., Silva, F.L.I.M., Salgado, I., Ferrarese-Filho, O., Rockel, P. and Kaiser, W.M. (2002). Nitric oxide and nitrate reductase in higher plants. *Physiol. Mol. Biol. Plants*, **8** : 11-17.

McIntosh, L. (1994). Molecular biology of the alternative oxidase. *Plant Physiol.*, **105** : 781–786.

Millar, A.H. and Day, D.A. (1996). Nitric oxide inhibits the cytochrome oxidase but not the alternative oxidase of plant mitochondria. *FEBS Lett.*, **398** : 155-158.

Mitller, R. and Lam, E. (1995). Identification, characterization, and purification of a tobacco endonuclease activity induced upon hypersensitive response cell death. *Plant Cell*, **7** : 1951-1962.

Mittler, R. and Lam, E. (1996). Sacrifice in the face of foes: pathogen-induced programmed cell death in plants. *Trends Microbiol.*, **4** : 10–15.

Modolo, L.V., Cunha, F.Q., Braga, M.R. and Salgado, I. (2002). Nitric oxide synthase-mediated phytoalexin accumulation in soybean cotyledons in response to the *Diaporthe phaseolorum* f. sp. *meridionalis* elicitor. *Plant Physiol.*, **130** : 1288-1297.

Moriya, R., Uehara, T. and Momura, Y. (2000). Mechanism of nitric oxide induced apoptosis in human neuroblastoma SH-SY5Y cells. *FEBS Lett.*, **484** : 253–260.

Murphy, M.P. (1999). Nitric oxide and cell death. *Biochim. Biophys. Acta*, **1411** : 401-414.

Nicotera, P., Brune, B. and Bagetta, G. (1997). Nitric oxide: inducer or suppressor of programmed cell death? *Trends Pharmacol. Sci.*, **18** : 189–190.

Ninnemann, H. and Maier, J. (1996). Indications for the occurrence of nitric oxide synthases in fungi and plants and the involvement in photoconidiation of *Neurospora crassa. Photochem. Photobiol.*, **64** : 393-398.

Noritake, T., Kawakita, K. and Doke, N. (1996). Nitric oxide induces phytoalexin accumulation in potato tuber tissues. *Plant Cell Physiol.*, **37** : 113-116.

Orzáez, D. and Granell, A. (1997). DNA fragmentation is regulated by ethylene during carpel senescence in *Pisum sativum. Plant J.*, **11** : 137-144.

Pedroso, M.C. and Durzan, D.J. (2000). Effect of different gravity environments on DNA fragmentation and cell death in *Kalanchoë* leaves. *Ann. Bot.,* **86** : 983–994.

Pedroso, M.C., Magalhaes, J.R. and Durzan, D. (2000a). Nitric oxide induce cell death *Taxus* cells. *Plant Sci.*, **157** : 175-180.

Pedroso, M.C., Magalhaes, J.R. and Durzan, D. (2000b). A nitric oxide burst precedes apoptosis in angiosperm and gymnosperm callus cells and foliar tissues. *J. Exp. Bot.*, **51** : 1027-1036.

Petit, P.X., Susin, S.A., Zamzami, N., Mignotte, B. and Kroemer, G. (1996). Mitochondria and programmed cell death: back to the future. *FEBS Lett.*, **396** : 7-13.

Poderoso, J.J., Carreras, M.C., Schöpfer, F., Lisdero, C.L., Riobó, N.A., Giulivi, C., Boveris, A.D., Boveris, A. and Cadenas, E. (1999a). The reaction of nitric oxide with ubiquinol: kinetic properties and biological significance. *Free Radical Biol. & Med.*, **26** : 925-935.

Poderoso, J.J., Lisdero, C., Schöpfer, F., Riobó, N., Carreras, M.C., Cadenas, E. and Boveris, A. (1999b). The regulation of mitochondrial oxygen uptake by redox reactions involving nitric oxide and ubiquinol. *Proc. Natl. Acad. Sci. USA,* **274** : 37709–37716.

Pollock, J.S, Förstermann, U., Mitchell, J.A., Warner, T.D., Schmidt, H.H., Nakane, M. and Murad, F. (1991). Purification and characterization of particulate endothelium-derived relaxing factor synthase from cultured and native bovine aortic endothelial cells. *Proc. Natl. Acad. Sci. USA*, **88** : 10480-10484.

Purvis, A.C. and Shewfelt, R.L. (1993). Does the alternative pathway ameliorate, chilling injury in sensitive plant tissues? *Physiol. Plant.*, **88** : 712-718.

Ribeiro, E.A., Cunha, F.Q., Tamashiro, W.M.S.C. and Martins, I.S. (1999). Growth phase-dependent subcellular localization of nitric oxide synthase in maize cells. *FEBS Lett.*, **445** : 283-286.

Riobó, N.A., Clement, E., Melani, M., Boveris, A., Cadenas, E., Moncada, S. and Poderoso, J.J. (2001). Nitric oxide inhibits mitochondrial NADH: ubiquinone reductase activity through peroxynitrite formation. *Biochem. J.*, **359** : 139–145.

Ryerson, D.E. and Heath, M.C. (1996). Cleavage of nuclear DNA into oligonucleosomal fragments during cell death induced by fungal infection or by abiotic treatments. *Plant Cell*, **8** : 393–402.

Salgado, I., Modolo, L.V., Ribeiro, J.N., Magalhães, J.R. and Tamashiro, W.M.S.C. (2002). Parallels between plants and animals in the production and molecular targets of nitric oxide. *Physiol. Mol. Biol. Plant*, **8** : 185-191.

Sarti, P., Lendaro, E., Ippoliti, R., Bellelli, A., Benedetti, P.A., and Brunori, M. (1999). Modulation of mitochondrial respiration by NO: investigation by single cell fluorescence microscopy. *FASEB J.*, **13** : 191-197.

Saviani, E.E., Orsi, C.H., Oliveira, J.F.P., Pinto-Maglio, C.A.F. and Salgado, I., (2002). Participation of the mitochondrial permeability transition pore in nitric oxide-induced plant cell death. *FEBS Lett.*, **510** : 136-140.

Schöpfer, F., Riobo, N., Carreras, M.C., Alvarez, B., Radi, R., Boveris, A., Cadenas, E. and Poderoso, J.J. (2000). Oxidation of ubiquinol by peroxynitrite: implications for protection of mitochondria against nitrosative damage. *Biochem. J.*, **349** : 35-42.

Shen, Y.Y., Wang, X.L. and Wilcken, D.E. (1998). Nitric oxide induces and inhibits programmed cell death through different pathways. *FEBS Lett.*, **433** : 1125–1131.

Siedow, J.N. and Umbach, A.L. (1995). Plant mitochondrial electron transfer and molecular biology. *The Plant Cell,* **7** : 821-831.

Simons, B.H., Millenaar, F.F., Mulder, L., Van Loon, L.C. and Lambers, H. (1999). Enhanced expression and activation of the alternative oxidase during infection of *Arabidopsis* with *Pseudomonas syringae* pv tomato. *Plant Physiol.*, **120** : 529–538.

Sleath, P.R., Hendrickson, R.C., Kronheim, S.R., March, C.J. and Black, R.A. (1990). Substrate specificity of the protease that processes human interleukin-1β. *J. Biol. Chem.*, **265** : 14526-14528.

Stein, J.C. and Hansen, G. (1999). Mannose induces an endonuclease responsible for DNA laddering in plant cells. *Plant Physiol.*, **121** : 71–79.

Stuehr, D.J. and Nathan, C.F. (1989). NO: a macrophage product responsible for cytostasis and respiratory inhibition in tumour target cells. *J. Exp. Med.*, **169** : 1543-1555.

Sun, Y.L., Zhao, Y., Hong, X. and Zhai, Z.H. (1999). Cytochrome *c* release and caspase activation during menadione-induced apoptosis in plants. *FEBS Lett.*, **462** : 317-321.

Susin, S.A., Lorenzo, H.K., Zamzami, N., Marzo, I., Snow, B.E., Brothers, G.M., Mangion, J., Jacotot, E., Constantini, P., Loeffler, M., Larochette, N., Goodlett, D.R., Aebersold, R., Siderovski, D.P., Penninger, J.M. and Kroemer, G. (1999). Molecular characterization of mitochondrial apoptosis-inducing factor. *Nature*, **397** : 441-446.

Tenhaken, R., Levine, A., Brlsson, L.F., Dlxon, R.A. and Lamb, C. (1995). Function of the oxidative burst in hypersensitive disease resistance. *Proc. Natl. Acad. Sci. USA*, **92** : 4158-4163.

Thornberry, N.A. and Lazebnik, Y. (1998). Caspases: enemies within. *Science*, **281** : 1312-1316.

Tomei, L.D. and Cope, F.O. (1994). Apoptosis II: the molecular basis of apoptosis in disease. Communications in: *Cell and Molecular Biology* (Cold Springer Harbor, NY: Cold Spring Harbor Laboratory Press).

Torres, J., Darley-Usmar, V. and Wilson, M.T. (1995). Inhibition of cytochrome *c* oxidase in turnover by NO: mechanisms and implications for control of respiration. *Biochem. J.*, **312** : 169-173.

Uritani, I. and Asahi, T. (1980). Respiration and related metabolic activity in wounded and infected tissues. In: *The Biochemistry of Plants,* (Eds. Stumpf, P.K. and Conn, E.E.) Academic Press, London, **v.2**, pp 463-485.

Vanlerberghe, G.C. and McIntosh, L. (1992). Lower temperature increases alternative pathway capacity and alternative oxidase protein in tobacco. *Plant Physiol.*, **100** : 115–119.

Vanlerberghe, G.C. and McIntosh, L. (1997). Alternative oxidase: from gene to function. *Annu Rev. Plant Physiol. Plant Mol. Biol.*, **48** : 703–734.

Villa, P., Kaufmann, S.H. and Earnshaw, W.C. (1997). Caspases and caspase inhibitors. *Trends Biochem. Sci.*, **22** : 388-393.

Viveira, H.L., Belzacq, A.S., Haouzi, D., Bernassola, F., Cohen, I., Jacotot, E., Ferri, K.F., El Hamel, C., Bartle, L.M., Melino, G., Brenner, C., Goldmacher, V. and Kroemer, G. (2001). The adenine nucleotide translocator: a target of nitric oxide, peroxynitrite, and 4-hydroxynonenal. *Oncogene*, **20** : 4305-4316.

Wang, H., Li, J., Bostock, R.M. and Gilchrist D.G. (1996a). Apoptosis: a functional paradigm for programmed plant cell death induced by a host-selective phytotoxin and invoked during development. *Plant Cell,* **8** : 375–391.

Wang, M., Oppedijk, B.J., Lu, X., Van Duijn, B. and Schilperoort, R.A. (1996b). Apoptosis in barley aleurone during germination and its inhibition by abscisic acid. *Plant Mol. Biol.,* **32** : 1125–1134.

Warner, T.D., Mitchell, J.A., Sheng, H. and Murand, F. (1994). Effects of cyclic GMP on smooth muscle relaxation. *Adv. Pharmacol.*, **26** : 171-194.

Woltering, E.J., van der Bent, A. and Hoeberichts, F.A. (2002). Do plant caspases exist? *Plant Physiol.*, **130** : 1764-1769.

Wyllie, A.H., Morris, R.G., Smith, A.L. and Dunlop, D. (1984). Chromatin cleavage in apoptosis: association with condensed chromatin morphology and dependence on macromolecular synthesis. *J. Pathol.*, **142** : 67-77.

Xue, D. and Horvitz, H.R. (1995). Inhibition of the *Caenorhabditis elegans* cell-death protease CED-3 by a CED-3 cleavage site in baculovirus p35 protein. *Nature*, **377** : 248-251.

Yamasaki, H., Sakihama, Y. and Takahashi, S. (1999). An alternative pathway for nitric oxide production in plants: new features of an old enzyme. *Trends Plant Sci.*, **4** : 128-129.

Yamasaki, H., Shimoji, H., Ohshiro, Y. and Sakihama, Y. (2001). Inhibitory effects of nitric oxide on oxidative phosphorylation in plant mitochondria. *Nitric Oxide: Biol. Chem.*, **5** : 261-270.

Zoratti, M. and Szabò, I. (1995). The mitochondrial permeability transition. *Biochim. Biophys. Acta*, **1241** : 139-176.

Zottini, M., Formentin, E., Scattolin, M., Carimi, F., Lo Schiavo, F. and Terzi, M. (2002). Nitric oxide affects plant mitochondrial functionality *in vivo*. *FEBS Lett.*, **515** : 75-78.

Chapter 5

PEROXISOMES AS A SOURCE OF NITRIC OXIDE

Francisco J. Corpas[1★], Juan B. Barroso[2], Ana M. León[1], Alfonso Carreras[2], Miguel Quiros[3], José M. Palma[1], Luisa M. Sandalio[1] and Luis A. del Río[1]

[1]*Departamento de Bioquímica, Biología Celular y Molecular de Plantas, Estación Experimental del Zaidín, CSIC, Granada;* [2]*Departamento de Bioquímica y Biología Molecular, Facultad de Ciencias Experimentales, Universidad de Jaén;* [3]*Departamento de Química Iuorgánica, Facultad de Gieucias, Universidad de Granada, Spain*
★Corresponding author : E-mail : javier.corpas@eez.csic.es; Fax : 34 958 129600

Summary

Peroxisomes are subcellular organelles with a simple morphology but with a complex antioxidative network. In addition to the key antioxidant enzyme catalase, plant peroxisomes contain superoxide dismutase, the four enzyme components of the ascorbate-glutathione cycle (ascorbate peroxidase, glutathione reductase, monodehydroascorbate reductase and dehydroascorbate reductase), and three NADP-dependent dehydrogenases. More recently, evidences have been reported showing the presence of nitric oxide synthase (NOS) activity and its reaction product, nitric oxide (NO) in peroxisomes. These findings open new questions in cell biology of reactive nitrogen and oxygen species (RNS and ROS) and suggest that peroxisomes could have a role in the transduction pathways of plant cells as a source of signal molecules, such as nitric oxide, hydrogen peroxide (H_2O_2) and possibly S-nitrosoglutathione (GSNO).

Keywords : Antioxidants, nitric oxide, nitric oxide synthase, peroxisomes, signaling.

In : Nitric Oxide Signaling in Higher Plants, 2004
(Eds Jose R. Magalhaes, Rana P. Singh and Leonidas P. Passos)
Studium Press, LLC, Houston, USA, pp 111-129

1. INTRODUCTION

Peroxisomes are subcellular organelles present in virtually all eukaryotic cells. They are small structures (0.1-2 μm diameter) bounded by a single membrane which have an essentially oxidative type of metabolism. They contain catalase and H_2O_2-producing flavin oxidases as basic enzymatic constituents. The simple morphology of peroxisomes does not reflect the complexity of their enzymatic composition (Tabak *et al.*, 1999; Baker and Graham, 2002). In plants, there are several types of peroxisomes which are specialized in certain metabolic functions. Glyoxysomes are specialized peroxisomes, occurring in the storage tissue of oilseeds, that contain the fatty acid β-oxidation and glyoxylate cycle enzymes to convert the seed reserve lipids into sugars which are used for germination and plant growth. Leaf peroxisomes are specialized peroxisomes present in photosynthetic tissues that carry out the major reactions of photorespiration. Another type of specialized peroxisomes are root-nodule peroxisomes from certain tropical legumes, in which the synthesis of allantoin, the major metabolite for nitrogen transport within these plants, is carried out (Huang *et al.*, 1883; Baker and Graham, 2002).

During the last two decades, the existence of a complex antioxidant network inside peroxisomes including different superoxide dismutases (SODs), the enzyme components of the ascorbate-glutathione cycle and several NADP-dehydrogenases, has been demonstrated (for reviews see: Corpas *et al.*, 2001; del Río *et al.*, 2002). In addition to the well known generation of H_2O_2 in peroxisomes, the production of superoxide radicals ($O_2^{\cdot -}$), another reactive oxygen species (ROS), has also been proved (del Río *et al.*, 2002). All these findings evidence the oxidative type of metabolism of these organelles.

The gaseous free radical nitric oxide ($NO^{\cdot}$) is a widespread intracellular and intercellular messenger with a broad spectrum of regulatory functions in many physiological processes (Moncada *et al.*, 1991; Wendehenne *et al.*, 2001). In animal systems, a considerable attention has been focussed on this molecule and the enzyme responsible for its endogenous production, nitric oxide synthase (NOS; EC 1.14.13.39) (Hemmens and Mayer, 1998; Alderton *et al.*, 2001). In plant systems, comparatively much less is known about NO and the source of its production (Wendehenne *et al.*, 2001; Neill *et al.* 2002b, 2003; Wojtaszek, 2000). In the last years, endogenous NO was described to be involved in several plant physiological processes, such as ethylene emission (Leshem and Haramaty, 1996), response to drought (Leshem, 1996), disease resistance (Delledonne *et al.*, 1998, 2001; Durner *et al.*, 1998; Clarke *et al.*, 2000), growth and cell proliferation (Ribeiro *et al.*, 1999), maturation and senescence (Leshem *et al.*, 1998), apoptosis/programmed cell death (Magalhaes *et al.*, 1999; Pedroso and Durzan, 2000; Pedroso *et al.*, 2000a; Clarke *et al.*, 2000), and stomatal closure (García-Mata and Lamatina, 2002; Neill *et al.*, 2002). Furthermore, the application of exogenous NO to plants has been used as a valuable tool to study how this molecule affects or modulates some physiological and biochemical processes. For example, exogenous NO inhibits the activities of tobacco aconitase (Navarre *et al.*, 2000), catalase and ascorbate peroxidase (Clark *et al.*, 2000), participates in cell wall lignification (Ferrer and Ros Barceló, 1999), induces stomatal closure in several plants species (García-Mata and Lamattina, 2001), increases respiration through the alternative oxidase pathway (Huang *et al.*, 2002), induces cell death (Pedroso *et al.*, 2000b; Saviani *et al.*, 2002), mediates the accumulation of ferritin (Murgía *et al.*, 2002), negatively modulates wound signaling (Orozco-Cárdenas and Ryan, 2002), and is required for root organogenesis (Pagnussat *et al.*, 2002).

However, very little is known so far on how NO is produced in plants (Leshem, 2000). NO is a component of the nitrogen cycle and the enzyme nitrate reductase has been demonstrated to be a source of NO (Yamasaki *et al*, 1999; Yamasaki and Sakihama, 2000; Rockel *et al.*, 2002; see Chapter 8 in this volume). There is a growing body of evidence which shows the presence of NOS activity and animal NOS immunorelated proteins in different plant tissues and situations (Wendehenne *et al.*, 2001). However, neither the plant NOS has been purified to homogeneity nor the cDNA encoding this protein has been isolated yet.

This chapter analyses the potentiality of plant peroxisomes as a cellular source of NO. This is supported by two different evidence : i) the identification of nitric oxide as a product of an enzymatic reaction in isolated peroxisomes; and ii) the biochemical characterization of NOS activity of plant peroxisomes and its subcellular localization in the matrix of these oxidative organelles.

2. NITRIC OXIDE IS A COMPONENT OF PEROXISOMAL METABOLISM

The production of NO in plant tissues is well established, but to our knowledge the localization of NO in specific cell compartments has not been studied yet. Although the presence of this molecule in peroxisomes appears to be logical, considering the existence of a NOS-like activity in these organelles (see point 3 of this chapter), due to the actual controversy on the occurrence of this enzyme in plant tissues, it is important to demonstrate unequivocally the presence of NO in subcellular compartments. Consequently, the presence of NO in isolated pea leaf peroxisomes has been studied using two different approaches: spin trapping electron paramagnetic resonance (EPR) spectroscopy, and fluorometric analysis with 4,5-diaminofluorescein diacetate (DAF-2 DA).

2.1. Nitric oxide detection in isolated leaf peroxisomes by EPR spectroscopy

EPR spin trapping techniques have been demonstrated to be accurate and specific methods to determine the direct formation of NO in biological systems (Kotake *et al.*, 1996; Nagano and Yamaguchi, 2002) and are considered a clear evidence of the NO presence in these systems.

EPR measurements were performed using the trap $Fe(MGD)_2$ that reacts with NO forming a stable complex, NO-$Fe(MGD)_2$, with a characteristic three-line EPR spectrum. To standardize the EPR method, a commercial pure nitric oxide synthase from animal origin (nNOS from Calbiochem©) was used as positive control (Fig. 1). The EPR spectrum obtained with this commercial nNOS has the characteristic triplet signal of NO-$Fe(MGD)_2$ (g = 2.05 and a_N = 12.8 G) reported by Xia *et al.* (1997, 2000). When the peroxisomal fractions of pea leaves were assayed, similar three-line signals with the same values for g and a_N were obtained (Fig. 1).

This EPR technique has also been used to detect the presence of NO in homogenates of soybean embryonic axes (Caro and Puntarulo, 1999). The relatively weak triplet signal of the $NO^{\cdot}$-spin adduct detected in isolated peroxisomes could be due to a low enzymatic generation of the signal molecule $NO^{\cdot}$, as it has been reported in animal systems for the constitutive NOSs (nNOS and eNOS) (Joshi *et al.*, 1999). However, another possibility for the low intensity of

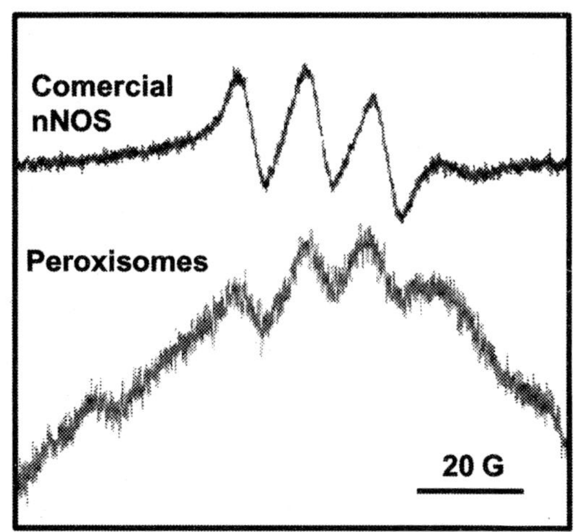

Figure 1 : EPR spectra of the NO-spin adduct of the $Fe(MGD)_2$ complex. For the detection of NO·, samples were added to a reaction mixture containing the substrate and all the cofactors of the NOS reaction, plus the NO-spin trap $Fe(MGD)_2$, and were incubated for 1 h at 37ºC. Then, samples were analyzed by EPR. Pure nNOS from Calbiochem® was used. Peroxisomes (270 μg of protein) were isolated from pea leaves by differential and density-gradient centrifugation. EPR parameters of the spectrum were g = 2.05 and a_N = 12.8 G. Representative spectra from 3-5 independent measurements are shown. The spectrum of peroxisomes was obtained by the accumulation of 5 recordings.

the EPR signal derives from the well-known affinity of $NO^{\cdot}$ for heme proteins (Sharma *et al.*, 1987; Kanner *et al.*, 1992). The heme-enzyme catalase is very abundant in peroxisomes, and $NO^{\cdot}$ could form complexes with its heme-groups. This way, catalase could compete with the spin trap for the $NO^{\cdot}$ generated in the peroxisomal NOS reaction, producing a lower NO-$Fe(MGD)_2$ EPR signal. On the other hand, peroxisomes have $O_2^{\cdot-}$-generating systems in their matrices and membranes (del Río *et al.*, 1989; López-Huertas *et al.*, 1999), and the spin adduct NO-$Fe(MGD)_2$ can react with $O_2^{\cdot-}$ radicals yielding EPR-silent species (Nagano and Yoshimura, 2002). This could equally lead to an underestimation of the endogenous $NO^{\cdot}$ present in peroxisomes.

2.2. Nitric oxide detection in isolated leaf peroxisomes by fluorometric analysis

The second approach to examine the presence of NO in pea leaf peroxisomes was fluorometric analysis with DAF-2 DA as fluorophore. This fluorescent probe has become a common technique to detect NO in animal and plant systems (Nakatsubo *et al.*, 1998; Nagano and Yoshimura, 2002). The method was developed by Nagano´s group (Koyima *et al.*, 1998) and the detection of NO consists in the incubation of samples with the membrane-permeable DAF-2 DA which is subsequently hydrolysed by cytosolic estereases releasing DAF-2. At physiological pH, DAF-2 is relatively non-fluorescent, but in the presence of NO and oxygen, DAF-2 triazole (DAF-2T), a bright green fluorescent product, is formed. DAF-2 does not react either with stable oxidized forms of NO, such NO_2^-, NO_3^-, or with other reactive oxygen/nitrogen species such as $O_2^{\cdot -}$, H_2O_2, and $ONOO^-$. Another advantage of using DAF-2 for cellular imaging is its visible excitation wavelength that is less damaging to cells. Therefore this probe has been used to obtain a real-time bio-imaging of NO with fine temporal and spatial resolution in plant cells (Foissner *et al.*, 2000; Pedroso *et al.*, 2000; García-Mata and Lamattina, 2002; Neill *et al.*, 2002). The detection limit of DAF-2T is less that 2-5 nM and shows a linear correlation up to around 1000 nM (Nakatsubo *et al.*, 1998). For these reasons, we decided to analyse the presence of NO in isolated peroxisomes by this method.

Figure 2A shows a linear correlation between fluorescence intensity and the amount of peroxisomal protein indicating a direct relationship between the amounts of NO and peroxisomal protein. To corroborate these results several additional assays were carried out: i) the peroxisomal samples were preincubated with 2 mM aminoguanidine, a known inhibitor of NOS; ii) L-arginine, the substrate of NOS, was not added to the reaction mixtures; and iii) prior to the addition of DAF-2 DA, the peroxisomal samples were denatured by heating at 95ºC for 10 min. In these cases, reductions of 88, 65 and 66 % in the relative fluorescence compared to the control reaction, respectively, were determined (Fig. 2B). Therefore, it can be concluded that, at least, 65% of the $NO^{\cdot}$ detected by spectrofluorimetric analysis has an enzymatic origin.

3. CHARACTERIZATION OF PEROXISOMAL NITRIC OXIDE SYNTHASE ACTIVITY

Since 1996, there has been a growing number of reports showing the presence of a nitric oxide synthase activity in plants similar, to a certain extent, to mammalian NOS. Table 1 summarizes, in chronological order, the NOS activity described in several plant

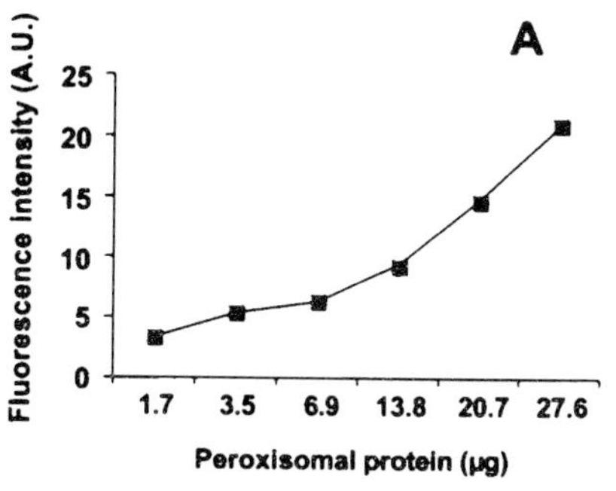

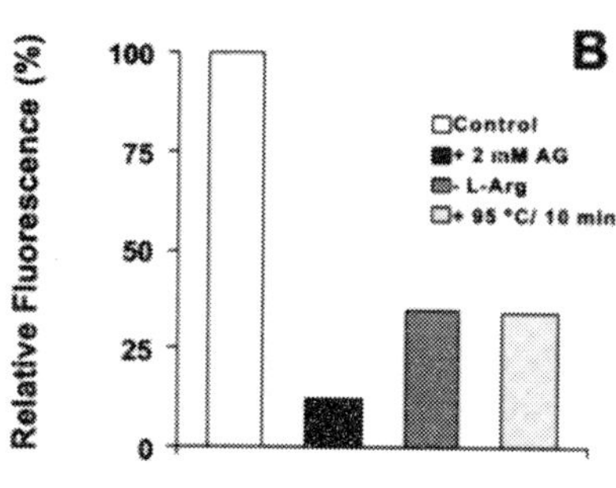

Figure 2 : Spectrofluorimetric detection of NO in pea leaf peroxisomes with DAF-2 DA. Peroxisomes freshly isolated from pea leaves were added to the reaction mixtures and the fluorescence measured. Values shown are means of three independent experiments. A, different volumes of peroxisomal fractions were added to the reaction mixture and the fluorescence produced was expressed as arbitrary units (A.U.). B, effect on the NO· generation of: i) preincubating peroxisomes with 2 mM aminoguanidine (AG); ii) removing L-arginine from the reaction mixture (L-Arg); and iii) denaturing the samples by heating (+ 95ºC/10 min). The amount of peroxisomal proteins in the reaction mixtures was 12 µg and fluorescence was expressed as percent of the control (relative fluorescence).

species. Without discarding the existence of other sources of NO different from the NOS activity, we will concentrate on the characterization of the peroxisomal enzyme that could be responsible for the generation of NO in these organelles.

3.1. Biochemical evidence of nitric oxide synthase activity in peroxisomes

The existence of a NOS-like activity in pea leaf peroxisomes was evidenced using two biochemical approaches. First, the measurement in isolated peroxisomes of NOS activity by monitoring the conversion of L-[^{3}H]arginine to L-[^{3}H]citrulline. This activity showed a linear correlation with the amount of peroxisomal protein (Fig. 3A) and was increased during the first 30 min of incubation, and then the NOS activity remained steady (Fig. 3B). The specific activity was 5.61 nmol of L-[^{3}H]citrulline · mg^{-1} protein · min^{-1} and was Ca^{2+}-dependent, being NADPH strictly necessary (Table 2). The incubation of the peroxisomal fractions with several inhibitors of mammalian NOS reduced the production of L-[^{3}H]citrulline between 59-100%. (Table 2). Additionally, the incubation of peroxisomes with an antibody against murine iNOS produced a 88% reduction of the NOS activity with an antibody dilution of 1/200 (Fig. 4A).

Table 1. Nitric oxide synthase activity reported in some plant species.

Species/Tissue or cell type	Assays	Reference
Pisum sativum/Leaves	Gaseous NO emission sensitive to NOS inhibitors	Leshem & Haramaty (1996)
Lupinus albus/Roots and nodules	Arginine-citrulline assay	Cueto *et al.* (1996)
Mucuna hassjoo	Arginine-citrulline assay	Ninnemann & Maier (1996)
Glycine max/ *Pseudomonas synringae*-infected cell suspensions	NO production sensitive to NOS inhibitors	Delledonne *et al.* (1998)
Nicotiana tabacum/TMV-infected leaves	Arginine-citrulline assay	Durner *et al.* (1998)
Glycine max/Embryonic axes	NADPH-diaphorase activity	Caro & Puntarulo (1999)
Zea mays/Root tips and young leaves	Arginine-citrulline assay	Ribeiro *et al.* (1999)
Pisum sativum/Leaf peroxisomes	Arginine-citrulline assay	Barroso *et al.* (1999)
Taxus brevifolia/Callus	NO production sensitive to NOS inhibitors	Pedroso *et al.* (2000)
Nicotiana tabacum/Leaf epidermal cells	NO production sensitive to NOS inhibitors	Foissner *et al.* (2000)
Nicotiana tabacum/Cell cultures *Arabidopsis*/Cell cultures *Petrosilenum crispum*/ Cell cultures	NO production sensitive to NOS inhibitors	Tun *et al.* (2001)
Glycine max/Cotyledons	Arginine-citrulline assay	Modolo *et al.* (2002)

The second approach to study the NOS activity of pea leaf peroxisomes consisted in Western blot analysis using a polyclonal antibody against the peptide PT387 (Ac-Cys-residues 1131 to 1144) from the C terminus of the deduced amino acid sequence of murine iNOS (Uttenthal *et al.*, 1998). This showed the presence in

Table 2. Biochemical properties of the nitric oxide synthase (NOS) activity of pea leaf peroxisomes. The different percentages of inhibition of the peroxisomal NOS activity produced by each inhibitor are indicated in brackets

Specific activity	5.61 ± 0.54 nmol L-[^{3}H]citrulline · mg^{-1} protein · min^{-1}	
Subunit molecular mass	130 kDa	
Cofactors	Ca^{2+} NADPH (strict dependency)	
Inhibitors	Aminoguanidine	(100 %)
	L-NAME	(90 %)
	L-NMMA	(88 %)
	Thiocitrulline	(80 %)
	Diphenyliodonium	(60 %)
	7-nitroindazole	(59 %)
	L-*N*5-(1-Imenoethyl)-ornithine	(59 %)

L-NAME, N^G-nitro-L-arginine methyl ester; L-NMMA, N^G-monomethyl-L-arginine

peroxisomes of an immunoreactive polypeptide band of about 130 kDa (Fig. 4B, lane 4). Immunoreactive bands with a similar mobility were obtained with the hepatic iNOS from LPS-induced rat and

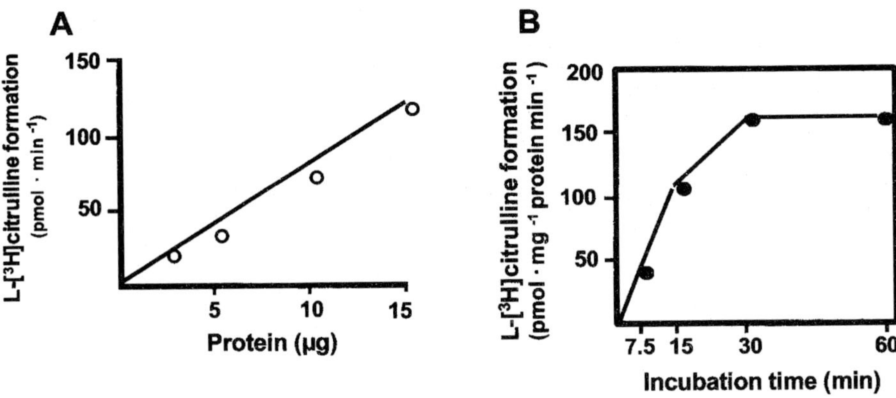

Figure 3 : Protein- and time-dependent formation of L-[^{3}H]citrulline from L-[^{3}H] arginine in peroxisomal fractions. A, increasing amounts of peroxisomal protein samples were added to the standard assay mixtures and the NOS activity was assayed at a concentration of 1 mM Arginine. B, peroxisomal samples (15 µg protein) were added to the standard assay mixture and the NOS activity was determined.

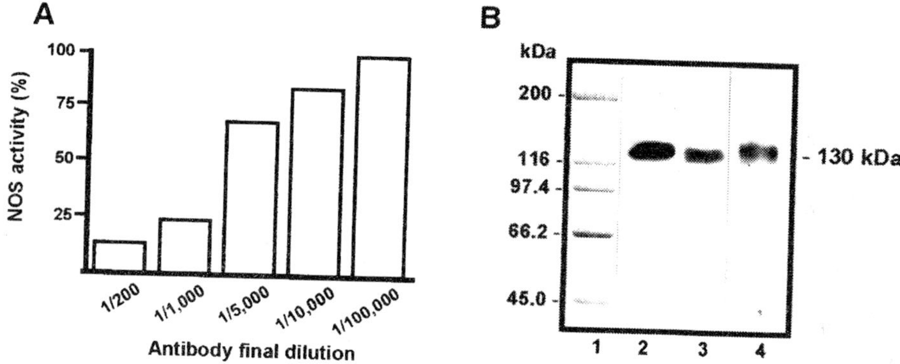

Figure 4 : Enzymatic and Western blot analysis of NOS in peroxisomal fractions with a polyclonal antibody against murine iNOS. A, effect of the antibody against iNOS on the peroxisomal NOS activity. Peroxisomal fractions were incubated at 25 ºC for 2 h with different dilutions of the antibody against murine iNOS. Then, the NOS activity of samples was assayed using an Arginine concentration of 1 mM. B, immunoblot analysis of peroxisomal fractions with the polyclonal antibody to murine iNOS. 1, molecular mass markers. 2, crude extracts of lipopolysaccharide-pretreated rat liver (30 μg of protein). 3, murine macrophage control lysate (30 μg protein). 4, pea peroxisomal fraction (50 μg protein). A dilution of 1/2500 was used for lanes 2 and 3, and 1/500 for lane 4. Reproduced with permission from *J. Biol. Chem.* 274: 36729-36733 (1999).

murine macrophage lysates, used as positive controls (Fig. 4B, lanes 2 and 3).

All these biochemical data evidence the existence of NOS activity in peroxisomes. In its low activity and Ca^{2+}-dependency it was similar to the constitutive animal NOS (neuronal and endothelial NOSs). However, the peroxisomal NOS was related to the inducible isoform (iNOS) since its subunit molecular mass was in the range reported for these mammalian enzymes, and was inhibited by the antibody against iNOS. Therefore, the peroxisomal NOS activity has its own biochemical peculiarities.

3.2. Immunolocalization of nitric oxide synthase in peroxisomes

To probe the localization of NOS in peroxisomes, two different approaches were used: i) confocal laser immunofluorescence microscopy with antibodies against iNOS and catalase, a characteristic marker enzyme of peroxisomes; and ii) immunogold electron microscopy analysis with an antibody against murine iNOS.

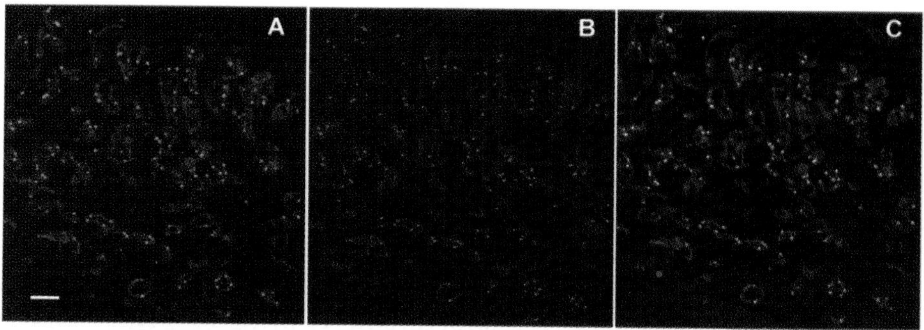

Figure 5 : Colocalization of NOS and catalase in pea leaves. Representative images illustrating the scanning confocal laser immunofluorescent detection of NOS and catalase in spongy mesophyll cells of pea leaf sections (60-µm thick). (A) Cy2-streptavidin immunofluorescence punctuates (green) attributable to anti-iNOS. (B) Cy3 immunofluorescence punctuates (red) attributable to anti-catalase. (C) Colocalized immunofluorescence punctuates (yellow). Scale bar = 10 µm.

Figure 5 (panel A) shows spongy mesophyll cells of a pea leaf section with the green punctuate immunofluorescence pattern of Cy2-streptavidin, corresponding to the antibody against iNOS. The panel B shows the same leaf section with the red immunofluorescence pattern of Cy3 attributable to the antibody against catalase. Peroxisomes appeared as small distinct fluorescent spots within non-vacuolar cytoplasmic regions of spongy mesophyll cells. Figure 5C shows the colocalized immunofluorescence punctuate patterns of Cy2-streptavidin and Cy3. The nearly complete overlapping of the two punctuate patterns indicated that NOS and catalase were localized in the same cell compartment, the peroxisome. Figure 6 shows representative electron micrographs of thin sections of pea leaves

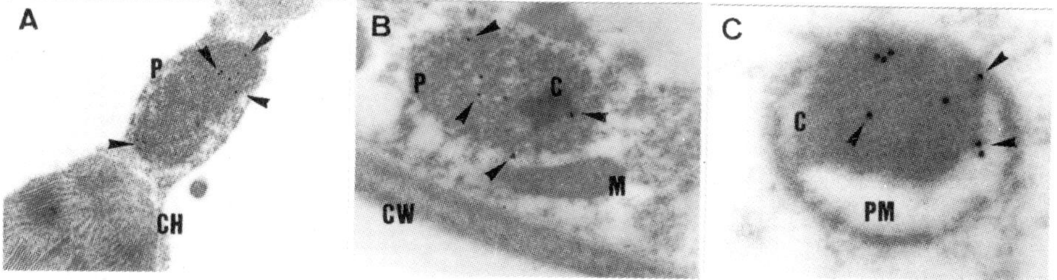

Figure 6 : Immunogold electron microscopy localization of NOS in peroxisomes from three plant species. The electron micrographs are representative of thin sections of pea leaves (A), olive leaves (B), and sunflower hypocotyls showing specific immunolocalization of NOS. Magnification is x 10,000 in panel A, x 16,000 in panel B, and x 63,000 in panel C.

(panel A), olive leaves (panel B) and sunflower hypocotyls (panel C). In these three plant species, the immunolocalization with the antibody against murine iNOS showed that the gold particles were localized inside peroxisomes.

Very recently, the peroxisomal localization of iNOS in animal cell peroxisomes has been reported, in organelles from cultured rat hepatocytes, and this iNOS was associated with the up-regulation and decreased expression of catalase (Stolz *et al.*, 2002). These results suggest that NOS could be a characteristic enzyme of peroxisomes.

4. CONCLUSIONS AND FUTURE PROSPECTS

An important feature of eukaryotic cells is the compartmentation of proteins within membrane-bound organelles which is essential to coordinate, regulate and integrate the different metabolic pathways (Igamberdiev and Lea, 2002). Considering that a characteristic property of peroxisomes is the participation in metabolic pathways with other cell compartments (Huang *et al.*, 1983; Baker and Graham, 2002; Igamberdiev and Lea, 2002), such as chloroplasts and mitochondria, the presence of NOS and $NO^{\cdot}$ in peroxisomes can have an impact on the metabolic regulation of these and other cell organelles. The high diffusion coefficient of $NO^{\cdot}$ in biological membranes (Denicola *et al.*, 1996) makes it that can act as a signal molecule in another organelles of the same cell or even in other cells. Nitric oxide could also have a role in the activity regulation of different peroxisomal heme/flavin enzymes involved in the metabolism of ROS, such as catalase and ascorbate peroxidase (Brunelli *et al.*, 2001; Clark *et al.*, 2000) or xanthine oxidase (Sakuma *et al.*, 1997). In plant cells, $NO^{\cdot}$ produced in peroxisomes could act on plant mitochondria, where apparently NOS is not present (Barroso *et al.*, 1999), and modulate the activity of the alternative oxidase (Huang *et al.*, 2002). Additionally, $NO^{\cdot}$ could also regulate the activity of cytosolic aconitase (Navarre *et al.*, 2000).

The presence of $NO^{\cdot}$ in peroxisomes suggests different interactions of this molecule with other components of the metabolism of ROS (Fig. 7). Besides catalase, several antioxidative enzyme systems have been demonstrated in plant peroxisomes, including different superoxide dismutases, the four enzymes of the ascorbate-glutathione cycle plus ascorbate and glutathione, and three NADP-dependent dehydrogenases (del Río *et al.*, 2002). The NADP-dehydrogenases

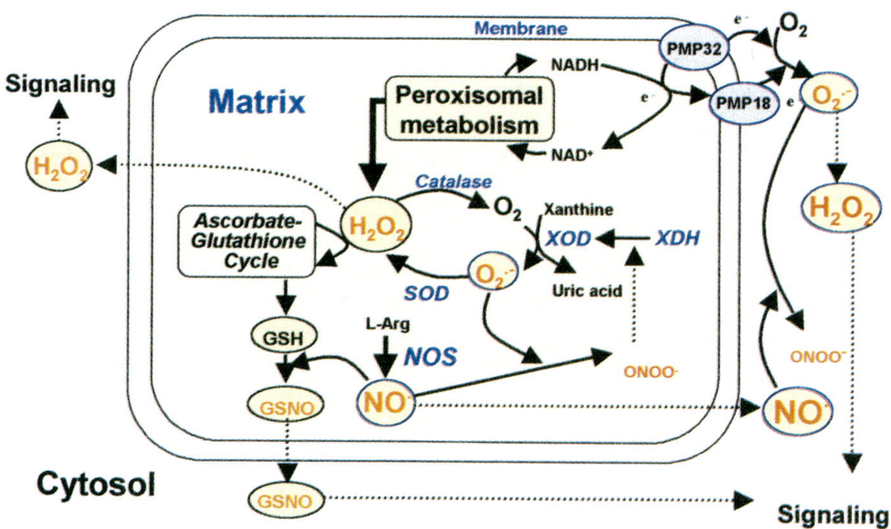

Figure 7 : Hypothetical model for the function of NOS/NO in leaf peroxisomes and the role of these organelles in the generation of the signal molecules hydrogen peroxide (H_2O_2), superoxide radicals ($O_2^{\cdot -}$), nitric oxide ($NO^{\cdot}$), and *S*-nitrosoglutathione (GSNO). PMP32, 32-kDa peroxisomal membrane polypeptide. PMP18, 18-kDa peroxisomal membrane polypeptide. L-Arg, L-arginine. $ONOO^-$, peroxynitrite. GSH, reduced glutathione. (*Journal of Experimental Botany* **53**, 1267; 2002. Copyright 2002 Society for Experimental Biology, UK).

found in the peroxisomal matrix (Corpas *et al.*, 1998, 1999) could provide the necessary NADPH for the $NO^{\cdot}$-producing NOS reaction. But inside peroxisomes $NO^{\cdot}$ could react with $O_2^{\cdot -}$ radicals generated in the peroxisomal matrix by xanthine oxidase (del Río *et al.*, 1989, 2002) to form peroxynitrite ($ONOO^-$), a powerful oxidizing reactive nitrogen species (RNS). On the other hand, $NO^{\cdot}$ can react with reduced glutathione (GSH), present in peroxisomes (Jiménez *et al.*, 1997), to generate *S*-nitrosoglutathione (GSNO), another reactive nitrogen species. This compound is a powerful inducer of defence genes (Durner *et al.*, 1998) and can also function as a long-distance signal molecule, transporting glutathione-bound $NO^{\cdot}$ (Durner and Klessig, 1999) and can be metabolized to other bioactive nitrogen oxides (Gaston, 1999).

The occurrence of $NO^{\cdot}$ and NOS in peroxisomes indicates that these organelles can be considered as cellular compartments with the capacity to generate and release into the cytosol important signal molecules such as $NO^{\cdot}$ and probably GSNO, apart from $O_2^{\cdot -}$ and

H_2O_2. These molecules could be used in different signaling pathways (Bolwell, 1999; Neill *et al.*, 2002, 2003; Lamattina *et al.*, 2003) and contribute to a more integrated communication among cell compartments particularly in response to biotic and abiotic stresses. In plants the cellular population of peroxisomes can proliferate during senescence and under different stress conditions (del Río *et al.*, 1998, 2002; López-Huertas *et al.*, 2000; Baker and Graham, 2002), and this has prompted to propose that plant peroxisomes could act as subcellular indicators or sensors of plant stress by releasing the signaling molecules $NO^{\cdot}$, $O_2^{\cdot -}$ and H_2O_2 to the cytosol and triggering a specific gene expression (Corpas *et al.*, 2001).

Further research is necessary to achieve the purification and characterization of the peroxisomal protein responsible for the NOS activity and the cloning of the corresponding gene. This will supply information on NO biosynthesis and its regulation, and will allow to design molecular strategies directed to improve the tolerance of plants to different biotic and abiotic stresses.

ACKNOWLEDGEMENTS

A.M.L. acknowledges a Ph.D. fellowship (F.P.I.) from the Ministry of Education and Science, Spain. This work was supported by the DGESIC, Ministry of Education and Science (grant PB98-0493-01), the European Union (contract HPRN-CT-2000-00094) and the *Junta de Andalucía* (groups CVI 0192 and CVI 0286). We are grateful to Prof. José Rodrigo, *Instituto de Neurobiología "Ramón y Cajal"*, CSIC, Madrid, for his generous donation of the antibody against iNOS, and Dr. Susana Puntarulo, School of Pharmacy and Biochemistry, University of Buenos Aires, Argentina, for her helpful advices about the spin trapping EPR method used. The EPR and confocal laser microscopy analyses, were carried out at the Centre of Scientific Instrumentation of the University of Granada and the Technical Services of the University of Jaén, respectively.

REFERENCES

A-H-Mackerness, S., John, C.F., Jordan, B. and Thomas, B. (2001) Early signaling components in ultraviolet-B responses: distinct roles for different reactive oxygen species and nitric oxide. *FEBS Lett.*, **489** : 237-242.

Alderton, W.K., Cooper, C.E., Knowles, R.G. (2001) Nitric oxide synthases: structure, function and inhibition. *Biochem J.*, **357** : 593-615.

Barroso, J.B., Corpas, F.J., Carreras, A., Sandalio, L.M., Valderrama R., Palma, J.M., Lupiáñez, J.A. and del Río, L.A. (1999) Localization of nitric oxide synthase in plant peroxisomes. *J. Biol. Chem.,* **274** : 36729-36733.

Baker, A. and Graham, I. (2002) Plant peroxisomes. Biochemistry, Cell Biology and Biotechnological Applications. Kluwer Academic Publishers, Dordrecht.

Beevers, H. (1979) Microbodies in higher plants. *Annu. Rev. Plant Physiol.,* **30** : 159-193.

Beligni, M.V. and Lamattina, L. (1999a) Nitric oxide counteracts cytotoxic processes mediated by reactive oxygen species in plant tissues. *Planta,* **208** : 337-344.

Beligni, M.V. and Lamattina, L. (1999b) Nitric oxide protects against cellular damage produced by methylviologen herbicides in potato plants. *Nitric Oxide* **3**, 199-208.

Bolwell, G.P. (1999) Role of active oxygen species and NO in plant defence responses. *Curr. Opin. Plant Biol.,* **2** : 287-294.

Brunelli, L., Yermilov, V. and Beckman J.S. (2001) Modulation of catalase peroxidatic and catalatic activity by nitric oxide. *Free Radic. Biol. Med.,* **30** : 709-714.

Caro, A. and Puntarulo, S. (1999) Nitric oxide generation by soybean embryonic axes. Possible effect on mitochondrial function. *Free Radic. Res.,* **31** : S205-212.

Clark, D., Durner J., Navarre, D.A., Klessig, D.F. (2000) Nitric oxide inhibition of tobacco catalase and ascorbate peroxidase. *Mol. Plant Microbe Interact.,* **13** : 1380-1384.

Corpas, F.J., Barroso, J.B and del Río, L.A. (2001) Peroxisomes as a source of reactive oxygen species and nitric oxide signal molecules in plant cells. *Trends Plant Sci.,* **6** : 145-150.

Corpas, F.J., Barroso, J.B., Sandalio, L.M., Distefano, S., Palma, J.M., Lupiáñez, J.A. and del Río, L.A. (1998) A dehydrogenase-mediated recycling system of NADPH in plant peroxisomes. *Biochem J.,* **330** : 777-784.

Corpas, F.J., Barroso, J.B., Sandalio, L.M., Palma, J.M., Lupiáñez, J.A. and del Río, L.A. (1999) Peroxisomal NADP-dependent isocitrate dehydrogenase. Characterization and activity regulation during natural senescence. *Plant Physiol.,* **121** : 921-928.

Cueto, M., Hernández-Perea, O., Martín, R., Ventura, M.L. Rodrigo, J., Lamas, S. and Golvano, M.P. (1996) Presence of nitric oxide synthase activity in rotos and nodules of *Lupinus albus. FEBS Lett.,* **398** : 159-164.

Delledonne, M., Xia, Y., Dixon, R.A. and Lamb, C. (1998) Nitric oxide functions as a signal in plant disease resistance. *Nature,* **394** : 585-858.

Delledonne, M., Zeier, J., Marocco, A. and Lamb, C. (2001) Signal interactions between nitric oxide and reactive oxygen intermediates in the plant hypersensitive disease resistance response. *Proc. Natl. Acad. Sci. USA.,* **98** : 13454-13459.

del Río, L.A., Corpas, F.J., Sandalio, L.M., Palma, J.M., Gómez, M. and Barroso, J.B. (2002) Reactive oxygen species, antioxidant systems and nitric oxide in peroxisomes. *J. Exp Bot.,* **53** : 1255-1272.

del Río, L.A., Fernández, V.M., Rupérez, F.L., Sandalio, L.M. and Palma J.M. (1989) NADH induces the generation of superoxide radicals in leaf peroxisomes. *Plant Physiol.,* **89** : 728-731.

Denicola, A., Souza, J.M., Radi, R., Lissi, E. (1996) Nitric oxide diffusion in membranes determined by fluorescence quenching. *Arch. Biochem. Biophys.* **328** : 208-212.

Durner. J and Klessig, D.F. (1999) Nitric oxide as a signal in plants. *Curr. Opin. Plant Biol.,* **2** : 369-374.

Durner, J., Wendehenne, D. and Klessig, D.F. (1998) Defense gene induction in tobacco by nitric oxide, cyclic GMP, and cyclic ADP-ribose. *Proc. Natl. Acad. Sci. USA,* **95** : 10328-33.

Elfering, S.L., Sarkela, T.M. and Giulivi, C. (2002) Biochemistry of Mitochondrial Nitric-oxide Synthase. *J. Biol. Chem.,* **277** : 38079-38086.

Esteban, F.J., Jiménez, A., Fernández, A.P., Delmoral, M.L., Sánchez-López, A.M., Hernández, R., Garrosa,M., Pedrosa, J.A., Rodrigo, J. and Peinado, M.A. (2001) Neuronal nitric oxide synthase immunoreactivity in the guinea-pig liver: distribution and colocalization with neuropeptide Y and calcitonin gene-related peptide. *Liver,* **21** : 374-379.

García-Mata, C. and Lamattina, L. (2001) Nitric oxide induce stomatal closure and enhances the adaptive plant responses against drought stress. *Plant Physiol.,* **126** : 1196-1204.

García-Mata, C. and Lamattina, L. (2002) Nitric oxide and abscisic acid cross talk in guard cells. *Plant Physiol.,* **128** : 790-792.

Gaston, B. (1999) Nitric oxide and thiol groups. *Biochim. Biophys. Acta,* **1411** : 385-400.

Ferrer, M.A. and Ros Barcelo, A. (1999) Differential effects of nitric oxide on peroxidase and H_2O_2 production by the xylem of *Zinnia elegans. Plant, Cell Environ.,* **22** : 891-897.

Foissner, I., Wendehenne, D., Langebartels, C. and Durner, J. (2000) In vivo imaging of an elicitor-induced nitric oxide burst in tobacco. *Plant J.,* **23** : 817-824.

Hemmens, B. and Mayer, B. (1998) Enzymology of nitric oxide synthases. *Methods Mol. Biol.,* **100** : 1-32.

Huang, X, Rad, U. and Durner, J. (2002) Nitric oxide induces transcriptional activation of the nitric-oxide-tolerant alternative oxidase in *Arabidopsis* suspension cells. *Planta,* **215** : 914-923.

Huang, A.H.C., Trelease, R.N. and Moore, T.S, Jr. (1983) Plant Peroxisomes. Academic Press, New York.

Igamberdiev, A.U. and Lea, P.J (2002) The role of peroxisomes in the integration of metabolism and evolutionary diversity of photosynthetic organisms. *Phytochemistry,* **60** : 651-674.

Joshi, M.S., Ponthier, J.L. and Lancaster J.R. Jr. (1999) Cellular antioxidant and pro-oxidant actions of nitric oxide. *Free Radic. Biol. Med.,* **27** : 1357-1366.

Kanner, J., Harel, S. and Granit, R. (1992) Nitric oxide, an inhibitor of lipid oxidation by lipoxygenase, cyclooxygenase and hemoglobin. *Lipids,* **27** : 46-49.

Kojima, H., Nakatsubo, N., Kikuchi, K., Kawahara, S., Kirino, Y., Nagoshi, H., Hirata, Y. and Nagano, T. (1998) Detection and imaging of nitric oxide with novel fluorescent indicators:diaminofluoresceins. *Anal. Chem.,* **70** : 2446-2453.

Kotake, Y., Tanigawa, T., Tanigawa, M., Ueno, Y., Allen, D.R. and Lai, C.S. (1996) Continuos monitoring of cellular nitric oxide generation by spin trapping with an iron-dithiocarbamate complex. *Biochim. Biophys. Acta,* **1289** : 362-368.

Lamattina, L., Garcia-Mata, C., Graziano, M. and Pagnussat, G. (2003). Nitric oxide : the versatility of an extensive signal molecule. *Annu. Rev. Plant Biol.,* **54** : 109-136.

Leshem, Y.Y. (1996) Nitric oxide in biological systems. *Plant Growth Regul.,* **18** : 155-159.

Leshem, Y.Y. (2000) Nitric oxide in plants. Occurrence, function and use. Kluwer Academic Publishers, Dordrecht.

Leshem, Y.Y. and Haramaty, E. (1996) The characterization and contrasting effects of the nitric oxide free radical in vegetative stress and senescence of *Pisum sativum* Linn. Foliage. *J. Plant Physiol.,* **148** : 258-263

Leshem, Y.Y., Wills, R.B.H. and Veng-Va Ku, V. (1998) Evidence for the function of the free radical gas-nitric oxide (NO) as an endogenous maturation and senescence regulating factor in higher plants. *Plant Physiol. Biochem.,* **36** : 825-833.

López-Huertas, E., Corpas, F.J., Sandalio, L.M. and del Río, L.A.. (1999) Characterization of membrane polypeptides from pea leaf peroxisomes involved in superoxide radical generation. *Biochem J.,* **337** : 531-536.

López-Huertas, E., Charlton, W.L., Johnson, B., Graham, I.A. and Baker, A. (2000) Stress induces peroxisome biogenesis genes. *EMBO J.,* **19** : 6770-6777.

López-Huertas, E., Sandalio, L.M. and del Río, L.A. (1995) Integral membrane polypeptides of pea leaf peroxisomes: characterization and response to plant stress. *Plant Physiol. Biochem.,* **33** : 295-302.

Lum, H.K., Burtt, Y.K.C. and Lo, S.C.L. (2002) Hydrogen peroxide induces a rapid production of nitric oxide in mung bean (*Phaseolus aureus*). *Nitric Oxide,* **6** : 205-213.

Magalhaes, J.R., Pedroso, MC. and Durzan, D. (1999) Nitric oxide, apoptosis and plant stresses. *Physiol. Mol. Biol. Plants,* **5** : 115-125.

Minorsky, P.V. (2002) Peroxisomes: organelles of diverse function. *Plant Physiol.,* **130** : 517-518

Mittler, R. (2002) Oxidative stress, antioxidants and stress tolerance. *Trends Plant Sci.,* **7** : 405-410.

Modolo, L.V., Cunha, F.Q., Braga, M.R. and Salgado, I. (2002) Nitric oxide synthase-mediated phytoalexin accumulation in soybean cotyledons in response to the *Diaporthe phaseolorum* f. sp. *Meridionalis* Elicitor. *Plant Physiol.,* **130** : 1288-1297.

Moncada, S., Palmer, R.M.J. and Higgs, E..A. (1991) Nitric oxide: physiology, pathophysiology and phamarcology. *Pharmacol. Rev.,* **43** : 109-142.

Murgia, I, Delledonne, M. and Soave, C. (2002) Nitric oxide mediates iron-induced ferritin accumulation in Arabidopsis. *Plant J.,* **30** : 521-528.

Nagano, T. (1999) Practical methods for detection of nitric oxide. *Luminescence.,* **14** : 283-290.

Nagano, T. and Yoshimura,T. (2002) Bioimaging of nitric oxide. *Chem. Rev.,* **102** : 1235-1269.

Nakatsubo, N., Kojima, H., Kikuchi, K., Nagoshi, H., Hirata, Y., Maeda, D., Imai, Y., Irimura, T. and Nagano, T. (1998) Direct evidence of nitric oxide production from bovine aortic endothelial cells using new fluorescence indicators: diaminofluoresceins. *FEBS Lett.,* **427** : 263-266.

Navarre, D.A., Wendehenne, D., Durner, J., Noad, R. and Klessig, D.F. (2000) Nitric oxide modulates the activity of tobacco aconitase. *Plant Physiol.*, **122** : 573-582.

Neill, S.J., Desikan, R., Clarke, A. and Hancock, J.T. (2002a) Nitric oxide is a novel component of abscisic acid signaling in stomatal guard cells. *Plant Physiol.*, **128** : 13-16.

Neill, S.J., Desikan, R., Clarke, A., Hurst, R.D. and Hancock, J.T. (2002b) Hydrogen peroxide and nitric oxide as signaling molecules in plants. *J. Exp. Bot.*, **53** : 1237-1247.

Neill, S.J., Desikan, R. and Hancock, J.T. (2003). Nitric oxide signalling in plants. *New Phytol.*, **159** : 11-35.

Ninnemann, H. and Maier, J. (1996) Indications for the occurrence of nitric oxide synthases in fungi and plants and the involvement in photoconidiation of *Neurospora crassa. Photochem. Photobiol.*, **64** : 393-98.

Orozco-Cardenas, M.L. and Ryan, C.A. (2002) Nitric oxide negatively modulates wound signaling in tomato plants. *Plant Physiol.*, **130** : 487-93.

Pagnussat, G.C., Simontacchi, M., Puntarulo, S. and Lamattina, L. (2002) Nitric oxide is required for root organogenesis. *Plant Physiol.*, **129** : 954-956.

Pedroso, M.C. and Durzan, D.J. (2000) Effect of different gravity environments of DNA fragmentation and cell death in *Kalanchoë* leaves. *Ann. Bot.*, **86** : 983-994.

Pedroso, M.C., Magalhaes, J.R. and Durzan, D.J. (2000a). A nitric oxide burst precedes apoptosis in angiosperm and gymnosperm callus cells and foliar tissues. *J. Exp. Bot.*, **51** : 1027-1036.

Pedroso, M.C., Magalhaes, J.R., Durzan, D. (2000b) Nitric oxide induces cell death in *Taxus* cells. *Plant Sci.*, **157** : 173-180.

Ribeiro, E.A., Cunha, F.Q, Tamashiro, W.M.S.C. and Martins, I.S. (1999) Growth phase-dependent subcellular localization of nitric oxide synthase in maize cells. *FEBS Lett.*, **445** : 283-286.

Rockel, P., Strube, F., Rockel, A., Wildt, J. and Kaiser, W.M. (2002) Regulation of nitric oxide (NO) production by plant nitrate reductase *in vivo* and *in vitro*. *J. Exp. Bot.*, **53** : 103-110.

Sakuma, S., Fujimoto, Y., Sakamoto, Y., Uchiyama, T., Yoshioka, K., Nishida, H. and Fujita, T. (1997) Peroxynitrite induces the conversion of xanthine dehydrogenase to oxidase in rabbit liver. *Biochem. Biophys. Res. Commun.*, **230** : 476-479.

Saviani, E.E., Orsi, C.H., Oliveira, J.F.P., Pinto-Maglio, C.A.F. and Salgado, I. (2002) Participation of the mitochondrial permeability transition pore in nitric oxide induced plant cell death. *FEBS Lett.*, **510** : 136-140.

Sharma, V.S., Traylor, T.G., Gardiner, R. and Mizukami, H. (1987) Reaction of nitric oxide with heme proteins and model compounds of hemoglobin. *Biochemistry,* **26** : 3837-3842.

Stolz, D.B, Zamora, R, Vodovotz, Y, Loughran, P.A., Billiar, T.R., Kim, Y.M., Simmons, R.L., Watkins SC. (2002) Peroxisomal localization of inducible nitric oxide synthase in hepatocytes. *Hepatology,* **36** : 81-93.

Suzuki, N., Kojima, H., Urano, Y., Kikuchi, K., Hirata, Y. and Nagano, T. (2002) Orthogonality of calcium concentration and ability of 4,5-diaminofluorescein to detect NO. *J. Biol. Chem.*, **277** : 47-49.

Tabak, H.F., Braakman, I. And Distel, B. (1999) Peroxisomes: simple in function but complex in maintenance. *Trends Cell Biol.,* **9** : 447-453

Tu, N.N., Holk, A. and Scherer, G.F.E. (2001) Rapid increase of NO release in plant cell cultures induced by cytokinin. *FEBS Lett.*, **509** : 174-176.

Uttenthal, L.O., Alonso, D., Fernandez, A.P., Campbell, R.O., Moro, M.A., Leza, J.C., Lizasoain, I., Esteban, F.J., Barroso, J.B., Valderrama, R., Pedrosa, J.A., Peinado, M.A., Serrano, J., Richart, A., Bentura, M.L., Santacana, M., Martinez-Murillo, R., Rodrigo, J. (1998) Neuronal and inducible nitric oxide synthase and nitrotyrosine immunoreactivities in the cerebral cortex of the aging rat. *Microsc. Res. Tech.,* **43** : 75-88.

Van Bel, A.J.E., Ehlers, K. and Knoblauch, M. (2002) Sieve elements caught in the act. *Trends Plant Sci.,* **7** : 126-132.

Walz, C., Juenger, M., Schad, M. and Kelr, J. (2002) Evidence for the presence and activity of a complete antioxidant defence system in mature sieve tubes. *Plant J.*, **31** : 189-197.

Wendehenne, D., Pugin, A., Klessig, D.F. and Durner, J. (2001) Nitric oxide: comparative synthesis and signaling in animal and plant cells. *Trends Plant Sci,* **6** : 177-83.

Wojtaszek, P. (2000) Nitric oxide in plants. To NO or not to NO. *Phytochemistry,* **54** : 1-4.

Xia, Y., Cardounel, A.J., Vanin, A.F., Zweier, J.L. (2000) Electron paramagnetic resonance spectroscopy with N-methyl-D-glucamine dithiocarbamate iron complexes distinguishes nitric oxide and nitroxyl anion in a redox-dependent manner: applications in identifying nitrogen monoxide products from nitric oxide synthase. *Free Radic. Biol. Med.,* **29** : 793-797.

Xia, Y. and Zweier, J.L. (1997) Direct measurement of nitric oxide generation from nitric oxide synthase. *Proc. Natl. Acad. Sci. USA*, **94** : 12705-12710.

Yamaguchi, J. and Nishimura, M. (1984) Purification of glyoxysomal catalase and immunochemical comparison of glyxysomal and leaf peroxisomal catalase in germinating pumpkin cotyledons. *Plant Physiol.,* **262** : 261-267.

Yamasaki, H. and Sakihama, Y. (2000) Simultaneous production of nitric oxide and peroxynitrite by plant nitrate reductase: in vitro evidence for the NR-dependent formation of active nitrogen species. *FEBS Lett.,* **468** : 89-92

Yamasaki, H., Sakihama, Y. and Takahashi, S. (1999) An alternative pathway for nitric oxide production in plants: new features of an old enzyme. *Trends Plant Sci.,* **4** : 128-129.

Yang, T. and Poovaiah, B.W. (2002) Hydrogen peroxide homeostasis: activation of plant catalase by calcium/calmodulin. *Proc. Natl. Acad. Sci. USA*, **99** : 4097-4102.

Chapter 6

NITRIC OXIDE AS A MEDIATOR OF ABSCISIC ACID SIGNALING IN GUARD CELLS

Steven Neill*, Radhika Desikan, Jo Bright and John Hancock

Centre for Research in Plant Science, Faculty of Applied Sciences, University of the West of England, Bristol, Coldharbour Lane, Bristol BS16 1QY, UK
*Corresponding author : E-mail : Steven.Neill@uwe.ac.uk

Summary

Recent data indicate that nitric oxide (NO) is an endogenous mediator of stomatal closure induced by abscisic acid (ABA). Incubation of epidermal peels from several species, including Arabidopsis thaliana, Pisum sativum *and* Vicia faba, *with various nitric oxide donors, induces stomatal closure in a dose-dependent manner. Such stomatal closure is attenuated by co-incubation with NO scavengers. ABA-induced stomatal closure is also inhibited substantially by co-incubation with these NO scavengers, indicating a requirement for endogenous NO generated in response to ABA. ABA-induced NO synthesis in guard cells has been demonstrated using microscopy and a cell-permeable NO-sensitive fluorescent dye. In* A. thaliana *guard cells, NO is generated from nitrite via nitrate reductase: nitrite stimulates NO synthesis and stomatal closure, and stomata of the NR-deficient* nia1, nia2 *mutant do not synthesise NO in response to ABA. Pharmacological and genetic data obtained thus far indicate that NO signaling in guard cells involves synthesis and action of cyclic GMP and cyclic ADP ribose, in addition to elevated cytosolic*

In : Nitric Oxide Signaling in Higher Plants, 2004
(Eds Jose R. Magalhaes, Rana P. Singh and Leonidas P. Passos)
Studium Press, LLC, Houston, USA, pp 131-148

calcium and the action of protein kinases and protein phosphatases.

Keywords : abscisic acid, calcium, cyclic ADP ribose, cyclic GMP, guard cells, nitrate reductase, nitric oxide, nitric oxide synthase, reversible protein phosphorylation, signalling, stomata

1. INTRODUCTION

Water availability is one of the major environmental determinants of plant growth and crop yield as drought stress causes cellular dehydration that leads to a range of negative responses. Leaf expansion is particularly sensitive to dehydration and ceases when the cellular pressure potential falls below the critical value required for cell wall expansion. Other cellular processes such as photosynthesis and general metabolism are also impaired as desiccation develops, so it is not surprising that mechanisms have evolved that allow plants to combat water deficit (Bray, 1997). A key response to reduced soil water content is stomatal closure, initiated by the synthesis, re-distribution and transport of the endogenous anti-transpirant abscisic acid (ABA). ABA cellular re-distribution and *de novo* synthesis are triggered at pressure potentials approaching zero and transport of ABA in the transpiration stream to the guard cells elicits stomatal closure. ABA-induced stomatal closure is mediated via a number of inter-connecting intracellular signaling pathways, with several intermediates, including nitric oxide (NO), having only recently been identified. In this chapter, we outline data implicating the involvement of NO during ABA-induced stomatal closure and discuss some of the research questions relating to NO synthesis and action in guard cells.

2. ABA AND STOMATAL MOVEMENTS

Stomata operate as homeostatic sensory and effector systems that integrate various stimuli, including light, CO_2 concentrations, water potential and hormonal signals, to optimise Water Use Efficiency (WUE), a measure of the efficiency with which CO_2 influx is facilitated at the expense of water loss. Stomatal opening and closing are brought about via osmotic influx and efflux of water across the guard cell plasma membrane and tonoplast that follow the influx and efflux of K^+ and Cl^- ions (Schroeder *et al.*, 2001a,b). ABA activates stomatal closure via a complex web of potentially inter-linked pathways, with elevation of cytosolic calcium, achieved via both influx from the extracellular milieu and efflux from intracellular stores, representing an essential common step regulating subsequent ion channel activity. Potential calcium-mobilising second messengers mediating ABA-induced stomatal closure include inositol trisphosphate (IP_3), inositol hexakisphosphate (IP_6), cyclic ADP ribose (cADPR) and sphingosine-1-phosphate (S1P) (see Hetherington, 2001; Schroeder *et al.*, 2001b). Other components of guard cell ABA signal cascades include trimeric and monomeric G proteins, protein kinases and phosphatases, phospholipase D, syntaxins, hydrogen peroxide (H_2O_2), actin rearrangements, membrane trafficking, cytosolic alkalinisation, gene expression, and RNA processing and metabolism (Hetherington, 2001; Lemichez *et al.*, 2001; Li *et al.*, 2002; Schroeder *et al.*, 2001a; Wang *et al.* 2001). It is clear that ABA signalling in guard cells, perhaps like most intracellular signalling, operates via a network-based system with inter-linked “ABA signalling modules”. Such a network structure may be needed to provide robustness and flexibility, and be essential for the interpretation and processing of multiple signals (Hetherington, 2001). Recent data indicate that NO is yet another component that is somehow integrated into this ABA guard cell signaling network.

3. NO-MEDIATION OF ABA-INDUCED STOMATAL CLOSURE

It is now clear that NO is an important endogenous signaling molecule in plants, involved in several developmental and physiological processes and responses to the environment (Neill *et al.*, 2003 and chapters in this book). NO is a short-lived gas, so most experiments have utilised NO donors in order to determine the effects of exogenous NO. It should be recognised, however, that different NO donors might generate different molecular species of NO and that the donors themselves, or donor residues, may be biologically active (see Neill *et al.*, 2003). Thus, similar responses to different NO donors and a lack of response to depleted NO donors provide more confidence that the responses really are related to NO *per se*. In addition, inhibition of NO responses with NO scavengers, that remove NO before it induces biological responses, such as PTIO (2-phenyl-4,4,5,5-tetramethylimidazolinone-1-oxyl 3-oxide) or carboxy-PTIO (cPTIO) provides further evidence that any observed effects are linked to NO.

In 2001 Garcia-Mata and Lamattina showed that exogenous NO generated from SNP (sodium nitroprusside) or SNAP (*S*-nitroso-*N*-acetylpenicillamine) initiated stomatal closure in epidermal peels of *Vicia faba* (see Fig. 1). SNP also induced stomatal closure in epidermal peels of *Tradescantia* sp and *Salpichroa organifolia*; incubation with cPTIO attenuated the effects of SNP on stomatal closure in all three species. These effects of NO on stomatal closure have been similarly demonstrated in *Pisum sativum* and *Arabidopsis thaliana,* in which SNP and another NO donor *S*-nitrosoglutathione (GSNO) also induced stomatal closure (Fig. 1; Desikan *et al.*, 2002; Neill *et al.*, 2002b). Although NO is known to induce programmed cell death in other systems, it did not reduce the viability of guard cells and its effects were fully reversible (Neill *et al.*, 2002b). NO responses are dose-dependent, with stomatal apertures declining with increasing concentrations of NO donor. However, the NO-closing response is diminished at NO concentrations above a certain threshold (Fig. 1). Detailed kinetic analyses with *P. sativum* epidermal peels revealed that high concentrations of SNP induced an initial stomatal closure, but that the stomata then re-opened (Desikan *et al.*, data not shown). The mechanisms underlying these differential responses are not yet known, but they may well be of some physiological significance, especially if stimuli other than ABA, for example drought stress *per se*, induce NO synthesis and release, and thus exposure of guard cells to elevated amounts of NO *in planta*. The apparent discrepancy between the recent data of Sakihama *et al.* (2003), who reported that SNAP induced stomatal opening in *Vicia faba*, and those of Garcia-Mata and Lamattina (2001), working with the same species, might be explained by this concentration effect: Sakihama *et al.* (2003) used SNAP at 5 mM, whereas the highest concentration used by Garcia-Mata and Lamattina was 500 μM, ten times lower. Garcia-Mata and Lamattina (2001) observed that SNP reduced the transpiration rate of detached wheat leaves when applied via the transpiration stream, a response also seen in *Arabidopsis*, barley, tomato and wheat (Clarke *et al.*, unpublished).

Other studies have demonstrated that guard cells themselves can synthesise NO, and that NO is required for stomatal closure initiated by ABA. Removal of NO by co-incubation with the NO scavengers

PTIO or cPTIO attenuates ABA-induced stomatal closure in *P. sativum*, *V.faba* and *A. thaliana* (Fig. 1; Desikan *et al.*, 2002; Garcia-Mata and Lamattina, 2002; Neill *et al.*, 2002b). Inhibition of ABA effects by PTIO is always substantial but not always total (Desikan *et al.*, unpublished), possibly reflecting inherent flexibility in ABA

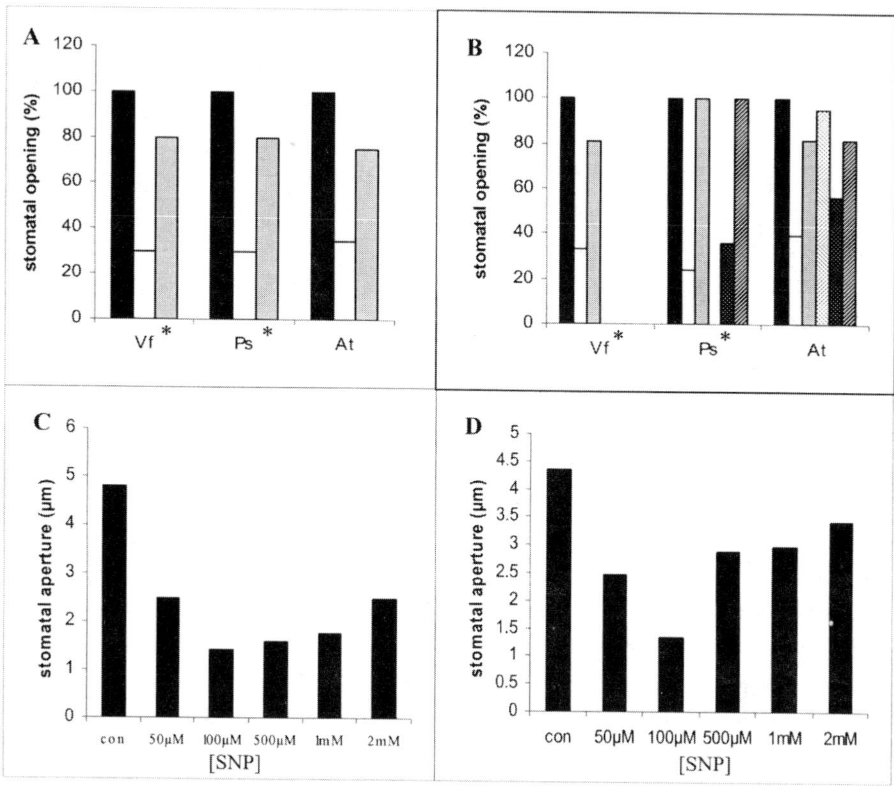

Figure 1 : Nitric oxide mediates ABA-induced stomatal closure.
(A) Stomatal closure induced by ABA is inhibited by co-incubation with an NO-scavenger in *V. faba* (Vf), *P. sativum* (Ps) and *A. thaliana* (At) ■ = control; □= ABA (10 µM); □= ABA + cPTIO (200 µM, Vf); ABA + PTIO (200 µM, Ps and At). (B) Nitric oxide provided by NO donors induces stomatal closure that is attenuated by co-incubation with an NO-scavenger in *V. faba* (Vf), *P. sativum* (Ps) and *A. thaliana* (At) ■ = control; □= SNP (200 µM Vf, 100 µM Ps, 50 µM At); = SNP + cPTIO (Vf 200 µM); SNP + PTIO (Ps, At 200 µM); □= as per Figure 1B GSNO (100 µM); ■= GSNO + PTIO (200 µM); ▨= light-inactivated SNP (50 µM). (C, D) Dose-dependence of SNP-induced stomatal cloure in *P. sativum* (C) *and A. thaliana* (D).
Figure includes data (*) redrawn from Garcia-Mata and Lamattina (2001, 2002) and Neill *et al.* (2002b) with permission from American Society of Plant Biologists.

guard cell signalling (see above). In order to determine if ABA induces NO synthesis in guard cells, a fluorescence-based microscopic NO assay has been used. This employs the cell-permeable NO sensitive dye diaminofluorescein diacetate (DAF-2 DA). DAF-2 DA is hydrolysed in cells to non-fluorescent DAF-2, which can be converted to the fluorescent product triazofluorescein (DAF-2T) by reaction with NO, and thus function as an intracellular NO monitor (Kojima *et al.*, 1998). Using this assay, it is clear that ABA does induce a rapid and substantial increase in NO production in stomatal guard cells of *P. sativum* , *V. faba* and *A. thaliana* (Fig. 2). By quantifying pixel intensity from such fluorescence measurements, it is possible to obtain an estimate of relative NO content. Significantly, in *A. thaliana*, a maximum response is achieved within 10 min, before any reductions in stomatal aperture are apparent, kinetics in keeping with a role for NO in mediating ABA responses (Desikan *et al.*, 2002, Neill *et al.*, 2003). In all three species, fluorescence is observed throughout the cytoplasm and, at least in *P. sativum* and *A. thaliana*, intense fluorescence is often observed associated with the chloroplasts. It is clearly important to verify that this assay really does report NO concentrations and not non-specific fluorescence. ABA-induced fluorescence is largely abolished by co-incubation with PTIO or cPTIO (Desikan *et al.*, 2002; Garcia-Mata and Lamattina, 2002; Neill *et al.*, 2002b) demonstrating that the observed increases in fluorescence are a result of NO production. In addition, the non-specific negative probe 4-AF DA was used with *V. faba* and *A. thaliana*. In neither case did ABA induce an increase in guard cell fluorescence (Desikan *et al.*, 2002; Garcia-Mata and Lamattina, 2002). It has been reported that DAF-2 DA-mediated fluorescence is amplified in the presence of Ca^{2+} ions, although still dependent on the presence of NO (Broillet *et al.*, 2001). This is an important point, because it is well known that ABA and other stimuli increase the cytosolic Ca^{2+} content of guard cells. However, treatment with EGTA-AM, a membrane-permeable form of the Ca^{2+} chelator EGTA (Wu *et al.*, 1997), did not reduce ABA-induced DAF-2 DA fluorescence in *P. sativum* (Neill *et al.*, 2002b). The correlations between NO content and stomatal aperture following treatment with ABA and an NO scavenger, coupled with the fact that NO itself induces stomatal closure, provide excellent evidence that NO truly is an endogenous mediator of ABA-induced stomatal closure.

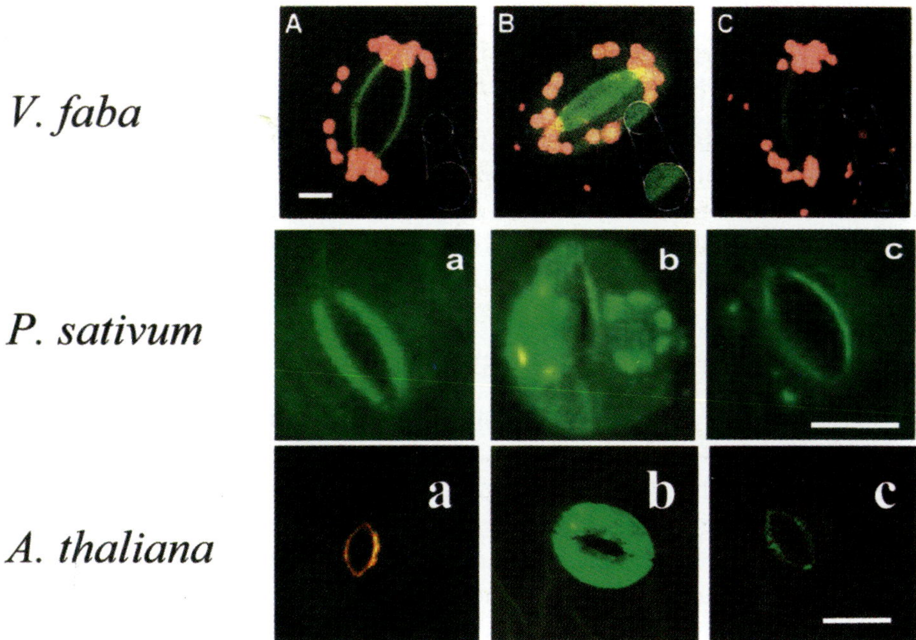

Figure 2 : ABA-induced NO synthesis visualised by DAF-2 DA fluorescence, in guard cells of *V. faba*, *P. sativum* and *A. thaliana*.
A, a : control.
B, b : ABA (*V. faba* 10 µM, 30 min; *P. sativum* (10 µM, 30 min); *A. thaliana* (50 µM, 30 min).
C, c : ABA + cPTIO (*V. faba* 200 µM) or ABA + PTIO (*P. sativum*, *A. thaliana* 200 µM).
Bars = 5 µm (V. *faba*), 7 µm (*P. sativum*), 14 µm (*A. thaliana*).
Data for *V. faba* and *P. sativum* reproduced from Garcia-Mata and Lamattina (2001) and Neill *et al.* (2002b) respectively, with permission from American Society of Plant Biologists.

The DAF-2 DA assay only indicates relative changes in NO content, not actual values. It will be important therefore to quantify the actual steady-state concentrations of NO achieved within guard cells. Whatever the biosynthetic source of guard cell NO, it will have to be sufficient to account for the amounts of NO turned over. Neill *et al.* (2002b) used the haemoglobin assay, that measures extracellular NO, to estimate an NO production rate of 2 nmol g^{-1} min^{-1} in ABA-treated *P. sativum* epidermal peels. It must be noted though, that this was from whole peels, not guard cells. Moreover, the relationship between NO release and intracellular NO content is not known, nor are the sites of NO action within guard cells.

In addition to its role in ABA-induced closure, a potential signalling role for NO during stomatal closure in the dark is suggested by the observation of Garcia-Mata and Lamattina (2001) that incubation with cPTIO partly prevented dark-induced stomatal closure in *V. faba*. Incubation in PTIO has a similar effect in *P. sativum* (Desikan *et al.*, unpublished).

4. NITRIC OXIDE SYNTHESIS IN GUARD CELLS

The biosynthetic origins of NO in plants have yet to be resolved, but both enzymatic and non-enzymatic sources have been suggested (see Stohr and Ullrich, 2002; Garcia-Mata and Lamattina, 2003; Neill *et al.*, 2003). In mammals, NO is synthesised largely from nitric oxide synthase (NOS), although other routes do exist (Neill *et al.*, 2003). NOS has not been identified unequivocally in plant tissues. NOS-like enzyme activity certainly exists (e.g. Barroso *et al.*, 1999), but to date no NOS-like DNA sequences have been identified or isolated. Interactions of plant proteins with antibodies raised against mammalian NOS have been used as evidence for the existence of plant NOS-like enzymes, but the lack of specificity is well known and has been demonstrated recently (Butt *et al.*, 2003). Similarly, inhibition of NO synthesis and responses with inhibitors of mammalian NOS have been used as evidence for NOS-like activity. Incubation of *P. sativum* epidermal peels with the mammalian NOS inhibitor L-NAME (N^G-nitro-L-Arg-methyl ester, an arginine analogue and inhibitor of mammalian NOS and plant NOS-like activity (Barroso *et al.*, 1999)) inhibited both ABA-induced stomatal closure and NO synthesis in pea (Neill *et al.*, 2002b). However, it is important to recognise that inhibitors can be notoriously non-specific and that NOS inhibitors might actually inhibit other NO-biosynthetic enzymes (Garcia-Mata and Lamattina, 2003).

It has long been known that nitrate reductase (NR) can synthesise NO from nitrite, NO_2^-, and increasing evidence indicates powerful physiological roles for this reaction (Garcia-Mata and Lamattina, 2003; Neill *et al.*, 2003 and chapter 8 in this book). Yamasaki and Sakihama (2000) characterised NO generation from NR *in vitro*, and

Rockel *et al.* (2002) showed that NR-mediated NO production occurs *in vivo* and is highly regulated by nitrite levels. Desikan *et al.* (2002) used a combined pharmacological, physiological and genetic approach to probe the requirement for NR-mediated NO synthesis during ABA-induced stomatal closure in *A. thaliana*. Firstly, incubation with nitrite induced stomatal closure in a dose-dependent manner. Secondly, treatment with the potential NOS inhibitor L-NAME did not inhibit ABA-induced stomatal closure or NO synthesis, whereas incubation with tungstate did. Tungstate is an inhibitor of NR that can replace molybdenum in the molybdenum cofactor essential for NR function (Notton and Hewitt, 1971; Mendel and Hansch, 2002). Tungstate does not inhibit stomatal closure induced by SNP, H_2O_2 or darkness and preliminary data indicate rapid inhibition of NO-generating NR activity *in vitro*, indicating that these effects of tungstate are physiologically relevant (Bright and Desikan, unpublished). Thirdly, treatment with either ABA or nitrite induced NO synthesis in wild type guard cells. Fourthly, in epidermal peels of the NR-deficient *nia1,nia2* double mutant (there are two NR genes in the *Arabidopsis* genome) neither nitrite nor ABA induced stomatal closure or NO synthesis (Desikan *et al.*, 2002). However, *nia1,nia2* stomata do respond to SNP, H_2O_2 or darkness by closure, indicating that there is not a general impairment of stomatal function in this mutant.

These data provide compelling evidence for a signalling role for NR in guard cells, to generate NO in response to ABA and thereby induce stomatal closure. Nevertheless, several key questions remain (and see Garcia-Mata and Lamattina, 2003). Firstly, it will be essential to determine which of the two NR enzymes are present and functional in guard cells (both genes are expressed in leaves (Crawford and Forde, 2002)). No doubt *Arabidopsis* gene chip experiments will soon be useful here. Using the already available and ever-increasing T-DNA and transposon insertion collections, as well as knock-down approaches, ideally with reduced NR expression restricted to guard cells, it should then be possible to determine whether one or both genes are required, and to assess fully the contribution of guard cell NR to whole plant water relations, as well as to ABA signalling in stomatal guard cells. ABA causes increased NO synthesis in guard cells, so it will be informative to assess the NO-generating NR activity of control and ABA-treated cells, as both the *NIA* genes and the NR protein respond to many signals (Crawford and Forde, 2002). What is the sub-cellular location of guard cell NR? The available literature indicates that NR is a cytosolic enzyme but, at least in some cases, NO levels seem to be particularly high in or around chloroplasts (see Fig. 2). NR-GFP fusions may reveal the sub-cellular localisation of NR. It will also be important to determine which cellular protein partners interact with NR. NR activity is known to be controlled via phosphorylation and dephosphorylation and interactions with 14-3-3 proteins, so yeast two-hybrid and proteomic identification of the potential regulatory partners and 'signalosomes' are attractive propositions.

What are the concentrations of nitrate and nitrite in the transpiration stream and in and around guard cells? Typically, the gross nitrite concentration in leaves of *Arabidopsis*, for example, is low, of the order of 10 μM (Bright, unpublished), yet higher exogenous concentrations than this are required to initiate stomatal closure. However, it may be that there are nitrite 'hot-spots' or, more likely perhaps, it is the rate of turnover of nitrate to nitrite and then on to NO, that is important. Nitrate can drive NO production *in vivo* (Yamasaki *et al.*, 2000) so it could be that steady state nitrite concentrations may not need to increase if NR activity does. NAD(P)H availability may also affect NR activity.

It is possible that alternative NO sources exist in guard cells. In mammals, the enzyme xanthine oxidoreductase (another flavin-molybdo-enzyme) can generate NO from nitrite (Godber *et al.*, 2000) and it may similarly catalyse this reaction in plants. Stohr and colleagues have described a new NO–generating enzyme activity in tobacco roots (see Stohr and Ullrich, 2002). This enzyme, termed nitrite:NO reductase (NI-NOR) is plasma membrane-located and has different properties to NR. It has been speculated that NI-NOR co-localises at the apoplastic side of the plasma membrane with a plasma membrane form of NR (PM-NR) that produces the nitrite subsequently utilised by NI-NOR (see Stohr and Ullrich, 2002). Thus, NI-NOR would be well-positioned to generate NO from apoplastic nitrite and function in a signaling role. It will be of interest to determine if similar enzymes are present in guard cells. NO can be generated non-enzymatically from nitrite at low pH in the presence of a reductant such as ascorbate or glutathione. Whether this can occur within guard cells (e.g. in chloroplasts) or in the leaf apoplast remains to be seen.

5. NITRIC OXIDE METABOLISM

Although NO is only a short-lived molecule, it is possible that mechanisms to sequester it exist. For example, GSNO is formed by the reaction of NO with glutathione, a ubiquitous metabolite present at high concentrations in plant cells, and may be phloem- or xylem-mobile. The enzyme GSNO reductase has recently been characterised as having a key role in degrading GSNO and thereby reducing nitrosative stress (Liu *et al.*, 2001). The same enzyme exists in plants and may well have a similar functional role (Sakamoto *et al.*, 2002). Haemoglobins are also known to operate as NO scavengers and again these proteins may have a similar function in plants (Hunt *et al.*, 2001).

6. NITRIC OXIDE SIGNAL TRANSDUCTION

Stomatal movements require modulation of the activity of various ion channels such as those for K^+, Cl^- and Ca^{2+}, located on both the plasma membrane and internal membranes such as the tonoplast. As NO induces stomatal closure, it must in some way alter the activity of these ion channels. This could be a direct effect, or may occur via the generation of second messengers and activation of intracellular signalling cascades. NO can alter protein conformation, and thus activity, by reaction with haem groups or via nitrosylation of thiol moieties (Wendehenne *et al.*, 2001; Neill *et al.*, 2003). Thus, it is possible that NO might directly effect changes in ion channel activity. *S*-nitrosylated proteins have been identified in mammalian cells (Jaffrey *et al.*, 2001) and it would certainly be of interest to apply similar proteomic analyses to NO-challenged guard cells.

7. NITRIC OXIDE AND CALCIUM

Increased cytosolic calcium induces stomatal closure and both influx from the cell exterior and release from intracellular stores are activated during ABA-induced stomatal closure. Calcium increases often oscillate, and the nature of these oscillations may reflect both the initial stimulus and the sources of the increased calcium (Schroeder *et al.*, 2001a,b). Calcium acts as a second messenger to activate various signal cascades, for example reversible protein phosphorylation, that culminate in the modulation of ion channel activity, although not all channels are regulated by Ca^{2+} (Hetherington, 2001). As yet, there are no data describing the effects of NO on guard cell Ca^{2+} elevations or ion channel activity, but removal of extracellular Ca^{2+} via the calcium chelator EGTA prevented SNP-induced stomatal closure in *V. faba* (Garcia-Mata and Lamattina, 2001). Undoubtedly, electrophysiological and calcium imaging experiments will be informative with regard to NO effects on ion channel activities.

8. NITRIC OXIDE AND CYCLIC ADP RIBOSE

Calcium elevations in guard cells are brought about via several mechanisms, including those involving synthesis and action of the second messenger cADPR (Fig. 3). In animals, cADPR is synthesised in response to NO by the enzyme ADP ribosyl cyclase, and functions by binding to cADPR-sensitive Ca^{2+} channels on internal membranes, thereby inducing intracellular Ca^{2+} release (see Wendehenne *et al.*, 2001; Neill *et al.*, 2003). As yet, neither cADPR nor its biosynthetic enzymes have been identified unequivocally in plants. However, there is already good pharmacological and physiological evidence that cADPR mediates ABA-induced calcium release in guard cells (Leckie *et al.*, 1998; MacRobbie, 2000). It is possible that these effects

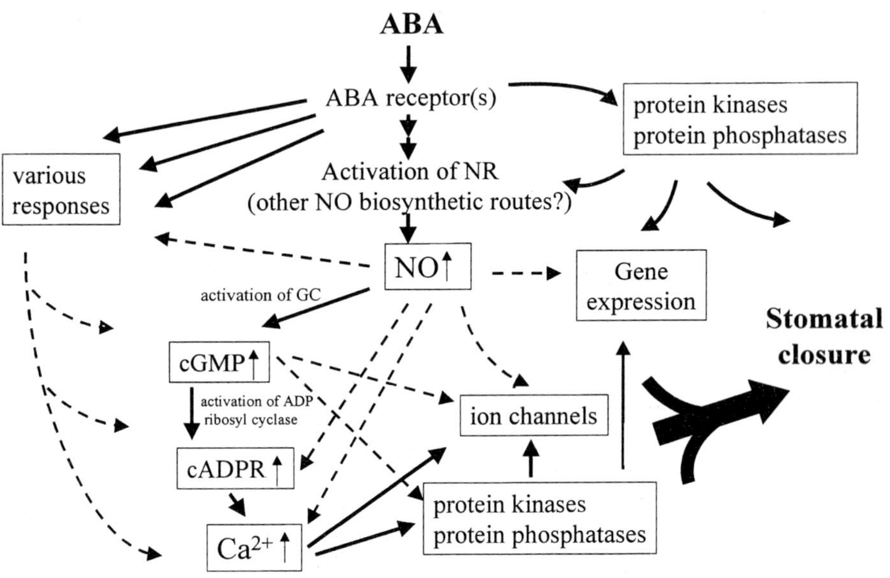

Figure 3 : Potential NO signaling in stomatal guard cells.
Solid arrows represent effects for which there is supporting evidence. Broken arrows represent potential effects for which there are not yet experimental data.

reflect ABA-induced NO synthesis and action in guard cells. Nicotinamide, an antagonist of cADPR production, inhibited stomatal closure induced by either ABA or NO in *P. sativum* (Neill *et al.*, 2002b). Quantification of cADPR, and cloning and manipulation of cADPR synthetic enzymes, will be required for more detailed analyses.

It remains to be determined whether or not NO signalling also involves other Ca^{2+}-releasing agents such as IP_3 and S1P. Inhibitors of IP_3 synthesis are available (Staxen *et al.*, 1999) and phospholipase C-deficient plants will also be valuable research tools (Hunt *et al.*, 2003).

9. NITRIC OXIDE AND CYCLIC GMP

A common response to NO is a transient increase in the second messenger cGMP, achieved via direct activation by NO of its biosynthetic enzyme guanylyl cyclase (GC) (Fig. 3; Wendehenne *et al.*, 2001; Neill *et al.*, 2003). Despite early controversy, the presence of cGMP in plants has now been demonstrated conclusively and there is good evidence for the involvement of cGMP in several NO-mediated processes (Minorsky, 2003; Neill *et al.*, 2003). Pharmacological data indicate the requirement for cGMP during ABA-induced stomatal closure in *P. sativum* (Neill *et al.*, 2002b) and *Arabidopsis* (Desikan *et al.*, unpublished). Co-incubation with ABA and ODQ (1*H*-(1,2,4)-oxadiazole-[4,3-*a*]-quinoxalin-1-one, an inhibitor of NO-sensitive guanylyl cyclase) inhibits stomatal closure induced by either ABA or NO. Inclusion of the cell-permeable cGMP analogue 8-Bromo-cGMP in the incubation medium negates the effects of ODQ, but incubation in 8-Bromo-cGMP alone has little effect, indicating that cGMP is required, but not sufficient for ABA and NO-induced stomatal closure. It is essential that these pharmacological data are extended by quantifying the guard cell cGMP content following stimulation, and by cloning and manipulating the enzymes of cGMP turnover. Several potential cGMP phosphodiesterase genes exist in the *Arabidopsis* genome and cGMP phosphodiesterase activity has been found in plants (see Neill *et al.*, 2003). Interrogation of plant databases with GC sequences from other eukaryotes does not reveal any obvious similarities. However, Ludidi and Gehring (2003) have recently reported the cloning of *AtGC1*, a gene encoding an enzyme with GC activity but a domain structure quite different to other GCs already characterised. This is an exciting development, and expression and functional analyses of this enzyme in relation to stomatal biology is an urgent research priority.

ADP ribosyl cyclase is a downstream target for cGMP in mammalian cells and possibly also in guard cells, inferring that calcium elevations may involve cGMP. Other cGMP targets include cGMP-gated ion channels. Cyclic nucleotide gated ion channels (CNGCs) have been cloned from plants (Trewavas *et al.*, 2003) and it will be interesting to assess their activity in guard cells.

10. REVERSIBLE PROTEIN PHOSPHORYLATION

Reversible phosphorylation is a central component of plant cell signalling, with guard cell signalling being no exception (Hetherington, 2001; Schroeder *et al.*, 2001a,b). ABA activates several protein kinases (PKs) in guard cells, and some ABA-insensitive mutants are mutated in genes encoding protein phosphatases (PPs) that include the two related PP2C enzymes ABI1 and ABI2 as well as a PP2A enzyme (Kwak *et al.*, 2002). Guard cells of the ABA-insensitive *abi1* and *abi2* mutants do not close in response to NO but do synthesise NO in response to ABA, indicating that the action of these two PP2C enzymes is downstream of NO synthesis (Desikan *et al.*, 2002). NR activation requires dephosphorylation, so it is possible that a PP enzyme mediates ABA responses by driving NO production (Fig. 3). In pea epidermal peels, ABA activates a PK with the characteristics of a mitogen-activated PK (MAPK) (Burnett *et al.*, 2000) and NO has been shown to activate a MAPK in other systems (Clarke *et al.* 2000, Kumar and Klessig, 2000). The MAPKK inhibitor PD98059 inhibits both ABA- and NO-induced stomatal closure and MAPK activation in pea epidermis, and NO activates a PK similar to that activated by ABA (Burnett *et al.*, 2000; Desikan *et al.* unpublished). Thus, MAPK activation may also be an integral part of the NO signal cascade in guard cells.

11. NO AND GENE EXPRESSION IN GUARD CELLS

It is clear that guard cells alter their spectrum of gene expression in response to ABA, and recent exciting data indicate a role for ABA in the modulation of RNA metabolism and processing (Hugouvieux *et al.*, 2001; Li *et al.*, 2002). It is also apparent that NO can alter gene expression (see Neill *et al.*, 2003); how this might affect guard cell functioning awaits to be discovered.

12. CONCLUSIONS AND FUTURE PROSPECTS

There can be no doubt that NO now ranks as an important endogenous signalling molecule in plants and that it is one of the key players in the game of ABA signalling in guard cells. Its effects are likely to impact on, and in turn be affected by, other signalling processes, in guard cells in particular, and in the plant in general. For example, both NO and H_2O_2 are made in other cells in tandem in response to various stresses (Neill *et al.*, 2002a), as they are in guard cells, but the interactions between NO, H_2O_2 and other signals remain to be unravelled. Similarly, NO can be synthesised by other cells, and it is possible that NO generated elsewhere in the plant, or in the environment (for example in the soil), will also affect stomatal apertures. It will be informative to determine the effects of gaseous NO, and potential NO precursors, on transpiration *in planta* as well as in epidermal peels.

There are now data indicating that guard cells generate NO via NR, and that downstream signalling may involve cGMP, cADPR, Ca^{2+} and reversible protein phosphorylation. Molecular genetic and biochemical approaches, linked to single cell studies and the use of various mutants, should help to clarify which signals are important, and how. Last, but not least, DNA microarray and proteomic analyses that identify guard cell genes and proteins will also aid in elucidating those mechanisms by which NO contributes to guard cell modulation of WUE. Given the importance of WUE in determining crop yields, such experiments should be well worth the effort.

ACKNOWLEDGEMENTS

Research in the authors' laboratory is supported by BBSRC and Defra.

REFERENCES

Barroso, J.B., Corpas, F.J., Carreras, Sandalio, L.M., Valderrama, R., Palma, J.M., Lupianez, J.A. and del Rio, L.A. (1999). Localization of nitric oxide synthase in plant peroxisomes. *J. Biol. Chem.*, **274** : 36729-36733.

Bray, E.A. (1997). Plant responses to water deficit. *Trends in Plant Sci.,* **2** : 48-54.

Broillet, M-C., Randin, O. and Chatton, J-Y. (2001). Photoactivation and calcium sensitivity of the fluorescent NO indicator 4,5-diaminofluorescein (DAF-2): implications for cellular NO imaging. *FEBS Lett.*, **491** : 227-232.

Burnett, E.C., Desikan, R., Moser, R. and Neill, S.J. (2000) ABA activation of an MBP kinase in *Pisum sativum* epidermal peels correlates with stomatal responses to ABA. *J. Exp. Bot.,* **51** : 197-205.

Butt, Y.K-C., Lum, J.H-K and Lo, S.C-L. (2003). Proteomic identification of plant proteins probed with mammalian nitric oxide synthase antibodies. *Planta,* **216** : 762-771.

Clarke, A., Desikan, R., Hurst, R.D., Hancock, J.T. and Neill, S.J. (2000). NO way back: nitric oxide and programmed cell death in *Arabidopsis thaliana* suspension cultures. *Plant J.*, **24** : 667-677.

Crawford, N.M. and Forde, B.G. (2002). Molecular and developmental biology of inorganic nitrogen nutrition. The Arabidopsis Book, eds. C.R. Somerville and E.M. Meyerowitz, American Society of Plant Biologists, Rockville, MD, doi/ 10.1199/tab.0009, http://www.aspb.org/publications/arabidopsis/.

Desikan, R. Griffiths, R., Hancock, J., and Neill, S. (2002). A new role for an old enzyme: nitrate reductase-mediated nitric oxide generation is required for abscisic acid-induced stomatal closure in *Arabidopsis thaliana. Proc. Natl. Acad. Sci. USA,* **99** : 16314-16318.

Garcia-Mata, C. and Lamattina, L. (2003). Abscisic acid, nitric oxide and stomatal closure – is nitrate reductase one of the missing links? *Trends Plant Sci.,* **8** : 20-26.

Garcia-Mata, C. and Lamattina, L. (2002). Nitric oxide and abscisic acid cross-talk in guard cells. *Plant Physiol.*, **128** : 790-792.

Garcia-Mata, C. and Lamattina, L. (2001). Nitric oxide induces stomatal closure and enhances the adaptive plant responses against drought stress. *Plant Physiol.*, **126** : 1196-1204.

Godber, B.L.J., Doel, J.J., Sapkota, G.P., Blake, D.R., Steven, C.R., Eisenthal, R. and Harrison, R. (2000). Reduction of nitrite to nitric oxide catalysed by xanthine oxidoreductase. *J. Biol. Chem.,* **275** : 7757-7763.

Hetherington, A.M. (2001). Guard cell signaling. *Cell,* **107** : 711-714.

Hugouvieux, V., Kwak, J.M. and Schroeder. J.I. (2001). A mRNA cap binding protein, ABH1, modulates early abscisic acid signal transduction in Arabidopsis. *Cell,* **106** : 477-487.

Hunt L, Mills LN, Pical C, Leckie CP, Aitken FL, Kopka J, Mueller-Roeber B, McAinsh MR, Hetherington AM & Gray JE. (2003). Phospholipase C is required for the control of stomatal aperture by ABA. *Plant J.*, **34** : 47-55.

Hunt, P. W., Watts, R.A., Trevaskis, B., Llewelyn, D.J., Burnell, J., Dennis, E.S. and Peacock, W.J. (2001). Expression and evolution of functionally distinct haemoglobin genes in plants. *Plant Molecular Biology,* **47** : 677-692.

Jaffrey, S.R., Erdjument-Bromage, H., Ferris, C.D., Tempst, P. and Snyder, S.H. (2001). Protein S-nitrosylation: a physiological signal for neuronal nitric oxide. *Nature Cell Biology,* **3** : 193-197.

Kojima, H., Nakatsubo N., Kikuchi, K., Kawahara, S., Kirin, Y., Nagoshi, H., Hirata, Y. and Nagano, T. (1998). Detection and imaging of nitric oxide with novel fluorescent indicators: diamiofluoresceins. *Anal. Chem.*, **70** : 2446-2453.

Kumar, D. and Klesig, D.F. (2000). Differential induction of tobacco MAP kinases by the defense signals nitric oxide, salicylic acid, ethylene and jasmomic acid. *Mol. Plant-Microbe Interact.,* **13** : 347-351.

Kwak, J.M., Moon, J.H., Murata, Y., Kuchitsu, K., Leonhardt, N., DeLong, A., and Schroeder, J.I. (2002). Disruption of a guard cell-expressed protein phosphatase 2A regulatory subunit, RCN1, confers abscisic acid insensitivity in *Arabidopsis. Plant Cell,* **14** : 2849-2861

Leckie, C.P., McAinsh, M.R., Allen, G.J., Sanders, D. and Hetherington, A.M. (1998). Abscisic acid-induced stomatal closure mediated by cyclic ADP-ribose. *Proc. Natl. Acad. Sci. USA,* **95** : 15387-15842.

Lemichez, E., Wu, Y., Sanchez, J.-P., Mettouchi, A., Mathur, J. and Chua, N.H. (2001). Inactivation of AtRac1 by abscisic acid is essential for stomatal closure. *Genes Dev.,* **15** : 1808-1816

Li, J., Kinoshita, T., Pandey, S., Ng, C, K.-Y., Gygi, S.P., Shimazaki, K. and Assman, S.A. (2002). Modulation of an RNA-binding protein by abscisic acid-activated protein kinase. *Nature,* **418** : 793-797.

Liu, L., Hausladen, L., Zeng, M., Que, L., Heitman, J. and Stamler, J.S. (2001). A metabolic enzyme for S-nitrosothiol conserved from bacteria to humans. *Nature,* **410** : 490-494.

Ludidi, N. and Gehring, C. (2003). Identification of a novel protein with guanylyl cyclase activity in *Arabidopsis thaliana. J. Biol. Chem.,* **278** : 6490-6494.

MacRobbie, E.A.C. (2002). ABA activates multiple Ca^{2+} fluxes in stomatal guard cells trigering vacuolar K^+ (Rb^+) release. *Proc. Natl. Acad. Sci. USA,* **97** : 12361-12368.

Mendel R.R. and Hansch, R. (2002). Molybdoenzymes and molybdenum cofactor in plants. *J. Exp. Bot.,* **53** : 1689-1698.

Minorsky, P.V. (2003). cGMP in plants. *Plant Physiol.,* **131** : 2003.

Neill, S.J., Desikan, R. and Hancock, J.T. (2003). Nitric oxide signalling in plants. *New Phytol.,* **159** : 11-35.

Neill, S.J., Desikan, R. and Hancock, J.T. (2002a). Hydrogen peroxide signalling. *Curr. Opin. Plant Biol.,* **5** : 388-395.

Neill, S.J., Desikan, R., Clarke, A. and Hancock, J.T. (2002b). Nitric oxide is a novel component of abscisic acid signalling in stomatal guard cells. *Plant Physiol.,* **128** : 13-16.

Notton, B.A. and Hewitt, E.J. (1971). The role of tungstate in the inhibition of nitrate reductase activity in spinach (*Spinacia oleracea* L.) leaves. *Biochem. Biophys. Res. Comm.,* **44** : 702-710.

Rockel, P. Strube, F., Rockel, A., Wildt, J. and Kaiser, W.M. (2002). Regulation of nitric oxide (NO) by plant nitrate reductase *in vivo* and *in vitro. J. Exp. Bot.,* **53** : 103-110.

Sakamoto, A., Ueda, M. and Morikawa, H. (2002). *Arabidopsis* glutathione-dependent formaldehyde dehydrogenase is an *S*-nitrososoglutathione reductase. *FEBS Letts.,* **515** : 20-24.

Sakihama, Y., Murakamai, S. and Yamasaki, H. (2003). Involvement of nitric oxide in the mechanism for stomatal opening in *Vica faba* leaves. *Biol. Plant.,* **45** : 117-117.

Schroeder, J.I., Kwak, J.M. and Allen, G.J. (2001a). Guard cell abscisic acid signalling and engineering drought hardiness in plants. *Nature* **410**: 327-330.

Schroeder, J.I., Allen, G.J. Hugouvieux, V., Kwak, J.M. and Waner, D. (2001b). *Annu. Rev. Plant Physiol. Plant Mol. Biol.,* **52** : 627-658.

Staxen, I., Pical, C., Montgomery L.T., Gray, J.E., Hetherington, A.M. & McAinsh, M.R. (1999) Abscisic acid induces oscillations in guard cell cytosolic free calcium that involve phosphoinositide-specific phospholipase C. *Proc. Natl. Acad. Sci USA,* **96** : 1779-1784.

Stohr, C. and Ullrich, W.R. (2002). Generation and possible roles of NO in plant roots and their apoplastic space. *J. Exp. Bot.,* **53** : 2293-2303.

Trewavas, A.J., Rodrigues, C. Rato, C. and Malho, M. (2002). Cyclic nucleotides: the current dilemma! *Curr. Opin. Plant Biol.,* **5** : 425-4209.

Wang, X-Q., Ullah, H., Jones, A.M. and Assman, S.A. (2001). G protein regulation of ion channels and abscisic acid signalling in *Arabidopsis* guard cells. *Science,* **292** : 2070-2072.

Wendehenne, D., Pugin, A., Klessig, D.F. and Durner, J. (2001). Nitric oxide: comparative synthesis and signalling in animal and plant cells. *Trends Plant Sci.,* **6** : 177-183.

Wu, Y., Kuzma, J., Marechal., E., Graeff, R., Lee, H.C., Foster, R. and Chua, N. H. (1997). Abscisic acid signaling through cyclic ADP-ribose in plants. *Science,* **278** : 2126-2130.

Yamasaki. H. and Sakihama, Y. (2000). Simultaneous production of nitric oxide and peroxynitrite by plant nitrate reductase: *in vitro* evidence for the NR-dependent formation of active nitrogen species. *FEBS. Lett.,* **468** : 89-92.

Chapter 7

NITRIC OXIDE AS A POTENTIAL SECOND MESSENGER IN CYTOKININ SIGNALING

Günther F.E. Scherer

Universität Hannover, Institut für Zierpflanzenbau, Baumschule und Pflanzenzüchtung, Abt. Spezielle Ertragsphysiologie, Herrenhäuser Str. 2, D-30419 Hannover, Germany

E-mail : scherer@zier.uni-hannover.de; Fax : +49-511-762-3608

Summary

Cytokinin induces rapid NO biosynthesis within 3 min of addition of hormone to several cell cultures. This effect is cytokinin-specific in that auxin, abscisic acid, gibberellic acid, and ACC do not increase NO release from these cells. Arginine-like inhibitors and the NO scavenger PTIO inhibit this NO release. Previously, it was shown that such inhibitors blocked the cytokinin-induced accumulation of betalains in Amaranthus*, a classical cytokinin response, and that NO donors induced betalain biosynthesis. In order to narrow down the source of NO which might be derived from nitrate reductase or from a different enzyme we challenged a mutant nia1/nia2 Arabidopsis line devoid of nitrate reductase with zeatin and it responded with NO production and showed exaggerated morphological responses to cytokinin, however, in general similar to the wild type. As a newly available cytokinin biotest we used* Arabidopsis *plants transformed with the promoter-gene construct ARR5-GUS. Again, NO and cytokinin are inducers and arginine-based inhibitors are blockers of GUS activity. The source enzyme for the nitrate reductase-*

In : Nitric Oxide Signaling in Higher Plants, 2004
(Eds Jose R. Magalhaes, Rana P. Singh and Leonidas P. Passos)
Studium Press, LLC, Houston, USA, pp 149-166

independent NO biosynthesis remains presently unclear. We speculate that it is rapidly up-regulated by cytokinin and NO may act upon downtream processes in the manner of a second messenger.

Keywords : Cytokinin, nitrate reductase, nitric oxide, NO synthase, second messenger

1. INTRODUCTION

The gaseous nitric oxide (NO) was discovered in plants - not in animals - but at first NO was misunderstood as a byproduct of nitrogen metabolism (Stutte and Weiland, 1978; Klepper 1979, Stutte *et al.*, 1979; Harper, 1981). In animals NO was discovered as a vasodilatory factor 1987 and its importance as a secondary messenger in medicine was clear from the start and this discovery later won the Nobel prize (Palmer *et al.*, 1987). Since then the importance in medicine of NO goes on to be documented in numerous publications and fields of medicine (Marletta, 1993, Marletta, 1994, Kerwin *et al.*, 1995, Mayer and Hermes, 1997). The hypothesis that cellular defense mechanisms in plants and animals might share partial similarities led to the first highly visible publications on NO functions in pathogen defence (Delledone *et al.*, 1998, Durner *et al.*, 1998, Delledone *et al.*, 2001) although the idea of NO as a signal substance in plants was proposed earlier (Leshem, 1996; Leshem and Hamarty, 1996; Cueto *et al.*, 1996; Ninnemann and Mayer, 1996, Leshem *et al.*, 1998). In the meantime this idea is generally accepted and we assume that NO could be both a chemical weapon in plant and a signal substance or second messenger fulfilling partially homologous functions to the functions of NO in animals (Durner and Klessig, 1999; Klessig *et al.*, 2000; Wojtaszek, 2000, Wendehenne *et al.*, 2001). It is clear already, however, that NO biology in plants will not be a copy of NO biology in animal systems as the plant hormones implicated in NO biology, such as cytokinin (Scherer and Holk, 2000; Tun *et al.*, 2001) and abscisic acid (Mata and Lammatini, 2001; Desikan *et al.*, 2002; Neill *et al.*, 2002) do not have a counterpart in animal physiology. Another difference is that nitrate reductase, an enzyme absent in animals, produces significant amounts of NO (Harper, 1981; Kaiser *et al.*, 2002; Rockel *et al.*, 2002) whereas the search for a genetically homologous enzyme in plants, so far, failed (see below).

1.1. The source of nitric oxide in plants and the relationship to cytokinin action

Due to the high reactivity of the NO molecule and diffusability one can expect for NO that the regulation of its biosynthesis will be similarly important for an understanding of its function as, for instance, this is the case for the gaseous hormone ethylene.

A long-known process to generate NO is the anaerobic generation of NO from nitrite and nitrate in the soil in the course of denitrification which a number of bacteria use as an energy source. An enzyme homologous to cytochrome-c-oxidase is present in these bacteria which generates NO as a side product (Zumft, 1997; Guiffre *et al.*,, 1999, Watmough, 1999; Hendricks *et al.*, 2000; Vollack and Zumft, 2001). At this point, the physiological influence of bacterially synthesized NO in the soil under anaerobic conditions on plants is unclear. Indications of the presence of a cytochrome-oxidase-like nitrite-reductase in he plant genome are not apparent.

The animal NO synthase is molecularly well characterized and a number of isoforms are known. All isoforms use arginine as a substrate which releases NO and which uses NADPH and O_2 as cofactors (Kerwin *et al.*, 1995; Crane *et al.*, 1997; Mayer and Hemmes, 1997). Many inhibitors of the animal NO synthase are known most of which are arginine derivatives or partially similar to arginine. Some of these inhibitors habe been successfully used in experiments with plants (Cueto *et al.*, 1996; Ninnemann and Mayer, 1996; Noritake *et al.*, 1996; Delledone *et al.*, 1998; Durner *et al.*, 1998; Beligni and Lammatina, 1999; 2000; Scherer and Holk, 2000; Delledone *et al.*, 2001, Mata and Lammatina, 2001; Maier *et al.*, 2001, Tun *et al.*, 2001; Neill et., al 2002; Beligni *et al.*, 2002; Garcia-Mata *et al.*, 2002). Although this fostered an impression that there should be a homologous plant NO synthase this simple conclusion probably is probably not correct. When the completed Arabidopsis genome was searched by BLAST no protein homologous to the arginine-binding domain of the animal NO synthase was found but in a number of bacteria such as *Deinococcus radiodurans* (acc. nr. NP_296316) and *Bacillus subtilis* (acc. nr. NP_388644) (G. Scherer, unpublished). As such a well-preserved homology can be found across such divergent phyla it should be easily discovered by the same criterion in the plant genome if it were present.

The (or one) more easily found answer is the renessaince of our knowledge on nitrate reductase. Soon after the finding of NO exudation by plants (Stutte and Weiland, 1978), Harper (1981) found that nitrate reductase can generate NO by a side reaction so that NO in plants seemed to be explained as a minor metabolite in nitrogen metabolism and research on NO in plants for long years also was a side topic (Wellburn, 1998). NO needed to be rediscovered just as the role of nitrate reductase as a source of NO experienced the same renessaince. It is clear now that this "side reaction" is important in plant NO biology (Yamasaki and Sakihama 2000; Kaiser *et al.*, 2002; Rockel *et al.*, 2002; Desikan *et al.*, 2002; Neill *et al.*, 2002).

Nitrate reductase, however, does not explain the numerous reports on clear effects of arginine-based inhibitors on NO biosynthesis in plants and moreover, our own finding, that the nitrate-reductase-free *nia1/nia2* mutant of *Arabidopsis* produced cytocinin-induced NO (see below).

Clear knowledge on other potential enzymes of NO biosynthesis besides the nitrate reductase is still scant. Reports on the purification of plant NO synthesizing enzymes did not yet proceed far enough to answer questions of molecular identity (Cueto *et al.*, 1996; Barroso *et al.*, 1999; Lo *et al.*, 2000; Stöhr *et al.*, 2001) or subcellular localisation (Barroso *et al.*, 1999, Foissner *et al.*, 2000; Mata and Lammatini, 2001; Desikan *et al.*, 2002, Neill *et al.*, 2002) or on the number of different enzymes (Barroso *et al.*, 1999; Foissner *et al.*, 2000). Other reactions generating NO in plants are also known such as a reduction of nitrite to NO by xanthine oxidase or the conversion of NO_2 to NO by carotenoids and light (Wendehenne *et al.*, 2001). However, the possibility of rapid regulation by hormones or elicitors or effects of arginine-like inhibitors on these processes seems to be remote so that here the presence of an as yet to be discovered enzyme is favoured as a working hypothesis.

2. RAPID ACTIONS IN CYTOKININ SIGNAL TRANSDUCTION AND NITRIC OXIDE AS A POTENTIAL SECOND MESSENGER

A few years ago almost nothing about cytokinin signal transduction was known. The general outline of current hypotheses is given in several articles (Hwang and Sheen, 2001; Schmülling, 2001; Hutchison and Kieber, 2002) and shown as a simplified model in figure 1. The discovery of a cytokinin receptor CKI 1 (or CRE1=AHK4) (Kakimoto, 1998, Inoue *et al.*, 2001, Suzuki *et al.*, 2001, Ueguchi *et al.*, 2001, Yamada *et al.*, 2001) showed that it is a two-component receptor having a histidine kinase domain. Four homologous histidine kinases are known (AHK1-AHK4) but the function of the other three has to be demonstrated as yet. Most likely a phosphate residue is transferred by them to histidine phosphotransfer proteins called AHP1-AHP5. So far AHP1 and AHP2 have been partially investigated in this respect, the functions of AHP3 to AHP5 remain unclear (Suzuki *et al.*, 1998; 2001a,b; 2002; Hwang and Sheen, 2001). It is thought that the phosphorylated AHP proteins migrate into the nucleus or are phosphorylated in the nucleus where they act as cofactors to transcription factors, the B-type ARR genes, to regulate the rapid expression of the A-type ARR genes. The regulation of these A-type ARR genes is cycloheximide-independent which shows no protein biosynthesis or gene regulation prior to the

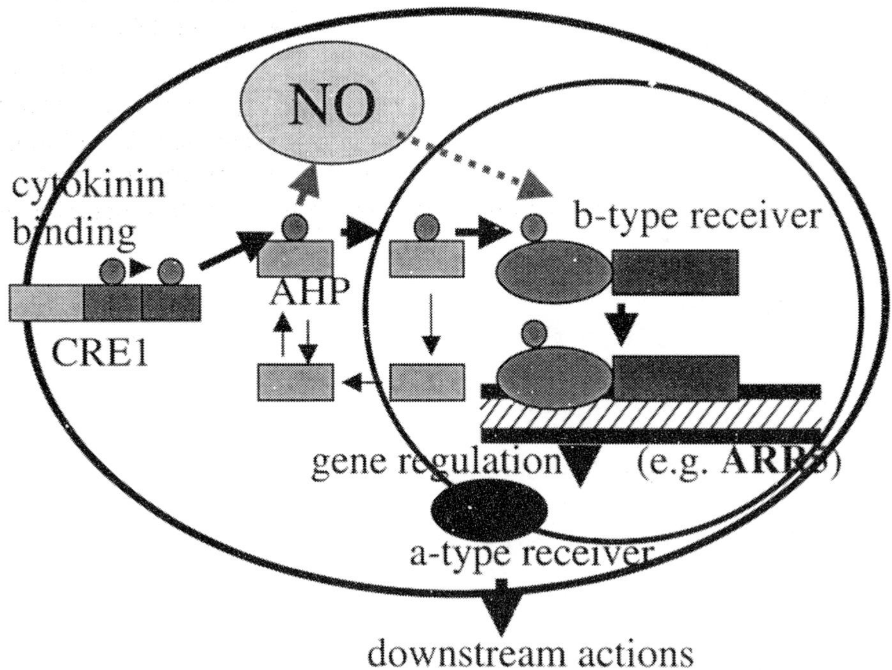

Figure 1 : Simplified model of cytokinin signal transduction showing the hypothetical function of nitric oxide (modified after Hutchison and Kieber, 2002).

expression of A-type ARR genes is necessary and that therefore A-type ARR genes are primary cytokinin-induced genes. It is important to realise that signal transduction operates in a cycloheximide-insensitive process with preexisting proteins and, therefore, gene regulation is a consequence of signal transduction.

The induction of several of the ten ARR genes is apparent already after 10 min (d'Agostino *et al.*, 2000; Kiba *et al.*, 1999) so that signal transduction reactions, if present, should be faster than that. B-type ARR genes are not regulated by cytokinin and the respective proteins are supposed to act as transcription factors or cofactors to other transcription factors and to regulate other cytokinin-regulated genes which may be called secondary genes. B-type ARR activity may be controlled by the proteasome (Smalle *et al.*, 2002). A summary of cytokinin-regulated genes was compiled (Schmülling *et al.*, 1997). Most of them are probably to be regarded as secondary cytokinin-regulated genes.

We found that NO biosynthesis is activated by cytokinin after 2-3 min in suspension-cultured cells (Tun *et al.*, 2001) so that NO biosynthesis could precede the regulation of primary ARR genes (Fig. 2). Other hormones tested were gibberellic acid, auxin, abscisic acid and the precursor to ethylene, 1-amino-1-carboxy-cyclopropane (ACC), all of which did not stimulate NO biosynthesis. The NO biosynthesis in cultured cells was inhibited by an inhibitor of animal NO synthase, 2-(aminoethyl) 2-thiopseudourea (AET), and by the NO scavenger 2-(4-carboxyphenyl)-4,4,5,5-tetramethylimidazoline-1-1-oxy-3-oxide (PTIO). In guard cells ABA induced NO biosynthesis via nitrate reductase up-regulation as the inhibitor characteristics showed (Desikan *et al.*, 2002; Neill *et al.*, 2002). Why abscisic acid was inactive in cultured cells is unclear. An explanation of this difference might be the differentiation status of these cell types which might not allow cell culture cells to respond in this way to abscisic acid with increased NO biosynthesis.

The inhibitor characteristics of rapid cytokinin-induced NO biosynthesis suggested that at least two sources of NO in plants are possible, nitrate reductase and an enzyme analogous to animal NO synthase. Therefore, we tested cytokinin-induced NO biosynthesis in seedlings of *nia1/nia2* mutants of *Arabidopsis* (Fig. 3) which are devoid of nitrate reductase activitity (Wilkinson and Crawford, 1991;

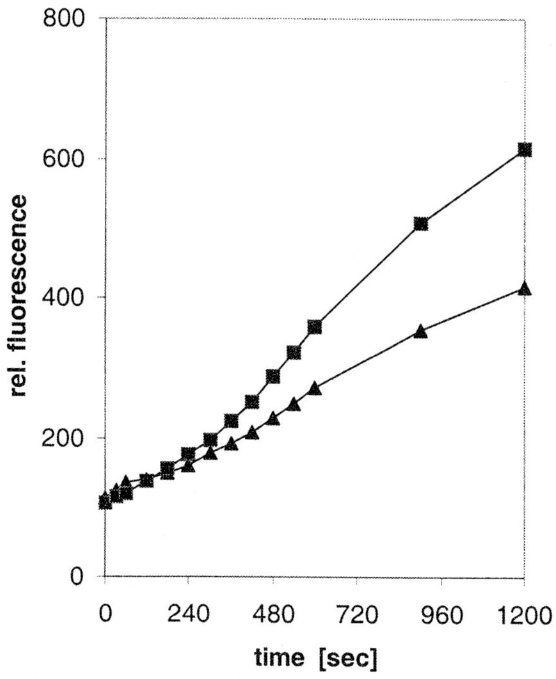

Figure 2 : Rapid induction of NO biosynthesis in tobacco BY2 cell culture cells by 5 μM benzyladenine purine (from: Tun *et al.*, 2001). Cells from a 5 d old culture were sieved and 200 mg wet weight was suspended in 1.7 mL MS medium (+2% sucrose) and 0.2 mL 50 mM K-phosphate buffer pH 7.4 was added to yield a final pH of 7.4. Cell-impermeable DAF2 was added at 5 μM concentration and hormone or solvent to measure the release of NO into the external medium. Samples (60 μL) were removed regularly and diluted into 3 mL water and fluorescence was read with 495 nm excitation and 515 nm emission in the fluorometer.

1993). The mutant seedlings clearly responded to cytokinin with an increase of NO biosynthesis and also showed the long-term physiological responses to cytokinin (see below). Hence, this suggests that cytokinin uses an enzyme different from nitrate reductase to rapidly generate NO. This view is further supported by the characteristics of inhibitors blocking cytokinin-induced NO biosynthesis. These are mimicking the guanidium group of arginine, AET and N^G-nitro-L-arginine methyl ester (L-NAME) (Tun *et al.*, 2001, and unpublished results), or the non-specific NO scavenger PTIO.

To be a second messenger for cytokinin action NO not only has to be generated rapidly by cytokinin but also should NO levels

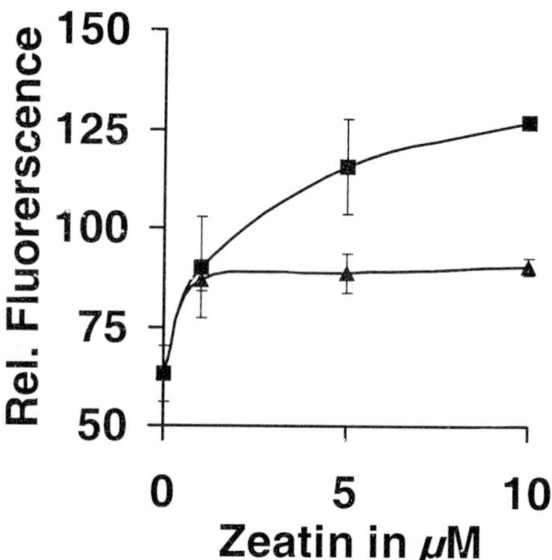

Figure 3 : Cytokinin-induced biosynthesis of NO in Arabidopsis seedlings. Squares : wild type seedlings; triangles: *nia1/nia2* mutant seedlings lacking nitrate reductase (N. Tun and G. Scherer, unpublished). Seedlings were grown at 26° in 12-well plates in the dark on a rotatory shaker in half strength MS medium plus 2% glucose. Ten seedlings were combined for one sample in 1 mL of the same medium and DAR-4M was added at 1 μM to measure the release of NO into the medium. After 30 min the fluorescence was read in the fluorometer at 495 nm excitation and at 515 nm emission. Background values were subtracted from medium controls without plants. Each data point was measured in triplicates. Wild type seedlings grow faster and the average fresh weight per sample was 6.9 mg, in *nia1/nia2* samples it was 4.1 mg.

influence downstream actions of cytokinin. We chose the most rapidly induced cytokinin-regulated gene ARR5 as a tool to investigate this. The promoter of the ARR5 gene linked to the *uid*A gene (GUS) and transformed into *Arabidopsis* as an indicator responds within 10 min to exogenous cytokinin (D'Agostino *et al.*, 2000) and this provides a convenient cytokinin biotest.

This promoter is activated by the chemical NO donor (+)-S-nitroso-N-acetylpenicillamine (SNAP) and by cytokinin but not by auxin. The inhibitors AET, L-NAME and PTIO inhibited the up-regulation of GUS activity by 1 hour cytokinin treatment so that NO is both sufficient and necessary for the regulation of the primary cytokinin-induced ARR5 gene (N. Tun and G. Scherer, unpublished results). Moreover, NO is a potential second messenger for cytokinin but not for auxin. When inhibitors of auxin action were used on the ARR5 promoter they did not inhibit cytokinin induction. These auxin inhibitors are phospholipase A inhibitors (Scherer and Arnold 1997; Paul *et al.*, 1998; Holk *et al.*, 2002) which, in turn, all block auxin activation of the auxin-inducible DR5 promoter (G. Scherer, unpublished results). As expected the auxin-inducible DR5 promoter is not affected by NO inhibitors. Hence, hormone specificity and inhibitor specificity, NO inhibitors for cytokinin action and phospholipase A inhibitors for auxin action, is evident. We are currently further exploiting this convenient system to obtain a more detailed model of cytokinin signal transduction.

Such a more detailed model for pathogen-induced signal transduction has been proposed. (after Durner *et al.*, 1998; Klessig *et al.*, 2000) :

Elicitor → Receptor → NO → cGMP → cADPR → Ca^{2+} → Gene Activation

Inhibitor studies suggested that, in analogy to animal models of signal transduction, NO could stimulate a guanylate cyclase. The increase in cGMP may lead to increased synthesis of another second messenger, cADPR (cyclic adenosine diphosphoriboside), which binds to calcium channels in the ER, thus increasing the cytosolic calcium levels. For most of the steps pharmacological evidence was provided in plants although it relies heavily on assumed homologies only for some of which are components identified in plant systems. A cADPR-sensitive calcium channel on the vacuole was proved (Allen *et al.*, 1995) and an increase due to NO of cGMP in spruce needles was found (Pfeiffer *et al.*, 1994). Moreover, for phytochrome action cGMP was suggested as a second messenger (Neuhaus *et al.*, 1993; Bowler *et al.*, 1994) and phytochrome and cytokinin have many overlapping responses (Piatelli *et al.* 1969; 1971; Bracale *et al.*, 1988; Cohen *et al.*, 1988; Kubo und Kakimoto 2000). It is possible that a similar model can be developed for cytokinin signaling but, presently, this must be regarded entirely as speculation.

The ARR5-GUS cytokinin biotest was also used to demonstrate an involvement of phospholipase D in cytokinin signaling (Romanov *et al.*, 2002). It was shown that only n-butanol but not isobutanol and also a large number of other inhibitors did not inhibit the cytokinin activation of ARR5-GUS but protein phosphorylation inhibitors also were effective. It is conceivable that both a NO synthesizing enzyme and a phospholipase D are activated prior to ARR5 gene transcription which might be parallel or consecutive events.

3. DOWNSTREAM AND SLOWER COMPLEX PHYSIOLOGICAL RESPONSES TO CYTOKININ AND ROLE OF NO

In contrast to this rapid ARR5-GUS biotest physiological responses to cytokinin are slower and probably require a whole network of signal transduction components to be initiated and conducted, thereby regulating whole sets of genes (Schmülling *et al.*, 1997). We investigated several physiological reactions to cytokinin and used a pharmacological approach similar to the experiments conducted with the ARR5-GUS system in order to investigate the role of NO in these slower physiological responses. We expect that NO is an important upstream second messenger so that inhibiting its function could inhibit many downstream responses but probably not all.

Our studies actually began with the classical and sensitive cytokinin test, the accumulation of the red pigment betalain in dark-grown *Amaranthus* seedlings (Scherer and Holk, 2000). After 3 days, the cytokinin benzylamino purine strongly stimulated the biosynthesis of betalaine in dark-grown seedlings of *Amaranthus caudatus*. When we tested the effect of several known inhibitors of animal NO synthase (Marletta 1994; Kerwin *et al.*, 1995) the order of potency of inhibition of these compounds was N^G-nitro-L-arginine (L-NNA) (50% inhibition at ca 0.35 mM), L-N^5-(iminoethyl)ornithine (L-NIO) (50% inhibition at ca 10 mM), N^G-nitro-L-arginine methyl ester (L-NAME) (50% inhibition at ca 20 mM), N^G-monomethyl-L-arginine acetate (L-NMMA) (50% inhibition at ca 10 mM), and methylisothiourea (M-ITU) (50% inhibition at ca 20 mM). All inhibitors can be considered as arginine derivatives modified at the guanidium group. In animal cell assays the methyl ester L-NAME is more effective than L-NNA and is hydrolysed inside the cell to L-NNA, the actual inhibitor, whereas in this plant assay the opposite was observed and, besides, the two nitro-derivatives were the most effective. More recently, we repeated this test with seedlings grown in wells as a hydroponic culture instead of growing them on filter paper. This increased the sensitivity to inhibitors greatly so that 100 µM of A*Et al*ready showed strong inhibition (N. Tun and G. Scherer, unpublished results). Probably, the uptake of inhibitors by the transpiration stream was limiting the inhibitor sensitivity in the previous results. Others observed 50% inhibition of plant NO synthesising activity isolated from root nodules by ca. 1 mM L-

NMMA (Cueto *et al.*, 1996) and in assays with plant NO synthesising activity isolated from tobacco leaves by ca 0.25 mM L-NMMA (Durner *et al.*, 1998) which is again far less sensitive compared to assays with animal NO synthase where these inhibitors have a K_i in the micromolar range. The generally lower sensitivity of the animal NO synthase inhibitors in plant experiments indicates as yet unknown differences of the animal NO synthases and the unknown NO synthesising plant enzyme.

We also tried to mimic cytokinin action in *Amaranthus* seedlings by applying the NO-releasing compounds diethylamine NONOate (DEA/NONO), 2,2'-(hydroxynitrosohydrazino) *bis*-ethanamine (NOC-18), and (+)-S-nitroso-N-acetylpenicillamine (SNAP) or NO itself. NO and all NO-releasing compounds, DEA/NONO, NOC18, and SNAP had to be applied several times in order to stimulate betalaine accumulation and about 20-25 % of the usual cytokinin response was obtained by the donors (Scherer and Holk, 2000). An explanation of the weak phenocopy of cytokinin-induced betalain accumulation could be that in this system cytokinin action may not depend on NO alone as a signalling intermediate, in which case only partial phenocopies of cytokinin action by NO or NO donors should be obtained or that the method of applying these compounds may have been only partially effective.

Another well-known physiological response to cytokinin is the deetiolation of *Arabidopsis* seedlings (Chory *et al.*, 1994). When cytokinins were applied to etiolated seedlings stunted growth of the hypocotyl and the root, opening of the hypocotyl hook, and the development of secondary leaves in the dark were observed. Animal NO synthase inhibitors reverted these responses partially both in the *Amaranthus* and *Arabidopsis* seedlings (N. Tun and G. Scherer, unpublished). An enhanced deetiolation in *Arabidopsis* seedlings was also observed by Beligni and Lammatina (2000).

As the *nia1/nia2* mutant in Arabidopsis should, by its absence, reveal the role of nitrate reductase-generated NO or, alternatively, the role of an additional NO synthesising enzyme in cytokinin action we also started to investigate the cytokinin responses in light-grown *nia1/nia2* seedlings. Remarkably, the response to cytokinin of *nia1/nia2* and of wild type seedlings is only basically similar but stronger in the mutant and occurred at lower hormone concentrations (Fig. 4).

Figure 4 : Response of wild type Arabidopsis seedlings and of *nia1/nia2* mutant seedlings to zeatin. Seedlings were grown at 25° in the light for 10 days on half strength MS medium plus 2% sucrose in 1% agar. Left half from left to right: wild type (control); wild type + 0.1 μM zeatin; wild type + 0.5 μM zeatin; wild type +1.0 μm zeatin. Right half: *nia1/nia2* (control); *nia1/nia2* + 0.1 μm zeatin; *nia1/nia2* + 0.5 μM zeatin; *nia1/nia2* + 1.0 μM zeatin. Bar = 1 cm.

Root growth was first increased by low hormone concentrations in both lines but with much stronger effect in *nia1/nia2* seedlings and, at higher hormone concentrations above 1 μM, roots become very short (not shown). Hypocotyls become shorter with increasing hormone concentrations in both lines but cotyledon enlargement and petiole elongation was much more prominent in the *nia1/nia2* line than in the wild type. At concentration of 1 μM or more the development of secondary leaves was halted in the mutant. Anthocyanin biosynthesis was induced in both lines but much stronger and at lower concentrations in the *nia1/nia2* line. Taken together, the mutant plants with one source of NO missing responded to cytokinin with an increase of NO biosynthesis (Fig. 2). The exaggerated phenotype might be explained either by a change in sensitivity to hormone or by the assumption that a component downstream of NO was strongly enhanced in sensitivity towards in the mutant line. Other explanations presently cannot be excluded.

This indicates that NO has a function in many downstream physiological responses to cytokinin and it could be a quite important and central mediator of cytokinin action close to the receptor. Therefore, in our modification of the currently popular model of cytokinin signal transduction (Fig. 1) we placed NO downstream of the AHP proteins as these AHP proteins are thought to be directly phosphorylated by the receptor histidine kinase with no intermediate in between (Hwang and Sheen, 2001; Schmülling, 2001; Hutchison and Kieber, 2002).

4. CONCLUSIONS AND FUTURE PROSPECTS

NO can be added now to the first rough draft of cytokinin signal transduction (see Fig. 1). As exemplified by the model for NO involvement in pathogen signal transduction where a good number of additional components have been identified by Durner and collegues (Durner *et al.*, 1998; Klessig *et al.*, 2000) one should expect for cytokinin the same complexity of a signal transduction network as for other hormones. So we are still at the beginning. For a better understanding the plant NO synthesising enzyme(s) besides nitrate reductase needs to be identified. The interaction of the two (major) sources of NO in the plant is just about visible in the comparison of the *nia1/nia2* mutant to the wild type when they respond to cytokinin and will provide a push in understanding the role of NO in cytokinin functions. The ARR5-GUS system will provide a better understanding of the rapid events after cytokinin receptor stimulation. Since nitrate and cytokinin and red light are known to be mutually enhancing their actions and each one of them is a signal on its own (Crawford, 1995; Fankhauser, 2002) this will be an extremely exciting field in plant physiology.

ACKNOWLEDGEMENT

Ni Ni Tun provided some experimental data prior to publication for this article. Dr. Joe Kieber (Chapel Hill, USA) gave us the ARR5-GUS *Arabidopsis* line.

REFERENCES

Allen, G.J., Muir, S.R., and Sanders, D., (1995). Release of Ca^{2+} from individual plant vacuoles by both $InsP_3$ and cyclic ADP ribose. *Science,* **268** : 735-737.

Barroso, J.B., Corpas, F.J., Carreras, A., Sandalio, L.M., Valderrama, R., Palma, J.M., Lupianez, J.A., and del Rio, L.A. (1999). Localization of nitric-oxide synthase in plant peroxisomes. *J. Biol. Chem.,* **274** : 36729-36733.

Beligni, M.V. and Lammattina, L. (1999). Nitric oxide counteracts cytotoxic processes mediated by reactive oxygen species in plant tissues. *Planta,* **208** : 337-344.

Beligni, M.V. and Lammatina, L. (2000). Nitric oxide stimulates seed germination and de-etiolation, and inhibits hypocotyl elongation, three light-inducible responses in plants. *Planta,* **210** : 215-221.

Bowler, C., Neuhaus, G., Yamagata, H., and Chua, N.H. (1994). Cyclic GMP and calcium mediate phytochrome phototransduction. *Cell,* **77** : 73-81.

Bracale, M., Longo, G.P., Rossi, G. and Longo, C.P. (1988). Early changes in morphology and polypeptide pattern of plastids from watermelon cotyledons induced by benzyladinine or light are very similar. *Physiol. Plant.,* **72** : 94-100.

Chory, J., Reinecke, D., Sim, S., Washburn, T. and Brenner, M. (1994). A role for cytokinins in de-etiolation in Arabidopsis (det mutants have an altered response to cytokinins). *Plant Physiol.,* **104** : 339-347.

Cohen, L., Arzee, T. and Zilberstein, A. (1988). Mimicry by cytokinin of phytochrome-regulated inhibition of chloroplast development in etiolated cucumber seedlings. *Physiol. Plant.*, **72** : 57-64.

Cueto, M., Hernádez-Perera, O., Martin, R., Bentura, M.L., Rodrigo, J., Lamas, S. and Golvano, M.P. (1996). Presence of nitric oxide synthase activity in roots and nodules of *Lupinus albus*. *FEBS Lett.,* **398** : 159-164.

Crane, B.R., Arvai, A.S., Gachhui, R., Wu, C., Ghosh, D.K., Getzoff, E.D., Stuehr, D.J. and Tainer, J.A. (1997). The structure of nitric oxide synthase oxygenase domain and inhibitor complexes. *Science,* **278** : 425-431.

Crawford, N.M. (1995). Nitrate: nutrient and signal for plant growth. *Plant Cell,* **7** : 859-868

D'Agostino, I.B., Deruere, J. and Kieber, J.J. (2000). Characterization of the response of the Arabidopsis response regulator gene family to cytokinin. *Plant Physiol.*, **124** : 1706-1717.

Delledone, M., Xia, Y., Dixon, R.A. and Lamb, C. (1998). Nitric oxide functions as a signal in plant disease resistance. *Nature,* **394** : 585-588.

Delledonne, M., Zeier, J., Marocco, A. and Lamb, C. (2001). Signal interactions between nitric oxide and reactive oxygen intermediates in the plant hypersensitive disease resistance response. *Proc. Natl. Acad. Sci. USA*, **98** : 13454-13459.

Desikan, R., Griffiths, R., Hancock, J. and Neill, S. (2002). A new role for an old enzyme: nitrate reductase-mediated nitric oxide generation is required for abscisic acid-induced stomatal closure in *Arabidospsis thaliana. Proc. Natl. Acad. Sci. USA,* **99** : 16314-16318.

Durner, J., Gow, A.J., Stamler, J.S. and Glazebrook, J. (1999). Ancient origins of nitric oxide signaling in biological systems. *Proc. Natl. Acad. Sci. USA,* **96** : 14206-14207.

Durner, J. and Klessig, D.F. (1999). Nitric oxide as a signal in plants. *Curr. Opin. Plant Biol.,* **2** : 369-374.

Durner, J., Wendehenne, D. and Klessig, D.F. (1998). Defense gene induction in tobacco by nitric oxide, cyclic GMP, and cyclic ADP-ribose. *Proc. Natl. Acad. Sci. USA,* **95** : 10328-10233.

Fankhauser, C. (2000). Light perception in plants: cytokinins and red light join forces to keep phytochrome B active. *Trends in Plant Sci.*, **7** : 143-145.

Foissner, I., Wendehenne, D., Langebartels, C. and Durner, J. (2000). In vivo imaging of an elicitor-induced nitric oxide burst in tobacco. *Plant J.,* **23** : 817-824.

Garcia-Mata, C. and Lamattina, L. (2002). Nitric oxide and abscisic acid cross talk in guard cells. *Plant Physiol.*, **128** : 790-792.

Giuffre, A., Stubauer, G., Sarti, P., Brunori, M., Zumft, W.G., Buse, G. and Soulimane, T. (1999). The heme-copper oxidases of *Thermus thermophilus* catalyze the reduction of nitric oxide: evolutionary implications. *Proc. Natl. Acad. Sci. USA,* **96** : 14718-14723.

Harper, J.E. (1981). Evolution of nitrogen oxide(s) during in vivo nitrate reductase assay of soybean leaves. *Plant Physiol.*, **68** : 1488-1493.

Hendriks, J., Oubrie, A., Castresana, J., Urbani, A., Gemeinhardt, S. and Saraste, M. (2000). Nitric oxide reductases in bacteria. *Biochim. Biophys. Acta.*, **1459** : 266-273.

Holk, A., Rietz, S., Zahn, M., Paul, R.U., Quader, H. and Scherer, G.F.E. (2002). Molecular identification of cytosolic, patatin-related phospholipases A from *Arabidopsis* with potential functions in plant signal transduction. *Plant Physiol.*, **130** : 90-101.

Hutchison, C.E. and Kieber, J.J. (2002). Cytokinin signaling in Arabidopsis. *Plant Cell,* **14S** : 47-59.

Hwang, I. and Sheen, J. (2001). Two-component circuitry in Arabidopsis cytokinin signal transduction. *Nature,* **413** : 383-389.

Inoue, T., Higuchi, M., Hashimoto, Y., Seki, M., Kobayashi, M., Kato, T., Tabata, S., Shinozaki, K. and Kakimoto, T. (2001). Identification of CRE1 as a cytokinin receptor from Arabidopsis. *Nature,* **409** : 1060-1063.

Kaiser, W.M., Weiner, H., Kandlbinder, A., Tsai, C.B., Rockel, P., Sonoda, M. and Planchet, E. (2002). Modulation nitrate reductase: some new insights, an unusual case and a potential important side reaction. *J .Exp. Bot.,* **53** : 875-882.

Kakimoto, T. (1998). Cytokinin signaling. *Curr. Opin. Plant Biol.,* **1** : 399-403.

Kerwin, J.F., Lancaster, J.R., jr. and Feldman, P.L. (1995). Nitric oxide: A new paradigm for second messengers. *J. Med. Chem.,* **38** : 4343-4362.

Kiba, T., Taniguchi, M., Imamura, A., Ueguchi, C., Mizuno, T. and Sugiyama, T. (1999). Differential expression of genes for response regulators in response to cytokinins and nitrate in *Arabidopsis thaliana. Plant Cell Physiol.*, **40** : 767–771.

Klepper, L. (1979). Nitric oxide (NO) and nitrogen dioxide (NO_2) emissions from herbicide-treated soybean plants. *Atmos. Environ.*, **13** : 537-542.

Klessig, D.F., Durner, J.F., Noad, R., Navarre, D.A., Wendehenne, D., Kumar, D., Zhu, J.M., Shah, J., Zhang, S., Kachroo, P., Trifa, Y., Pontier, D., Lam, E. and Silva, H. (2000). Nitric oxide and salicylic acid signaling in plant defense. *Proc. Natl. Acad. Sci. USA,* **97** : 8849-8855.

Kubo, M. and Kakimoto, T. (2000). The *CYTOKININ-HYPERSENSITIVE* genes of Arabidopsis negatively regulate the cytokinin-signaling pathway for cell division and chloroplast development. *Plant J.,* **23** : 385–394.

Leshem, Y.Y. (1996). Nitric oxide in biological systems. *Plant Growth Regul.,* **18** : 155-159.

Leshem, Y.Y. and Haramaty, E. (1996). Plant aging: the emission of NO and ethylene and effect of NO-releasing compounds on growth of pea (*Pisum sativum*) foliage. *J. Plant Physiol.*, **148** : 258-263.

Leshem, Y.Y., Wills, R.B.H. and Ku, V.V.V. (1998). Evidence for the function of the free radical gas nitric oxide (NO) as an endogenous maturation and senescence regulating factor in higher plants. *Plant Physiol. Biochem.*, **36** : 825-833.

Lo, S.C., Butt, Y.K. and Chan, Y.S. (2000). False nitric oxide synthase immunoreactivity in Asparagus bean (*Vigna sesquipdalis*) *Nitric Oxide*, **4** : 175.

Maier, J., Hecker, R., Rockel, P. and Ninnemann, H. (2001). Role of Nitric Oxide Synthase in the Light-Induced Development of Sporangiophores in *Phycomyces blakesleeanus*. *Plant Physiol.*, **126** : 1323-1330.

Mata, C.G. and Lamattina, L. (2001). Nitric oxide induces stomatal closure and enhances the adaptive plant responses against drought stress. *Plant Physiol.*, **126** : 1196-1204.

Marletta, M.A. (1993). Nitric oxide synthase structure and mechanism. *J. Biol. Chem.*, **268** : 1231-1234.

Marletta, M.A. (1994). Approaches toward selective inhibition of nitric oxide synthase. *J. Med. Chem.*, **37** : 1899-1907.

Mayer, B. and Hemmes, B. (1997). Biosynthesis and action of nitric oxide in mammalian cells. *Tr. Bioch. Sci.*, **22** : 477-481.

Neill, S.J., Desikan, R., Clarke, A. and Hancock, J.T. (2002). Nitric oxide is a novel component of abscisic acid signaling in stomatal guard cells. *Plant Physiol.*, **128** : 13-16.

Neuhaus, G., Bowler, C., Kern, R. and Chua, N.H. (1993). Calcium/calmodulin-dependent and -independent phytochrome signal transduction pathways. *Cell,* **73** : 937-952.

Ninnemann, H. and Maier, J. (1996). Indications for the occurrence of nitric oxide synthases in fungi and plants and the involvement in photoconidiation of *Neurospra crassa*. *Photochem. Photobiol.*, **74** : 393-398.

Noritake, T., Kawakita, K. and Doke, N. (1996). Nitric oxide induces phytoalexin accumulation in potato tuber tissue. *Plant Cell Physiol.*, **37** : 113-116.

Palmer, R.M.J., Ferrige, A.G. and Moncada, S. (1987). Nitric oxide release accounts for the biological activity of endothelium-derived relaxing factor. *Nature,* **327** : 524-526.

Paul, R., Holk, A. and Scherer, G.F.E. (1998). Fatty acids and lysophospholipids as potential second messengers in auxin action. Rapid activation of phospholipase A_2 activity by auxin in suspension-cultured parsley and soybean cells. *Plant J.*, **16** : 601-611.

Pfeiffer, S., Janystin, B., Jessner, G., Pichorner, H. and Ebermann, R. (1994). Gaseous nitric oxide stimulates guanosine-3‘,5‘-cyclic monophosphate (cGMP) formation in spruce needles. *Phytochemistry,* **36** : 259-262.

Piatelli, M., de Nicola, M.G. and Castrogiovanni, V. (1969). Photocontrol of amaranthin synthesis in *Amaranthus tricolor*. *Phytochemistry,* **8** : 731-736.

Piatelli, M., de Nicola, M.G. and Castrogionvanni, V. (1971). The effect of kinetin on amaranthin synthesis in *Amaranthus tricolor* in darkness. *Phytochemistry,* **10** : 289-293.

Rockel, P., Strube, F., Rockel, A., Wildt, J. and Kaiser, W.M. (2002). Regulation of nitric oxide (NO) production by plant nitrate reductase *in vivo* and *in vitro*. *J. Exp. Bot.*, **53** : 103-110.

Romanov, G.A., Kieber, J.J. and Schmülling, T. (2002). A rapid cytokinin response assay in Arabidopsis indicates a role for phospholipase D in cytokinin signalling. *FEBS Lett.*, **515** : 39-43.

Scherer, G.F.E. and Arnold, B. (1997). Auxin-induced growth is inhibited by phospholipase A_2 inhibitors. Implications for auxin-induced signal transduction. *Planta,* **202** : 462-469

Scherer, G.F.E. and Holk, A. (2000). NO donors mimick and NO inhibitors inhibit cytokinin action in betalaine accumulation in *Amaranthus caudatus* L. *Plant Growth Regul.,* **32** : 345-350.

Schmülling, T. (2001). Cream of cytokinin signalling: receptor identified. *Trends in Plant Sci.,* **6** : 281-284.

Schmülling, T., Schäfer, S. and Romanov, G. (1997). Cytokinins as regulators of gene expression. *Physiol. Plant.,* **100** : 505–519.

Smalle, J., Kurepa, J., Yang, P., Babiychuk, E., Kushnir, S., Durski, A. and Vierstra, R.D. (2002). Cytokinin growth responses in *Arabidopsis* involve the 26S proteasome subunit RPN12. *Plant Cell,* **14** : 17–32.

Stutte, C.A. and Weiland, R.T. (1978). Gaseous nitrogen loss and transpiration of several crop and weed species. *Crop Sci.,* **18** : 887-889.

Stöhr, C., Strube, F., Marx, G., Ullrich, W.R. and Rockel, P. (2001). A plasma membrane-bound enzyme of tobacco roots catalyses the formation of nitric oxide from nitrite. *Planta,* **212** : 835-841.

Stutte, C.A., Weiland, R.M.T. and Blem, A.R. (1979). Gaseous loss from soybean foliage. *Agron. J.,* **71** : 95-97.

Suzuki, T., Imamura, A., Ueguchi, C. and Mizuno, T. (1998). Histidine- containing phosphotransfer (HPt) signal transducers implicated in His-to-Asp phosphorelay in Arabidopsis. *Plant Cell Physiol.,* **39** : 1258–1268.

Suzuki, T., Miwa, K., Ishikawa, K., Yamada, H., Aiba, H. and Mizuno, T. (2001). The Arabidopsis sensor His-kinase, AHk4, can respond to cytokinins. *Plant Cell Physiol.,* **42** : 107-113.

Suzuki, T., Ishikawa, K., and Mizuno, T. (2002). An Arabidopsis histidine-containing phosphotransfer (HPt) factor implicated in phosphorelay signal transduction: Overexpression of AHP2 in plants results in hypersensitiveness to cytokinin. *Plant Cell Physiol.,* **43** : 123–129.

Suzuki, T., Sakurai, K., Ueguchi, C. and Mizuno, T. (2001). Two types of putative nuclear factors that physically interact with histidine-containing phosphotransfer (Hpt) domains, signaling mediators in His-to-Asp phosphorelay, in *Arabidopsis thaliana. Plant Cell Physiol.,* **42** : 37–45.

Tun, N.N., Holk, A. and Scherer, G.F.E. (2001). Rapid increase of nitric oxide (NO) release in plant cell cultures induced by cytokinin. *FEBS Lett.,* **509** : 174-176

Ueguchi, C., Sato, S., Kato, T. and Tabata, S. (2001). The AHK4 Gene Involved in the Cytokinin-Signaling Pathway as a Direct Receptor Molecule in Arabidopsis thaliana. *Plant Cell Physiol.,* **42** : 751-755.

Vollack, K.U. and Zumft, W.G. (2001). Nitric oxide signaling and transcriptional control of denitrification genes in *Pseudomonas stutzeri. J. Bacteriol.,* **183** : 2516-2526.

Watmough, N.J., Butland, G., Cheesman, M.R., Moir, J.W., Richardson, D.J. and Spiro, S. (1999). Nitric oxide in bacteria: synthesis and consumption. *Biochim. Biophys. Acta,* **1411** : 456-474.

Wellburn, A.R. (1998). Atmospheric nitrogenous compounds and ozone - is NO_x fixation by plants a possible solution? *New Phytol.,* **139** : 5-9.

Wendehenne, D., Pugin, A., Klessig, D.F. and Durner, J. (2001). Nitric oxide: comparative synthesis and signaling in animal and plant cells. *Trends in Plant Sci.*, **6** : 177-183.

Wilkinson, J.A. and Crawford, N.A. (1991). Identification of the *Arabidopsis CHL3* gene as the nitrate reductase structural gene *NIA2*. *Plant Cell,* **3** : 461-471.

Wilkinson, J.Q. and Crawford, N.M. (1993). Identification and characterization of a chlorate-resistant mutant of Arabidopsis with mutations both in *Nia1* and *Nia2* nitrate reductase structural genes. *Mol. Gen. Genet.*, **239** : 289-297.

Wojtaszek, P. (2000). Nitric oxide in plants. To NO or not to NO. *Phytochemistry,* **54** : 1-4.

Yamasaki, H. and Sakihama, Y. (2000). Simultaneous production of nitric oxide and peroxynitrite by plant nitrate reductase: *in vitro* evidence for the NR-dependent formation of active nitrogen species. *FEBS Lett.*, **468** : 89-92.

Yamada, H., Suzuki, T., Terada, K., Takei, K., Ishikawa, K., Miwa, K., Yamashino, T., and Mizuno, T. (2001). The Arabidopsis AHK4 histidine kinase is a cytokinin-binding receptor that transduces cytokinin signals across the membrane. *Plant Cell Physiol.,* **42** : 1017-1023.

Zumft, W.G. (1997). Cell biology and molecular basis of denitrification. *Microbiol. Mol. Biol. Rev.*, **61** : 533-616.

Chapter 8

NITRIC OXIDE PRODUCED BY NITRATE REDUCTASE IN HIGHER PLANTS

Jose R. Magalhaes*[1], Filomena, L.I.M. Silva[1], Ione Salgado[2], Osvaldo Ferrarese-Filho[3], Peter Rockel[4] and Werner M. Kaiser[5]

[1]*EMBRAPA - Gado de Leite – 36038-330 Juiz de Fora, MG, Brazil;* [2]*Dep. Bioquímica, IB, UNICAMP, 13083-970 Campinas, SP, Brazil;* [3]*Dep. Bioquímica - UEM - 87020-900 Maringá, PR - Brazil;* [4]*Forschungszentrum Jülich, Institut für Chemie der belasteten Atmosphäre, D-52425 Jülich, Germany;* [5]*Julius von Sachs Institut für Biowissenschaften, Lehrstuhl für Molekulare Pflanzenphysiologie und Biophysik, Julius von Sachs Platz, 2 D-97082, Würzburg -Germany*

*Corresponding author : E-mail : josemag@cnpgl.embrapa.br

Fax : (55-32)-3249-4852

Summary

Nitric oxide (NO) synthesis from the oxidation of L-arginine catalyzed by a family of NO synthases (NOS) in mammalians is well-known. However, NO synthesis has also been reported by NOS independent mechanisms in higher plants, algae, fungi and bacteria. A number of studies have indicated that plant cells possess a nitrite-dependent NO production pathway which can be distinguished from the NOS-mediated reactions. Using a nitrate reductase (NR) defective double mutant NIA1 NIA2, *it is demonstrated that the emission of NO in* Arabidopsis *originates from nitrate reductase. This chapter focuses on current understanding of the mechanism for the nitrite-dependent NO production in plants. Various methods for detecting NO production and emission from plants and plant tissues are summarized. The capacity of NR for NO production and the regulatory*

In : Nitric Oxide Signaling in Higher Plants, 2004

(Eds Jose R. Magalhaes, Rana P. Singh and Leonidas P. Passos)

Studium Press, LLC, Houston, USA, pp 167-181

properties of NR in context with observed complex patterns of NO production by plants in response to environmental factors and stress conditions is discussed.

Keywords : *Arabidopsis*, light, nitric oxide, nitrate reductase, *NIA*-mutant, plant stress, tobacco.

1. INTRODUCTION

Nitric oxide (NO) is a gaseous free radical, that reacts very rapidly with other atoms or molecules which contain unpaired electrons, having a very short half life-less than 10 seconds in the presence of oxygen (Lancaster Jr., 1992; MeBmer *et al.*, 1994). NO is produced as an air pollutant by industrial activities. It also plays a key role in atmospheric chemistry. Oxidation of NO by hydroxy peroxy radicals (HO_2) leads to the formation of hydroxyl radical (OH), nitrogen dioxide (NO_2) and to the photochemical production of ozone (O_3) in the troposphere. Thus, NO is important for atmospheric radical balance and for generation of photooxidants (Wildt *et al.*, 1997).

The NO is also produced by various members of the biosphere, including bacteria, fungi, plants and animals. Recently, NO has gained increasing attention as a general signaling compound in plant development and plant pathogen interactions. It is highly lipophilic and diffuses rapidly through membranes (Leshem, 1996). A major function in cells appears to be the activation of guanylate cyclase, which converts GTP to provide a local second messenger, cyclic guanosine monophosphate (cGMP) (Moncada, 1998). Compared to the vast amount of literature on NO in animals and humans, studies on the production and function of NO in plants are still scarce (Delledonne *et al.*, 1998; Durner *et al.*,1998; Kim *et al.*, 1998; Magalhaes *et al.*, 1999, 2000). Although the first reports on NO production by plants date back more than 25 years (Klepper, 1975) more insights into the role for NO as a signal in plant development and plant-pathogen interactions has directed new interest into plant NO research recently. Clearly, access to the multiple role of NO in plants and the environment requires detailed knowledge on the processes which are potentially involved in NO synthesis and its regulation.

In animals, NO is synthesized from L-arginine through a complex oxidation catalyzed by nitric oxide synthase (NOS, EC 1.14.13.39) family (Ignarro, 1996). The enzyme NOS catalyzes the NADPH- and oxygen-dependent oxygenation of L-arginine to NO and L-citrulline, in a reaction that requires at least six cofactors, including NADPH, FAD, FMN, tetrahydro-L-biopterin (H_4BPT), heme, calcium and calmodulin. In mammalian cells, several distinct NOS enzymes have been isolated, purified, cloned, and sequenced (Tzeng and Billar, 1996). Overall, NOS isoenzymes appear to be moderately to highly conserved proteins (Lin *et al.*, 1996; Marletta, 1999). It is worth to point it out that there are several possible reaction routes that may account for NO production independently of NOS, -NO pathway *in vivo*. This includes acidic degradation of tissue nitrite to NO, reaction between arginine and H_2O_2 to form NO, the anoxic reduction of nitrite to NO by xanthine oxidase and the reduction of nitrite to NO catalyzed by microbial nitrite reductase, (Zhang *et al.*, 1998). In plants, NO can also be produced in chloroplasts through light mediated conversion of NO_2 by carotenoids (Durner *et al.*, 1998).

In plants, NOS-like activities have also been reported (Ribeiro *et al.*, 1999). However, it has long been known that plants fertilized with nitrate may generate NO also via NAD(P)H-nitrate reductase (NR) as a by-product of nitrogen assimilation. Originally, this appeared restricted to a constitutive NR (Dean and Harper, 1988; Klepper, 1990). However, meanwhile evidence accumulated that inducible NR may also produce NO (Rockel *et al.*, 1996; Yamasaki *et al.*, 1999; Magalhaes *et al.*, 2000). This chapter will focus the knowledge on NO production via plant nitrate reductase.

2. BIOCHEMISTRY OF NITRIC OXIDE

The broader chemistry of NO involves an array of interrelated redox forms: nitrosonium cation (NO^+), nitric oxide ($NO^\cdot$), and nitroxyl anion (NO^-). This array is reminiscent of the redox forms of dioxigen- O_2, $O^{\cdot-}$, and $O_2{}^{2-}$ -the implications of which for O_2 biochemistry are well understood. Fundamental to the understanding of NO biochemistry is the interaction of established concepts in NO chemistry -the properties and reactivities of NO^+, $NO^\cdot$, and NO^- with current perspective on the biological functions of NO. It is notable that, by far the most widely studied of the nitroso-compounds, nitroprusside, has a NO^+ character, although under physiological conditions, NO can be interconverted among different redox forms with distinctive chemistries. Neutral $NO^{\cdot}$ has a single electron in its $2P$-π antibonding orbital. The charge neutrality of $NO^{\cdot}$ has been assumed to facilitate its free diffusibility in aqueous medium and across cell membranes. The reaction of $NO^{\cdot}$ with O_2 in both the gas phase and aqueous solution is a complicated process that is second order in $[NO^{\cdot}]$ $[K\ (NO^{\cdot})^2(O_2)]$; the biological half-life of $NO^{\cdot}$, generally assumed to be on the order of seconds, thus depends critically on its initial concentration (Stanler *et al.*, 1992).

3. NITRIC OXIDE PRODUCTION VIA NITRATE REDUCTASE IN ALGAE, FUNGI AND BACTERIA

The NO is formed via nitrate reductase (NR), not only in higher plants (Magalhaes *et al.*, 2000) but also in algae (Mallick *et al.*, 1999; Sakihama *et al.*, 2002), in fungi, (Takaya, 2002), and in bacteria, (Baumgärtner *et al.*, 1991; Zhang *et al.*, 1998).

In algae NO production was found in prokaryotic and eukaryotic algae like *Chlamydomonas reinhardtii* (Sakihama *et al.*, 2002), in *Scenedesmus, Anabaena doliolum,* and *Synechoccocus* (Mallick *et al.*, 1999). Experiments with inhibitors of photosynthesis (DCMU), of ATP synthesis, and of the uncoupler (2,4-DNP) and its analog arsenate clearly revealed that inhibition of plastidic nitrite assimilation is causally connected with NO emission, and a linear relationship between nitrite concentration in the culture medium and NO in the exhaust gas supports this view. A failure of *Scenedesmus*, grown in the medium substituted with W for Mo, to produce either NO/NO^-_2 in light or a light-off peak, and a resumption of these activities upon the addition of Mo proved that a functional nitrate reductase (NR) is necessary for the production of nitrite and NO by algae grown on nitrate as the nitrogen source. Moreover, the appearance of a NO peak immediately after nitrite supplementation under dark conditions in W-substituted cultures with or without glucose indicate role of NR in NO emission (Mallick *et al.*, 1999).

Nitric oxide (NO) production was also investigated in the unicellular green algae *Chlamydomonas reinhardtii* by amperometric techniques using an NO specific electrode. L- arginine, the substrate for NO synthase, did not induce NO production, and the NOS inhibitor N^P-nitro-l-arginine had no effect on the nitrite-dependent production of NO. A *Chlamydomonas* mutant lacking nitrate reductase (NR) activity, did not display any of the responses observed in the wild-type cells. The results provided direct *in vivo* evidence to confirm that NR is involved in the nitrite-dependent NO production in the green algae (Sakihama *et al.*, 2002).

In the fungus *Fusarium oxysporum* the pathway of respiratory nitrate denitrification is catalyzed by the sequential reactions of nitrate reductase and nitrite reductase. These enzymes are coupled with ATP generation through the respiratory chain and produce nitric oxide. Fungal nitric oxide reductase uses NADH as the direct electron donor in contrast to bacterial systems and thus might function in regeneration of NAD^+ and detoxification of the toxic radical, nitric oxide. Another pathway of nitrate dissimilation by fungi reduces nitrate to ammonium and couples the acetogenic reaction with substrate-level phosphorylation. This metabolic mechanism is also found in a variety of fungi and is called as ammonia fermentation (Takaya, 2002).

In bacteria, during oxidation of ammonium to nitrate, significant amounts of NO are released and nitrite is intermediately accumulated. Nitrification and NO release were detectable only in neutral pH. Addition of nitrapyrin inhibited both ammonium oxidation and NO production. NO was produced in rock samples containing *Nitrosomonas* or *Nitrosovibrio*, but was not produced in samples containing only *Nitrobacter*. Most of the NO production by corroding rocks was due to ammonium-oxidizing nitrifier bacteria (Baumgärtner *et al.*, 1991).

Under hypoxic conditions NO may also be formed by another NOS- and NR- independent mechanisms. Zhang *et al.* (1998) examined the reduction of nitrite to NO by xanthine oxidase (XO) under hypoxia, because the bacterial nitrate/nitrite reductases have structural similarity to XO. It was found that both purified and *in situ* xanthine oxidase catalyze the reduction of nitrite to NO, as demonstrated using a chemiluminescence NO meter. This redox reaction also requires NADH as an electron donor, and is oxygen independent. The inhibitory profiles suggest that reduction of nitrite takes place at the molybdenum center of XO whilst NADH is oxidized at the FAD center (Zhang *et al.*, 1998).

4. EVIDENCE THAT NITRATE REDUCTASE PRODUCES NITRIC OXIDE IN HIGHER PLANTS

The NO production by plant cells has been basically studied using the same methods as in animal research (Kojima *et al.*, 1998). We have recently shown that NO production in plant cells can be visualized by using the specific probe, diaminofluorescein diacetate (DAF-2DA), and that NO induces apoptosis in plants (Magalhaes *et al.*, 1999, 2000). NO production *in vitro* by enzyme solutions was most easily quantified by commercially available amperometric methods (Yamasaki, 2000), whereas NO emission into the atmosphere from plants or plant organs, as well as from enzyme solutions, has been followed by highly sensitive chemiluminescence detection (Wildt *et al.*, 1997; Rockel *et al.*, 2002).

Emission of NO from a variety of plant species was observed in a continuously stirred tank reactor, and the link between NO emission and CO_2 uptake ($3x10^{-6}$ mol of NO emitted per mol of CO_2 taken up) allowed estimation of the potential of the vegetation to evolve NO on a global scale as 0.23 Tg N yr^{-1} (Wildt *et al.*, 1997). Emission of NO from plants was found under specific conditions such as after herbicide treatment or under anaerobic conditions (Klepper, 1990).

Previous studies have indicated that plant cells possess a nitrite-dependent NO production pathway which can be distinguished from the NOS-mediated reaction. Nitrate reductase (NR, EC 1.6.6.1-3), as bifunctional enzyme, can reduce nitrate to nitrite: $NO_3^- + NAD(P)H + H^+ \rightarrow NO_2^- + NAD(P)^+ + H_2O$ (1); or at lower rates, it reduces nitrite to NO: $2\ NO_2^- + NAD(P)H + H^+ \rightarrow 2\ NO + NAD(P)^+\ 2\ OH^-$ (2). The NO production from plants was first observed by Klepper (1975) with soybean treated with photosynthetic inhibitor herbicides or other chemicals, as well as under dark anaerobic conditions (Klepper, 1990). Subsequently, NO_x production by plants was repeatedly confirmed and it was shown that a constitutive NAD(P)H nitrate reductase was responsible for evolution of NO_x (Harper, 1981; Ryan *et al.*, 1983; Dean and Harper, 1988; Klepper, 1990). More recently, electrochemical, fluorometric and chemiluminescence measurements *in vitro* showed that NO was also produced by inducible (purified) maize NR or by NR in crude desalted leaf extracts in the presence of nitrate or nitrite and NADH at pH 7 (Yamasaki, 2000; Yamasaki and Skihama, 2000; Rockel *et al.*, 2002). Further the NO production was inhibited by sodium azide, a known inhibitor of NR (Yamasaki, 2000). In addition, a plasma membrane bound nitrate reductase, coupled to a nitrite : NO reductase in membrane vesicles from tobacco roots has also been shown to produce NO (Stöhr *et al.*, 2001).

NO emission from non-infected leaves was completely insensitive to NOS inhibitors (Magalhaes *et al.*, 2000; Rockel *et al.*, 2002). Further evidence for a NR-dependent NO production came from the use of *NIA*-double mutants (Tobacco or *Arabidopsis*), which posses no NR activity. These plants did not produce NO, as measured with both, gas chromatography (Table 1) and with chemiluminescence detection (Rockel and Kaiser, unpublished). Experiments with ^{15}N-labelled nitrate as substrate for nitrate reduction showed that labelled NO_x was produced from $^{15}N\text{-}NO_3$ (Dean and Harper, 1986).

Table 1. NO emission in different genotypes of *Arabidopsis*, a double mutant *NIA1 NIA2*, nitrate reductase defective, and wild type. NO were measured in 25 day-old plants, with four replications, and the results are presented in $nL.gfw^{-1}.h^{-1}$ ± STD.

	WILD TYPE	MUTANT *NIA/NIA2*
NO ($nL.gfw^{-1}.h^{-1}$)	494 ± 57	000 ± 00

5. ESTIMATING THE CAPACITY OF NITRATE REDUCTASE FOR NITRIC OXIDE FORMATION

Crude leaf extracts (containing NR) or purified NR fed with NAD(P)H produced NO from nitrate, but only with a lag phase the duration of which depends on the experimental set up. If nitrate was replaced by nitrite, NO production started immediately (Yamasaki, 2000; Rockel *et al.*, 2002), indicating that nitrite was the actual substrate, and that NO production from nitrate required formation and accumulation of nitrite with time. NO production was not observed in the absence of NAD(P)H or of NR. It was completely inhibited by azide, which interrupts electron flow in NR at the cytochrome moiety (Yamasaki, 2000), indicating that chemical NO formation was negligible at physiological pH. The K_m for nitrite of NO formation by purified NR (maize) was relatively high (100 μM) compared to mean nitrite concentrations in illuminated leaves (10 μM), and nitrate was a competitive inhibitor (K_i = 50 μM). The maximum capacity of NR *in vitro* for NO formation from nitrite plus NAD(P)H was only about 1% of the nitrate-reducing capacity (up to 0.2 µmol g^{-1} fw^{-1} h^{-1} with spinach NR) when measured by chemoluminescence detection (Rockel *et al.*, 2002). Similar *in vitro* rates have been obtained by Yamasaki (2000) using amperometric detection.

Rates of NO emission into the gas phase by sunflower (*Helianthus annuus* L.) or by detached spinach leaves (*Spinacia oleracea* L.), measured by chemoluminescence were usually very low (Rockel *et al.*, 2002). In sunflower leaves, they were about 0.05 nmol g^{-1} fw^{-1} h^{-1} in the dark, and up to 0.5 nmol g^{-1} fw^{-1} h^{-1} in the light. Much higher rates of NO emission from leaves were achieved in the dark under anoxia, reaching up to 200 nmol.g^{-1} fw^{-1} h^{-1} (Rockel *et al.*, 2002). Intact Arabidopsis plants produced up to 20 nmol g^{-1} fw^{-1} h^{-1} NO in the light (Table 2). All together, in these more recent measurements, maximum NO production rates by NR *in vitro* or by plants and leaves were only a few percent of the NR capacity. These low rates contrast sharply with previous results from Klepper (1990), who found a 100-fold higher NO production rate, almost as high as the rate of nitrate reduction (up to 15 µmoles NO g^{-1} fw^{-1} h^{-1} under dark-anaerobic conditions). The reason for the big discrepancy is

Table 2. NO emission in *Arabidopsis*, as affected by light intensity and time exposure. Four-week-old plants were submitted to different light regimes by transferring the plants from 555 to 105 µmoles.m^{-2}.s^{-1}; from 555 µmoles.m^{-2}.s^{-1} to dark; from dark to 105 µmoles.m^{-2}.s^{-1} (in growth chamber) and transferring from dark to green house 555 µmoles.m^{-2}.s^{-1} or to full sun light 1500 µmoles.m^{-2}.s^{-1}. NO was measured at time zero, 2; 4; 6; 8 and 12 h with four replications. Results are presented in nL.gfw^{-1}. h^{-1} ± STD.

Light regime	hours				
	0	2	4	8	12
555 to 105 µmol.m^{-2}.s^{-1}	503 ± 62	187 ± 24	172 ± 23	89 ± 12	1.79 ± 0.20
555 µmol.m^{-2}.s^{-1} - Dark	503 ± 62	492 ± 47	477 ± 42	425 ± 41	232 ± 24
Dark - 105 µmol.m^{-2}.s^{-1}	264 ± 24	00 ± 00	00 ± 00	00 ± 00	00 ± 00
Dark - 555 µmol.m^{-2}.s^{-1}	264 ± 24	00 ± 00	278 ± 27	667 ± 57	543 ± 62

not clear, but in Klepper's measurement the very high rates may be based on an insufficient removal of other gaseous compounds emitted from leaves, which may also give rise to a chemoluminescence signal.

6. NITRIC OXIDE IN PLANT DEVELOPMENT AND ENVIRONMENTAL STRESSES

Stress-evoking factors, such as pathogen attack, wounding, mechanical perturbations or impedance, anaerobiosis, waterlogging and submergence, desiccation, chilling, freezing, salt stress, heavy metals, ozone, electrical currents, certain herbicides, induce a burst of stress ethylene (Kacperska, 1997), and this appears to be related to NO as signaling molecule in plants (Magalhaes *et al.*, 2000). As previously indicated, NO emission from plants is highly variable. Closer examination showed that NO emission varied greatly with developmental or growth stage of plants, light intensity and different stresses. Interestingly, an inverse trend for NO and ethylene emission was seen from the flower stage to the start of senescence (Magalhaes *et al.*, 2000). As plants entered senescence, NO decreased. The inverse correlation between ethylene and NO emission may suggest some interaction of the two synthesis pathways. However, the NR defective double mutant *NIA1 NIA2* did not emit NO but produced ethylene similarly to the wild type (Magalhaes *et al.*, 2000), suggesting that NO and ethylene biosynthesis are independent.

Drought and waterlogging drastically reduced NO emission (Magalhaes *et al.*, 2000). However, short term anoxia activates NR, leads to nitrite accumulation and high rates of NO emission from leaves (Rockel *et al.*, 2002) and roots (compare the chapter by Kaiser *et al.*). Thus, short term and long term hypoxia and anoxia may have different effects on NO production by roots. A decrease in NR has also been observed under drought stress (Garg *et al.*, 1998).

NO emission as function of light intensity was intricate and complicated. NO emission increased sharply during exposure to light, with greater light intensity. Although, significant NO emission was observed in dark before exposure to light. The emission dropped sharply to zero in the first two hours and then started to increase (Table 2). In the first two hours after sunrise, light intensity in the greenhouse was low (115 $\mu mol.m^{-2}.s^{-1}$ in average). This explains the NO emission drop to zero and is consistent with experiments in a growth chamber at 105 $\mu mol.m^{-2}.s^{-1}$ where NO was not produced at all.

Many replicated experiments showed significant diurnal variation in NO emission. For this reason, measurement of NO was best done at least three hours after plants were exposed to light (Table 2). This gave consistent and comparable results. Table 2 shows that transferring plants from dark to low light, 105 $\mu mol.m^{-2}.s^{-1}$ did not emit any NO during 12 h. By transferring from 555 to 105 $\mu mol\ m^{-2}\ s^{-1}$, a gradual decrease was observed in the first 4 h and then decreased to nearly zero by 12 h. When plants were transferred from 555 $\mu mol\ m^{-2}\ s^{-1}$ to darkness, NO emission declined gradually, but a considerable NO emission was observed by the end of 12 h. After transfering from light to dark, NR activity in leaves persisted during the dark phase for some time and then declined gradually likewise (Salalkar *et al.*, 1999). After re-exposure to light, NR increased rapidly in a pattern very similar to NO emission in this study, supporting the view that NO production rates parallel NR activity and eventually nitrite concentrations. It is also reasonable to explain the decrease of NO based on the decrease in NR activity with plant age (Anburaj and Francis, 1996; Lee *et al.*, 1998; Yu *et al.*, 1998). NR in lettuce showed peak activity 20 days after planting (Lee *et al.*, 1998), with a similar pattern of NO as observed by Magalhaes *et al.* (2000).

NR is a highly regulated enzyme (Campbell, 1999). It has a short half life of some hours and its induction requires nitrate and light (photosynthesis). The activity of the existing enzyme can be decreased within minutes by phosphorylation on a conserved serine residue in the hinge 1 region, and subsequent binding of a 14-3-3 dimer. Phosphorylation and inhibition depend on the presence of divalent cations. Reactivation is achieved by dephosphorylation. It also appears that the proteolytic degradation of NR is accelerated by its inactivation (for a recent review on NR modulation compare Kaiser *et al.*, 2002). *In vitro,* NR is inactivated by incubation with MgATP, eventually in the presence of PP2A inhibitors to prevent dephosphorlyation. Preincubation with ATP also decreases nitrite-dependent NO production, while reactivation by dephosphorylation increases NO production (Rockel *et al.*, 2002).

It has been mentioned above that light plus high CO_2 activate NR and that this also leads to higher rates of NO emission. That would suggest that NR modulation *in vivo* is paralleled by NO production. Indeed, artificial treatments that are all known to activate NR in the dark (where it is usually inactivated), all lead to high NO-emission (which is also usually low in the dark). Aside of anoxia, such conditions are: feeding of uncouplers, feeding of the 5'-AMP analogue AICAR (5-aminoimidazole-4-carboxyamide-1-ß-ribofuranosyl 5' – monophosphate) and others. If NR activation was prevented by the protein phosphatase inhibitor, okadaic acid, NO emission was blocked (Rockel *et al.*, 2002). Conclusively, NR modulation *in vivo* also modulates NO emission.

In incompatible plant pathogen interactions, one of the first responses after inoculation of the host is a burst of reactive oxygen species (ROS). Superoxide radicals or hydrogen peroxide react with NO very rapidly, leading at least in part to the formation of the highly toxic peroxonitrite, which readily nitrates aromatic amino acids *e.g.* tyrosine. Interestingly, NR is also able to catalyze at low rates the reduction of molecular oxygen to superoxide (Yamasaki, 2000). Simultaneous production of superoxide and NO will almost inevitably lead to the formation of peroxonitrite. Indeed, tobacco mutants with deleted nitrite reductase (NiR) but normal NR accumulate nitrite, emit high amounts of NO and have a high degree of tyrosine nitration (Morot-Gaudry-Talarmain *et al.*, 2002).

While the above considerations have been focused on NR dependent NO formation, NOS as a source for NO in plants cannot be neglected. There are good indications for the presence of an NOS-like activity in plants (Cueto *et al.*, 1996; Huang and Knopp, 1998; Jaffrey and Snyder, 1996; Ninnemann and Maier, 1996; Ribeiro *et al.*, 1999). However, neither the gene nor the protein responsible for NOS activity has yet been isolated from plants. The measurement of NOS activity is usually based on a test using ^{3}H-labelled arginine. With that test system, as well as by immunoblot analysis, NOS was detected in isolated leaf peroxisomes and in chloroplasts of *Pisum sativum* L., while no activity was found in isolated mitochondria (Barroso *et al.*, 1999). A final proof for the presence of NOS genes in plants is still lacking, in spite of a meanwhile completed sequencing of the *Arabidopsis* genome. To date, neither cDNA, nor any protein with sequence similarity to a known NOS have been found in plants.

7. CONCLUSIONS AND FUTURE PROSPECTS

In the past, NR has been studied extensively as a key enzyme of nitrogen metabolism, whereas its ability to produce NO has not been considered likewise. If NO plays an important role in various physiological processes, from plant growth to biotic and abiotic responses, the production of NO and its subcellular localization and its regulation should be a key elements in determining its role in these various processes. Whether NO comes from NR or from NOS should be functionally not relevant, if only the production time, concentration and localization are "correct". If NO in plants originates from two different pathways, via putative NOS and NR, each one should be regulated, and interaction between the pathways seems possible *e.g.* the NO formed by NOS could be produced in response to plant infection and during senescence, whereas NO production by NR may reflect NO signaling in response to nutrition, light intensity, and photosynthesis. Physiological effects of NO, suggests a potential key role for this free radical as signaling molecule in adaptive situation and NO emission may also be useful as an index for stress-evoking factors in plants. Further studies should be done on this issue.

REFERENCES

Anburaj, A. and Francis, K. (1996). Partitioning of NO_3^- reduction between roots and shoots of Italian millet and small millet as influenced by plant age. *Plant Physiol. Bioch.*, **23** : 145-147.

Barroso, J.B., Corpas, F.J., Carreras, A, Sandalio, L.M, Valderrama, R., Palma, J.M., Lupianez, J.A, Del Rio, L.A. (1999) Localization of nitric-oxide synthase in plant peroxisomes. *J Biol. Chem.*, **274** : 36729-36733.

Baumgärtner, M.; Sameluck, F.; Bock, E. and R. Conrad, R. (1991). Production of nitric oxide by ammonium-oxidizing bacteria colonizing building stones. *FEMS Microb. Lett.*, **85** : 95-100.

Campbell, W.H. (1999). Nitrate reductase structure, function and regulation: bridging the gap between biochemistry and physiology. *Ann. Rev. Plant Physiol. Plant Mol. Biol.*, **50** : 277-303

Cueto, M., Hernandez-Perera, O., Martin, R., Bentura, M.L., Rodrigo, J., Lamas, S. and Golvano, M.P. (1996). Presence of nitric oxide synthase activity in roots and nodules. *FEBS Lett.*, **398** : 159-164.

Dean, J.V, Harper, J.E. (1986). Nitric oxide and nitrous oxide production by soybean and winged bean during in vivo nitrate reductase assay. *Plant Physiol.*, **82** : 718-723

Dean, J.V. and Harper, J.E. (1988). The conversion of nitrite to nitrogen oxide(s) by the constitutive NAD(P)H-nitrate reductase enzyme from soybean. *Plant Physiol.*, **88** : 389-395.

Delledonne, M., Xia, Y., Dixon, R.A. and Lamb, C. (1998). Nitric oxide functions as a signal in plant disease resistance. *Nature*, **394** : 585-588.

Durner, J., Wendehenne, D. and Klessig, F. (1998). Defense gene induction in tobacco by nitric oxide, cyclic GMP, and cyclic ADP-ribose. *Proc. Natl. Acad. Sci. USA*, **95** : 10328-10333.

García-Mata, C and Lamattina, L (2003). Abscisic acid, nitric oxide and stomatal closure is nitrate reductase one of the missing links? *Trends in Plant Sci.*, **8** : 20-26.

Garg, B.K., Vxas, S.P., Kathju, S. and Lahiri, A.N. (1998). Influence of water deficit stress at various growth stages on some enzymes of nitrogen metabolism and yield in clusterbean genotypes. *Indian J. Plant Physiol.,* **3** : 214-218.

Harper, J.E. (1981). Evolution of nitrogen oxide(s) during in vivo nitrate reductase assay of soybean leaves. *Plant Physiol.,* **68** : 1488-1493.

Huang, J.S., Knopp, J.A. (1998). Involvement of nitric oxide in *Ralstonia solanacearum*-induced hypersensitive reaction in tobacco. In: *Bacterial Wilt Disease, Molecular and Ecological Aspects.* (Eds. Prior, P., Elphinstone, J. and Allen, C.), Berlin:INRA and Springer Editions, pp. 218-224.

Ignarro, L.J. (1996). Physiology and pathophysiology of nitric oxide. *Kidney Int.,* (Supplement) **55** : 52-5.

Jaffrey, S.R. and Snyder, S.H. (1996). PIN: an associated protein inhibitor of neuronal nitric oxide synthase. *Science,* **274** : 774-777.

Kacperska, A. (1997). Ethylene synthesis and a role in plant responses to different stressors. In : *Biology and Biotechnology of the Plant Hormone Ethylene* (Ed. Kanellis, A.K.), Kluwer Academic Publishers, The Netherlands, pp. 207-216.

Kaiser, W.M., Weiner, H., Kandlbinder, A., Tsai, C.B., Rockel, P., Sonoda, M. and Planchet E. (2002) Modulation of nitrite reductase: some new insights, an unusual case and a potentially important side reaction. *J. Exp. Bot.,* **53** : 875-882

Kim, O.K., Murakami, A., Nakamura, Y. and Ohigashi, H. (1998). Screening of edible Japanese plants for nitric oxide generation inhibitory activities in RAW 264.7 cells. *Cancer Lett.,* **125** : 199-207.

Klepper, L.A (1975). Evolution of nitrogen oxide gases from herbicide treated plant tissues. *WSSA Absts.*, **184** : 70

Klepper, L.A. (1990). Comparison between NO_x evolution mechanism of wild type and *nr1* mutant soybean leaves. *Plant Physiol.*, **93** : 26-32.

Kojima, H., Nakatsubo, N., Kikuch, K., Kawahara, S., Kirino, Y., Nagoshi, H., Hirata, Y. and Nagano, T. (1998). Detection and imaging of nitric oxide with novel fluorescent indicators: diaminofluoresceins. *Anal. Chem.*, **70** : 2446-2453.

Lancaster Jr., J.R. (1992). Nitric oxide in cells. *American Scient.,* **80** : 248-259.

Lee, E.H., Lee, B.Y., Lee, J.W., Kim, K., Kwon, Y.S. (1998). Nitrate content and activity of nitrate reductase and glutamine synthetase as affected by plant age, leaf position, time of day of leaf lettuce in hydroponics. *J. Korean Soc. Hort. Sci.,* **39** : 149-151.

Leshem, Y.Y. (1996). Nitric oxide in biological systems. *Plant Growth Regul.,* **18** : 155-159.

Magalhaes, J.R., Pedroso, M.C. and Durzan, D. (1999). Nitric oxide, aptoptosis and plant stress. *Physiol. Mol. Biol. Plants,* **5** : 115-125.

Magalhaes, J.R., Monte, D.C. and Durzan, D. (2000). Nitric oxide, and ethylene emission in *Arabidopsis thaliana. Physiol. Mol. Biol. Plants*, **6** : 117-127.

Mallick, N., Rai, L.C., Mohn, F.H. and Soeder, C.J. (1999). Studies on nitric oxide (NO) formation by the green alga scenedesmus obliquus and the diazotrophic cyanobacterium anabaena doliolum. *Chemosphere*, **39** : 1601-1610.

MeBmer, U.K., Ankarcrona, M., Nicotera, P. and Brune, B. (1994). *p53* expression in nitric oxide induced apoptosis. *FEBS Lett.*, **355** : 23-26.

Moncada, S. (1998). Nitric oxide as a biological mediator-A ten year perspective. *Nitric oxide: Basic Research and Clinic Application-Extensive Abstracts*, **1** : 6-11.

Morot-Gaudry-Talarmain, Y., Rockel, P., Moureaux, T., Quillere, I., Leydecker, M.T., Kaiser, W.M. and Morot-Gaudry, J.F. (2002). Nitrite accumulation and nitric oxide emission in relation to cellular signaling in nitrite reductase antisense tobacco. *Planta,* **215** : 708-715.

Ninnemann, H. and Maier, J. (1996). Indications for the occurrence of nitric oxide synthases in fungi and plants and the involvement in photoconidiation of *Neurospora crassa. Photochem. Photobiol.*, **64** : 393-398.

Ribeiro, E.A., Cunha, F.Q., Tamashiro, W.M.S.C. and Martins, I. S. (1999). Growth phase-dependent subcellular localization of nitric oxide synthase in maize cells. *FEBS Lett.*, **445** : 283-286.

Rockel, P., Rockel, A, Wildt, J. and Segschneider, H.J. (1996) Nitric oxide (NO) emission by higher plants. In : *Progress in Nitrogen Cycling Studies.* (eds.) Van Cleemput, O. *et al.* Kluwer Academic Publishers, the Netherlands, pp 603-606

Rockel, P., Struba, F., Rockel, A., Wildt, J. and Kaiser, W.M. (2002). Regulation of nitric oxide (NO) production by plant nitrate reductase *in vivo* and *in vitro*. *J. Exp. Bot.*, **53** : 1-8

Ryan, S.A, Nelson, R.S. and Harper, J.E. (1983). Soybean mutants lacking constitutive nitrate reductase activity. II. Nitrogen assimilation, chlorate resistance, and inheritance. *Plant Physiol.*, **72** : 510-514.

Sakihama, Y.; Nakamura, S. and Yamasaki· H. (2002). Nitric Oxide Production Mediated by Nitrate Reductase in the Green Alga *Chlamydomonas reinhardtii*: an Alternative NO Production Pathway in Photosynthetic Organisms. *Plant Cell Physiol.*, **43** : 290-297.

Salalkar, B.K., Shaiki, R.S., Naik, R.M., Munjal, S.U., Desai, B.B., Prikhsayat, S., Naik, M.S. and Singh, P. (1999). Changes in leaf nitrate reductase activity *in vivo* and *in vitro* during light-dark transmition. *J. Plant Biochem. Biotechnol.*, **8** : 37-40.

Stanler, J.S., Singel, D. and Loscalzo, J. (1992) Biochemistry of nitric oxide and its redox-activated forms. *Science,* **158** : 1898-1901.

Stöhr, C., Strube, F., Marx, G., Ullrich, W.R. and Rockel, P. (2001) A plasma membrane-bound enzyme of tobacco roots catalyzes the formation of nitric oxide from nitrite. *Planta,* **212** : 835-841.

Takaya, N. (2002). Dissimilatory nitrite reduction metabolisms and their control in fungi. *J. Biosci. Bioengin.*, **94** : 506-510.

Takahashi, S. and Yamasaki, H. (2002). Reversible inhibition of photophosphorylation in chloroplasts by nitric oxide. *FEBS Lett.*, **512** : 145-148.

Wildt, J., Kley, D., Rockel, A., Rockel, P., and Segschneider, H.J. (1997). Emission of NO from several higher plant species. *J. Geophys. Res.,* **102** : 5919-5927.

Wilkinson, J.Q. and Crawford, N.M. (1993). Identification and characterization of a chlorate-resistant mutant of *Arabidopsis thaliana* with mutations in both nitrate reductase structural genes *NIA1* and *NIA2*. *Mol. Gen. Genet.*, **239** : 289-297.

Yamasaki, H. (2000). Nitrite-dependent nitric oxide production pathway: implications for involvement of active nitrogen species in photoinhibition *in vivo*. In : *NO Philosophical Transactions of the Royal Society of London. Series-B. Biological Sciences*, **355** : 1477-1488.

Yamasaki, H., Sakihama, Y. And Takahashi, S. (1999). An alternative pathway for nitric oxide production in plants: new features of an old enzyme. *Trends in Plant Sci.,* **4** : 128-129.

Yamasaki, H. and Sakihama, Y. (2000). Simultaneous production of nitric oxide and peroxynitrite by plant nitrate reductase: *in vitro* evidence for the NR-dependent formation of active nitrogen species. *FEBS-Lett.* **468** : 89-92.

Yamasaki, H., Shimoji, H., Ohshiro, Y. and Yasuko Sakihama, Y. (2001). Inhibitory effects of nitric oxide on oxidative phosphorylation in plant mitochondria. *Nitric Oxide,* **5** : 261-270.

Yu, X.D., Sukumaran, S., and Marton, L. (1998). Induced nitrate reductase activity is correlated with increased *NIA1* transcription and mRNA levels. *Plant Physiol.*, **116** : 1091-1096.

Zhang, Z., Naughton, D., Winyard, P.G., Benjamin, N., Blake, D.R. and Symons, M.C.R. (1998). Generation of nitric oxide by a nitrite reductase activity of xanthine oxidase : a potential pathway for nitric oxide formation in the absence of nitric oxide synthase activity. *Biochem. Biophy. Res. Commun.*, **249** : 767-772.

Chapter 9

MEASUREMENT OF NITRIC OXIDE - EMISSION FROM ENZYME SOLUTIONS, CELLS SUSPENSIONS, LEAVES AND ROOTS BY GAS-PHASE CHEMILUMINESCENCE DETECTION

Elisabeth Planchet, Maria Stoimenova, Masatoshi Sonoda and Werner M. Kaiser*

Julius-von-Sachs-Institut für Biowissenschaften, Lehrstuhl für Molekulare Pflanzenphysiologie und Biophysik, University of Würzburg, Julius-von-Sachs-Platz 2, D-97082 Würzburg, Germany

*Corresponding author : E-mail : kaiser@botanik.uni-wuerzburg.de

Fax : (0)931-888-6158

Summary

NO production by enzyme solutions, cell suspensions, leaves and roots was followed on-line by measuring NO emission into purified air through chemiluminescence detection. The stability of NO in buffer solutions and in cell suspensions was also determined. In spite of the rather high gas flow required by the ozone-collision chemiluminescence analyzer (1,5 L min^{-1}), the method was sensitive enough to quantify NO emission from leaves, roots or cell suspensions containing 0.1 to 1g FW. Generally, in leaves and cell suspensions, NO emission was strictly dependent on nitrate reductase (NR) activity and on the concentration of the substrate, nitrite. In leaves, NO emission was low in the dark in air, 3 to 10 fold higher in the light, and more then 100-fold higher under anoxia in the dark. This complex

In : Nitric Oxide Signaling in Higher Plants, 2004

(Eds Jose R. Magalhaes, Rana P. Singh and Leonidas P. Passos)

Studium Press, LLC, Houston, USA, pp 183-198

pattern depends on the relative changes in NR activity and nitrite reductase (NiR) activity, which together determine the nitrite concentration in the cells. In NR-containing roots, NO emission in air was very low, but as in leaves, it increased dramatically under nitrogen. Unexpectedly, even NR-free roots were able to emit NO when fed with nitrite under anoxia, indicating the participation of some other enzyme, at least in roots. NO emission from a solution of purified NR (with nitrite and NADH as substrates) was slightly higher in nitrogen than in air and could be effectively quenched by low hydrogen peroxide concentrations, suggesting that in vivo, apparent NO emission could underestimate NO production due to a concomitant production of reactive oxygen species. Injection of a defined-but unnaturally large- amount of NO (380 pmoles) into an agitated cell suspension under air or nitrogen revealed that cells trapped 95% of the added NO in air and 90% in nitrogen.

Keywords : Chemiluminescence detection, nitric oxide, nitrate reductase, *Nicotiana tabacum.*

1. INTRODUCTION

Nitric oxide is an uncharged gaseous free-radical with moderate water solubility. The Henry-Coefficient is 1.9 mM in pure water at atmospheric pressure. The physical half-life of NO in solution depends on its concentration. At the low concentrations (nanomolar) of NO to be expected in biological system, the physical half life in pure water is in the range of hours (Henry *et al.*, 1997). However, living cells contain and produce many compounds that react rapidly with NO, e.g. reactive oxygen species (ROS), sulfhydryl groups and heme-compounds. Therefore the half-life of NO *in vivo* may be in the range of a few seconds (Lancaster, 1997), and much of the NO produced is not emitted but trapped within the cells.

In spite of its short biological half life, and based on a vast amount of literature on the formation and function of NO in animals, it has been suggested that NO is an important second messenger also in plants. To date, NO appears involved in controlling disease resistance, programmed cell death, growth, stomatal movement, senescence and development (Dangl *et al.*, 1996; Leshem and Haramaty, 1996; Delledonne *et al.*, 1998; Durner and Klessing, 1999; Beligni and

Lamattina, 2000; García-Mata and Lamattina, 2002; Neill *et al.*, 2002). For understanding the physiological role of NO in plants, it is important to clarify by what reactions, in what situation and how much NO is produced. Various direct and indirect methods exist to detect NO in tissues, cells, solutions and in the gas phase, which have been summarized recently (Lamattina *et al.* 2003). All of them have distinct advantages and disadvantages, and most of them are at best semi-quantitative and of limited specificity. Here we present data on the emission of NO into purified air by leaves, roots, cell suspensions and enzyme solutions, measured on-line by the ozone-collision chemiluminescence method (CLD 770 AL ppt, ECO PHYSICS AG, Zürich, Switzerland). This NO analyzer specifically detects and quantifies NO in the gas phase down to the ppt-concentration range. In principle, this type of NO analyzer quantifies photon emission generated when the excited NO_2, produced by the collision of NO with ozone, backs to the ground state. Although plants may produce and emit other organic volatiles that generate chemiluminescence upon reaction with ozone, these reactions are distinguished kinetically from NO because their reaction with ozone is much slower than that of NO and may also have a different emission spectra.

2. STABILITY OF NO IN PHYSIOLOGICAL BUFFERS AND IN CELL SUSPENSIONS

For an initial check of the stability of NO in physiological buffers and in cell suspensions, we carried out the following experiments (Fig. 1): A buffer solution (MES-KOH 50 mM, pH 5.5) was continuously flushed with NO (100 ppm in nitrogen) and was used as a source for a defined amount of NO. According to the Henry coefficient (1,9 mM at atmospheric pressure), 1 mL of this solution contained 190 pmoles NO. It should be noted that this NO concentration in the gas phase was at least 105 higher than NO concentrations produced by plants emitting NO into the atmosphere (see results below). Aliquots (2 mL) of that solution were rapidly injected into a vigorously stirred buffer solution in the headspace cuvette, which was flushed with purified air. Expectedly, after injection the NO rapidly escaped from the stirred solution into the gas phase, where it was measured. At 22°C, it took about 30 min until NO emission from the solution came to an end. Integration of the emission curve revealed that the recovery of the injected NO was almost complete (94%). However, when 2 mL of the NO-equilibrium solution were injected into a stirred tobacco cell suspension, only 4.7% of the injected NO could be recovered, indicating that a large fraction of the NO had rapidly reacted in the cells (Fig. 1). Interestingly, when the same amount of NO was injected into a cell suspension under a stream of nitrogen, about 10% of the injected NO was released into the gas phase (Fig. 2). Thus, while aerobic cells trapped about 95% of the added NO, anoxic cells trapped only 90%. In conclusion, the contribution of oxygen derived products to NO quenching was only 5% of the total quench.

When the cells were killed by brief boiling prior to the injection of NO, recovery of NO was still only 39% (Fig. 1). Thus, while part of the injected NO was trapped only by living cells, cellular

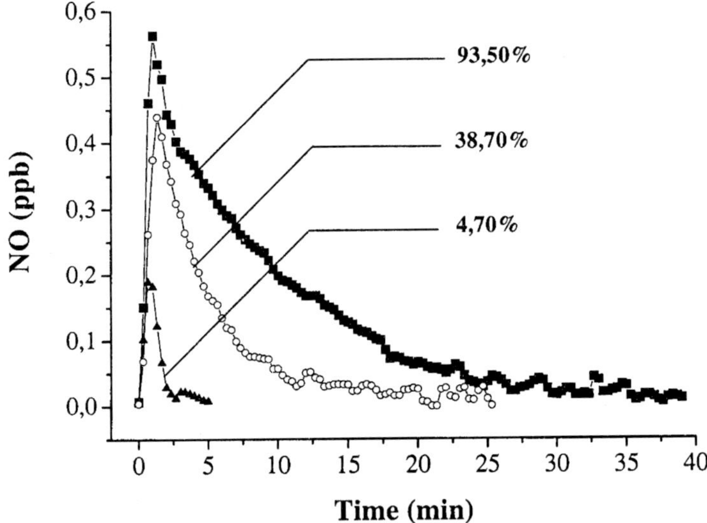

Figure 1 : NO recovery from a buffer solution (50 mM MES-KOH, pH 5.5) brought to equilibrium with 100 ppm NO in air. Two mL of the solution were injected into 1 mL of buffer solution (!), or into 1 mL of a cell suspension (Nicotiana tabacum cv. Xanthi) (7), or into 1 mL of a boiled cell suspensions (-) in a small Petri dish placed in a glas cuvette, mounted on a magnetic stirrer. The NO equilibrium concentration in the 1 mL solution flushed with 100 ppm NO in nitrogen was calculated from the Henry coefficient to be 190 pmol mL-1. As 2 mL were injected, the total amount of NO injected was 380 pmol. The NO-amount emitted from the liquid phase was measured by chemiluminescence detection in air. Numbers at the curves give the total amount of NO (obtained by integrating the area under the emission curve) as percentage of the theoretically added NO. The gas flow through the cuvette was 1,5 L min-1.

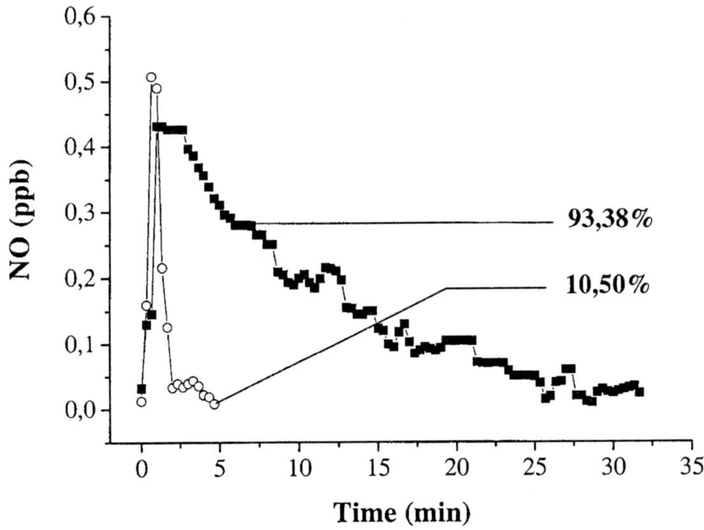

Figure 2 : Nitric oxide recovery after injection of NO-solution into buffer or into a cell suspension (as before), but under a stream of nitrogen. Similar experiment as in Fig. 1.

constituents where still present in the boiled extract which still reacted with 60 % of the injected NO.

The physical half life of NO was quantified in a slightly modified experiment (Fig. 3). Here, 2 mL of the above NO equilibrium solution were pulled into a syringe already containing 2 mL of air-saturated buffer. The syringe was immediately closed and left on the lab bench for up to one hour under gentle shaking, in order to allow NO to react with the dissolved oxygen. At T=0, T=30min or T=1h, the 4 mL aliquots were injected into the measuring cuvette under continuous stirring under a stream of nitrogen. Integration of the emission curves revealed that the amount of NO in the solution (in the syringe) half-saturated with air decreased only slowly, with an estimated half life of 42 min. Again we would like to point out that at the much lower NO concentrations produced by cells or plant organs, the physical (and probably also the biological) half life of NO should be even higher.

3. NO-EMISSION FROM A SOLUTION OF PURIFIED NITRATE REDUCTASE

Purified NR has been shown previously to immediately produce NO

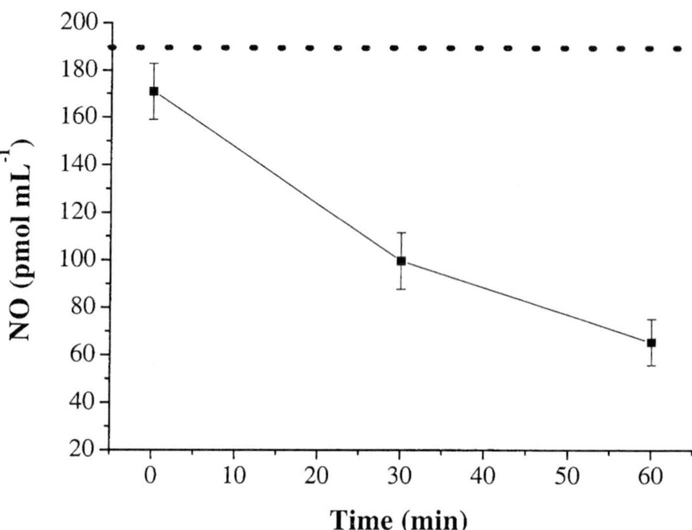

Figure 3 : Nitric oxide recovery from a buffer solution (50 mM MES-KOH, pH 5.5) brought to equilibrium with 100 ppm NO in an atmosphere of nitrogen. Two mL of the solution were pulled into a syringe containing already 2 mL air-saturated water. The syringe was closed, kept on the lab bench under gentle shaking and was injected at T=0 (!), T=30 min (-) or T=1h (7) into the cuvette under a stream of nitrogen. The dashed line gives the theoretical initial NO concentration (pmol mL-1). Other conditions as in Fig. 1.

when supplied with nitrite and NADH (Yamasaki *et al.*, 1999, Yamasaki and Sakihama, 2000; Rockel *et al.*, 2002). In a stirred buffer solution containing highly purified NR from maize (Sigma, Deisenhofen, Germany), NO emission from the aerated enzyme solution started immediately after nitrite addition, and increased continuously to reach a steady state, where the emission rate was 0,110 nmol min^{-1}. The NR activity contained in the solution was 13,8 nmol min^{-1} (measured in the presence of excess EDTA). Thus, the rate of NO emission in air was only about 0.8% of the NR activity in air, consistent with our previous results (Rockel *et al.*, 2002). In nitrogen, steady state NO-emission was slightly higher (0,91% of NRA), indicating either that a small part of the NO had reacted with dissolved oxygen, or with other products produced by NR. It was shown previously that NR can utilize molecular oxygen as an electron acceptor and that the product, $O_2^{\cdot -}$ (superoxide), is primarily generated via the MO-pterin center (Ruoff and Lillo, 1990; Barber and Kay, 1996). In addition, Yamasaki and Sakihama (2000) showed that NO, produced by purified NR in presence of NAD(P)H, rapidly reacts with superoxide to produce the extremely toxic peroxynitrite ($ONOO^-$) under aerobic conditions. In vitro studies have suggested that a reaction between gaseous NO and H_2O_2 produces singlet oxygen or hydroxyl radicals, a highly cytotoxic species (Noronha-Dutra *et al.*, 1993). Indeed, under our conditions NO production by NR was strongly inhibited by 100 μM hydrogen peroxide (Fig. 4). When catalase plus superoxide dismutase (SOD) were added together with NR, NO emission in air was as high as under nitrogen (not shown). Notably, normal NR activity (reduction of nitrate to nitrite) was not measurably affected by hydrogen peroxide (100 μM) (not shown). Thus, the differences in the NO emission rates were not due to effects of ROS on NR activity, but to trapping of NO by ROS.

One important outcome of this simple experiment is that if cells produce NO and reactive oxygen at the same time and in suitable quantities, as e.g. suggested for incompatible plant pathogen interactions, part of the NO will react immediately within the cell.

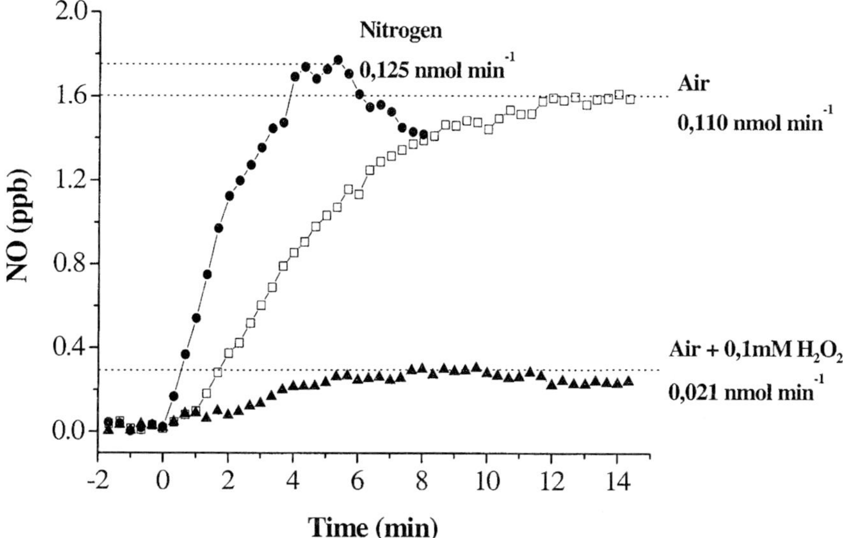

Figure 4 : Nitrite-dependent NO-emission from a solution of purified NR in vitro. At time zero, 200 μM NADH were added to 1 mL reaction buffer (100 mM HEPES-KOH, pH 7.6) containing commercially available purified maize NR and 200 μM nitrite in air ("), in nitrogen (,) or in air plus 100 μM H_2O_2 (7). Nitrate reductase activity was 13,8 nmol min-1 in air but this activity was not affected by H_2O_2. Other conditions as in Fig. 1.

However, at least in cell suspensions, ROS appeared not very relevant for NO quenching (compare fig. 2).

4. NITRIC OXIDE-EMISSION FROM CELL SUSPENSIONS

Cell suspensions are very efficient tools for physiological studies, and have been used e.g. for measuring NO production of cells in response to pathogens (Delledonne *et al.*, 1998; Clarke *et al.*, 2000). Using non-green nitrate-grown cell suspensions, we found that NO emission in air was negligible (Fig. 5). Under nitrogen, NO emission increased, reaching a steady state after 30 minutes. Addition of nitrite to the suspension more than doubled the rate of NO emission, indicating that it was nitrite limited. After return to aerobic conditions, the NO-emission rates (in the presence of nitrite) dropped to a slightly higher steady state than before (without nitrite addition). At this point we do not know why aerobic NO emission from cell suspensions was so low, even with added nitrite (similar observations were made with roots, see below). There are several possibilities: i) either NO quenching in air was very high (but see Figs. 1 and 2); ii)

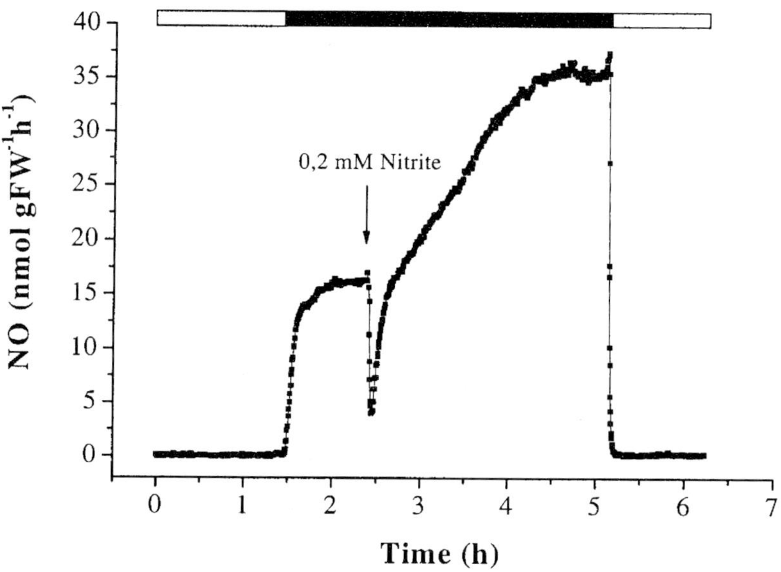

Figure 5 : NO-emission of nitrate grown cell suspensions (N. tabacum cv. Xanthi). The cells (0,8 g FW) were suspended in a total volume of 10 ml of the culture medium, in a petri dish. The dish was inclosed in the cuvette, which was mounted on a rotary shaker (100 rotations min^{-1}) and was flushed with air (white) or with nitrogen (black). KNO_2 (200 µM) was added to the cell suspension as indicated. Other conditions as in Fig. 1.

the NR activation state in air was very low; iii) the NR activation state was high, yet NR activity in aerobic cells was strongly limited by NADH.

Cells that were cultivated in the absence of nitrate contained no measurable NR activity and were not able to emit NO when fed with nitrite (not shown). Whether the very low NO emission by cells in air was due to quenching of NO by ROS, or to a low nitrite accumulation under aerobic conditions or both, is presently unknown.

5. NO-EMISSION PATTERN FROM DETACHED LEAVES

In the experiment shown in Figure 6, NO emission into the gas phase by detached tobacco leaves (Nicotiana tabacum var. Xanthi) was investigated. The leaves were harvested from plants grown hydroponically with nitrate as an N-source. Such leaves express NR, and a mean NR activity in extracts from illuminated leaves was 0,8 µmole gFW^{-1} h^{-1}. About the same extractable NR activity was obtained from tobacco leaves kept in darkness under nitrogen, were NR is also activated (Kaiser and Huber, 1997). During darkness in

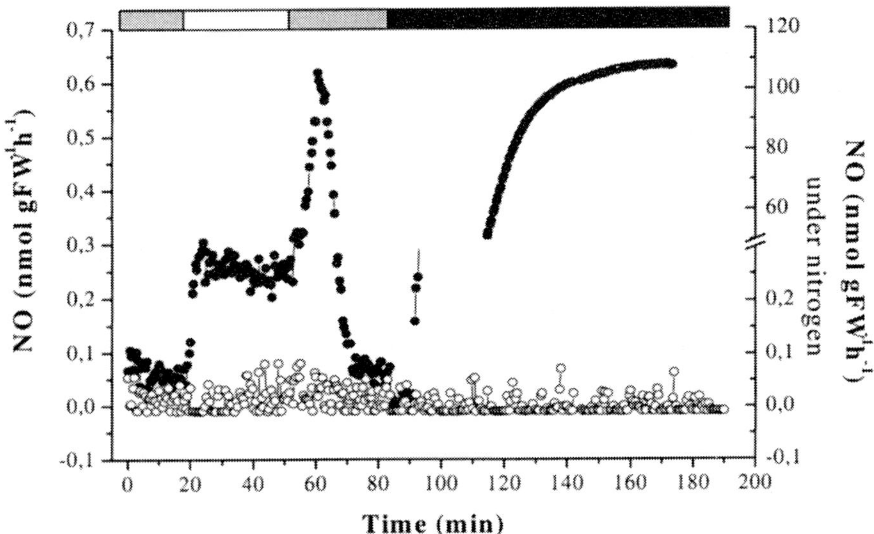

Figure 6 : NO-emission pattern of non-elicited tobacco leaves (*N. tabacum* cv. Xanthi) during a dark-light-dark-N_2 transient. NO-emission was measured from two detached leaves (about 4 g FW) of nitrate grown plants (,) or of ammonium-grown plants (-) in a glas cuvette flushed with air in dark (grey), with air in light (white) or with nitrogen in dark (black). Other conditions as in Figure 1.

air, detached tobacco leaves, with the petioles in nitrate solution (10 mM) emitted only small amounts of NO (Fig. 6) (also compare Rockel *et al.*, 1996, 2002). In the light, NO-emission increased to 0,1-0,5 nmol gFW^{-1} h^{-1}). When the light was switched off, NO release transiently increased but again settled at a very low steady-state value in the dark ("light-off peak"). Upon flushing the leaves in the dark with nitrogen, NO emission increased drastically, reaching a steady state of about 105 nmoles gFW^{-1} h^{-1}. Here again, the *in vivo* capacity of NR to produce NO (under nitrogen) was only a small fraction of NR activity, as in the in vitro experiment with purified NR (compare Fig. 4).

When we used leaves from plants that had been grown hydroponically on ammonium as the sole nitrogen source, extracts from these leaves contained practically no measurable NR activity, and the leaves produced no NO emission under all the above circumstances, just like the above-described cell suspensions (Fig. 5). Similar results were obtained with NR-free *NIA* double mutants (not shown here). These observations confirm the statement by Magalhaes *et al.* (this volume) that NO emission depends exclusively on NR, and that no other reaction in leaves participates in NO production under normal conditions.

The complex pattern of NO emissions from leaves has been traced back to two factors (compare Rockel *et al.* 2002): One factor is NR activity, which is low in the dark, higher in the light and also high in the dark under anoxia. A second factor is the cellular nitrite concentration, which depends on one hand on NR activity, and on the other hand on the rate of nitrite reduction in the plastids. Interestingly it was noticed that in leaves, the nitrite concentration was usually somewhat higher in the light than in the dark. Together with the light activation of NR, this explains the observed increase in NO emission in the light. In the dark under anoxia, NR is also largely activated, yet nitrite reduction is almost completely inhibited, leading to a high nitrite accumulation. In the dark, photosynthetic electron transport is zero. The oxidative pentose phosphate cycle could be a source for reductant in the plastids in the dark. However, the OPP-cycle requires a continuous supply of glucose-6-phosphate (G6P). Under anoxia (darkness), cellular ATP levels are low and G6P levels are also drastically decreased (Saint Ges *et al.*, 1991). Thus, the OPP-cycle is largely inoperative, explaining the accumulation of nitrite and the high NO emission in the dark under anoxia. The "light-off-peak" is the result of a transient "overshoot" of nitrate reduction over nitrite reduction. Photosynthetic electron transport stops immediately after darkening, yet the dark-inactivation of NR requires 5 to 15 min (Kaiser and Huber, 1994). Accordingly, there will be a transient nitrite accumulation after light off (also compare Riens and Heldt, 1992), which furnishes the transient NO-emission peak.

5.1. Nitric Oxide Emission from Roots

Roots of nitrate grown plants also contain NR and NiR activity. Accordingly, roots should be also able to produce NO. In order to test this, root segments (0,5 cm, including the root tips) were submerged in buffer solution (20 mM Hepes-KOH, pH 7,0) in a Petri dish, and enclosed in the measuring cuvette as before. No nitrate was added to the solution, since the roots contained about 50 mM stored nitrate. The cuvette was mounted on a shaker (150 rotations min-1), and was flushed with purified air or nitrogen, as indicated in Fig. 7 A, B and Fig. 8. Under these conditions, aerated root segments, like non-green suspension cells, gave only very little NO emission, which was hardly detectable. Under nitrogen, however, NO emission increased continuously, reaching a high steady state only after several hours. Upon aeration, NO emission rapidly approached zero, as before. The maximum (steady state) NO emission rates of nitrate fed WT tobacco roots in nitrogen were about 10 nmol g^{-1} FW h^{-1}, which corresponds to approximately 1,5 % of the extractable NRmax activity of these roots.

When aerated root segments, which produced almost no NO were supplied with 200 μM nitrite, they emitted NO already in air. This indicates that NO emission from roots, like in leaves, depends on the nitrite concentration. Root segments suspended in buffer solution continuously release nitrite into the surrounding medium (not shown). Therefore, unlike leaves, during nitrate reduction roots will not built up high nitrite concentrations, except under anoxia, where nitrite reduction ceases (explanations above).

Roots from ammonium-grown plants contained no or only residual NR activity, just like leaves or suspension cells. Expectedly, they were not able to emit NO when supplied with nitrate (Fig. 7B), even in nitrogen. However, when such NR-free roots were supplied with

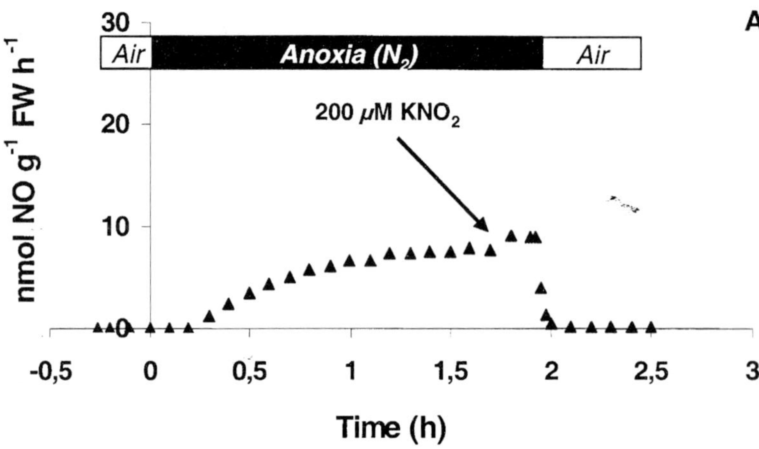

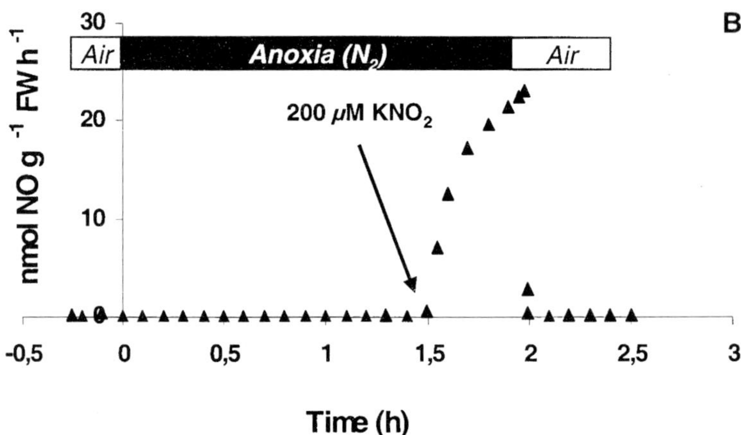

Figure 7 : Nitric oxide emission from tobacco (*N. tabacum* cv. Gat) root segments. (A) Root segments (1 g, 0,5 cm, including the root tips) from WT tobacco grown on 5 mM nitrate as sole nitrogen source were submerged in 10 ml 20 mM Hepes, pH 7,0 and subjected to aerobic treatment, prior to the onset of anoxia at time 0. These rootsegements contained about 50 mM stored nitrate. Gas exchange was aided by continuous shaking of the root segments and their surrounding medium at 150 rpm on a rotary shaker. Where indicated 200 μM KNO_2 was injected directly into the root surrounding medium. (B) Root segments (1 g, 0,5 cm, including the root tips) from WT tobacco grown on 5 mM NH_4Cl as sole nitrogen source were submerged in 10 ml 20 mM Hepes, pH 7,0 and subjected to aerobic treatment, prior to the onset of anoxia at time 0. These root segements contained no detectable storage nitrate and no NR activity. Gas exchange was aided by continuous shaking of the root segments and their surrounding medium at 150 rpm on a rotary shaker. Where indicated 200 μM KNO_2 was injected directly into the root surrounding medium. Other conditions as in Figure 7A.

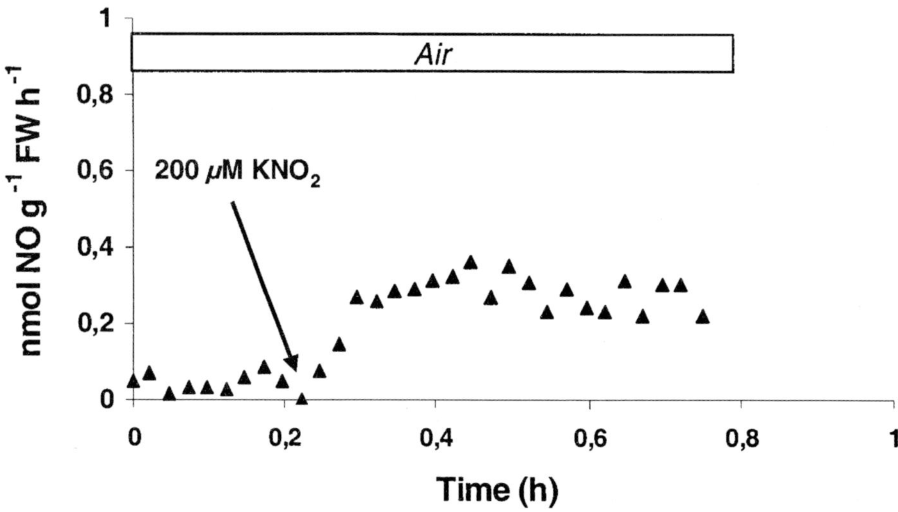

Figure 8 : Nitric oxide production of segments of WT tobacco (*N. tabacum* cv. Gat) roots grown on 5 mM nitrate as sole nitrogen source. Aerated root segments (1g, 0,5 cm, including the root tips) were submerged in 10 ml 20 mM Hepes, pH 7.0 and NO production was measured in the presence or absence of 200μM KNO_2 was recorded. Other conditions as in Figure 7.

200 μM nitrite, they emitted NO at remarkably high rates (Fig. 7B). It is at present not known by which reactions this NO is produced. Possible candidates are a nitrite:NO reductase (Stöhr *et al.* 2001), or, eventually, a xanthine oxidase/dehydrogenase (Millar *et al.* 1998; Li *et al.* 2003).

6. CONCLUSIONS AND FUTURE PROSPECTS

The above data show that chemiluminescence is able to detect even the low amounts of NO emitted from plant organs, cell suspensions and NR solutions into purified air. The advantage is that the method is specific for NO, non-destructive and suited for on-line measurements. However, as all other methods, it has some inherent disadvantages. These are: i) Only the fraction of NO which did not react inside the cells is detected. And ii) the method requires a continuous gas flow of 1,5 L min^{-1}, which does not allow much accumulation of NO in or around the tissues, because for efficient emission a large concentration gradient from inside the cells into the gas phase is created.

Clearly, it would be highly desirable to know how much of the

NO produced inside cells is emitted under a given set of conditions. Our experiments have shown that dissolved NO at nanomolar concentrations, despite its rather long physical half life at low concentrations, reacts rapidly with living cells. NO trapping may depend on many and variable physiological factors like e.g. the production of ROS. This problem is demonstrated by solutions of purified NR, where ROS are produced by NR in parallel with NO. In that simple system, NO emission in air was increased by 10 to 20% (occasionally 50%) in nitrogen or in the presence of excess catalase and superoxide dismutase. Further, trapping of dissolved NO added to cell suspensions was not much different in nitrogen than in air. Thus, quenching of NO should actually not depend much on reaction with ROS, but with other cellular constituents.

When cells or tissues produced and emitted NO, concentrations in the gas flow were in the range of 1 ppb, as an example. The equilibrium concentration of dissolved NO within such tissues would be about 1,9 pM, which is 105 times lower than in the experiments depicted in Figure 1. Even if we assume a concentration gradient by a factor of 10 or 100 to exist from inside cells into the gas phase, cellular NO concentrations would still be extremely low. It is not known how rapidly NO would react with cellular constituents at those low concentrations. However, the fact that maximum rates of NR-dependent NO emission from leaves (dark, anoxia) were also 1%, occasionally up to 15%) of the extractable NR activity (compare Rockel *et al.*, 2002), may be taken as an indication that at those low NO production rates NO quenching was actually negligible.

We have recently found that tobacco cell suspensions treated with the elicitor cryptogein or with the avirulent bacterium (*Pseudomonas syringae* pv. tomato) emitted NO at rather low rates (1 nmol gFW^{-1} h^{-1}) about 4 h after infection (Planchet *et al.*, unpublished). These rates were even lower than those observed above for NR-dependent NO emission. However, NO concentrations in cells required for the hypersensitive response have been reported to be 2 μM (Delledonne *et al.*, 1998; Delledonne *et al.*, 2001). According to the above data, such concentrations appear overestimated by several orders of magnitude.

ACKNOWLEDGEMENTS

This work was supported by the DFG (SFB 567) and Ka 456 13-1 and 15-1 to WM Kaiser. M. Sonoda was recipient of fellowships by AvH and JSPS. We greatfully acknowledge technical assistance of M. Lesch.

REFERENCES

Barber, M.J. and Kay, C.J. (1996). Superoxide production during reduction of molecular oxygen by assimilatory nitrate reductase. *Arch. Biochem. Biophys.*, **326** : 227-232.

Beligni, M.V. and Lamattina, L. (2000). Nitric oxide stimulates seed germination and de-etiolation and inhibits hypocotyls elongation, three light inducible responses in plants. *Planta*, **210** : 215-221.

Clarke, A., Desikan, R., Hurst, R.D., Hancock, J.T. and Neill, S.J. (2000). NO way back: nitric oxide and programmed cell death in *Arabidopsis thaliana* suspension cultures. *Plant J.*, **24** : 667-677.

Dangl, J.L., Dietrich, R.A. and Richberg, M.H. (1996). Death don't have no mercy: cell death programs in plant microbe interactions. *Plant Cell,* **8** : 1793-1807.

Delledonne, M., Xia, Y., Dixon, R.A. and Lamb, C. (1998). Nitric oxide functions as a signal in plant disease resistance. *Nature*, **394** : 585-588.

Delledonne, M., Zeier, J., Marocco, A. and Lamb, C. (2001). Signal interactions between nitric oxide and reactive oxygen intermediates in the plant hypersensitive disease resistance response. *Proc. Natl. Acad. Sci. USA,* **98** : 13454-13459.

Durner, J. and Klessig, D.F. (1999). Nitric oxide as a signal in plants. *Curr. Opin. Plant Biol.*, **2** : 369-374.

García-Mata, C. and Lamattina, L. (2002). Nitric oxide and abscissic acid cross talk in guard cells. *Plant Physiol.*, **128** : 790-792.

Henry, Y.A., Ducastel, B. and Guissani, A. (1997). Basic chemistry of nitric oxide and related nitrogen oxides. In: *Nitric Oxide Research from Chemistry to Biology* (Eds Henry, Y.A., Guissani A. and Ducastel, B.) Austin, TX: Landes Co. Biomed. Publ. pp15-46.

Kaiser, W.M. and Huber, S.C. (1994). Post-translational regulation of nitrate reductase in higher plants. *Plant Physiol.*, **106** : 817-821.

Kaiser, W.M. and Huber, S.C. (1997). Correlation between apparent activation state of nitrate reductase (NR), NR hysteresis and degradadtion of NR protein. *J. Exp. Bot.*, **48** : 1367-1374.

Lamattina, L., García-Mata, C., Graziano, M. and Pagnussat, G. (2003). Nitric oxide: the versatile of an extensive signal molecule. *Annu. Rev. Plant Biol.*, **54** : 109-36.

Lancaster, J.R. Jr. (1997).A tutorial on the diffusibility and reactivity of free nitric oxide. *Nitric oxide*, **1** : 18-30.

Leshem, Y.Y. and Haramaty, E. (1996). The characterization and contrasting effects of the nitric oxide free radical in vegetative stress and senescence of *Pisum sativum* L. foliage: *J. Plant Physiol.*, **148** : 258-263.

Li, H., Samouilov, A., Liu, X. and Zweier, J.L. (2003). Characterization of the magnitude and kinetics of xanthine oxidase-catalyzed nitrate reduction: evaluation of its role in nitrite and nitric oxide generation in anoxic tissues. *Biochemistry*, **42** : 1150-1159.

Millar, T.M., Stevens, C.R., Benjamin, N., Eisenthal, R., Harrison, R. and Blake, D. R. (1998). Xanthine oxidoreductase catalyses the reduction of nitrates and nitrite to nitric oxide under hypoxic conditions. *FEBS Lett.*, **427** : 225-228.

Neill, S.J., Desikan, R., Clarke, A. and Hancock, J.T. (2002). Nitric oxide is a novel component of abscissic acid signalling in stomatal gurad cells. *Plant Physiol.*, **128** : 13-16.

Noronha-Dutra, A., Epperlein, M. and Woolf, N. (1993). Reaction with nitric oxide with hydrogen peroxide to produce potentially cytotoxic singlet oxygen as a model for nitric oxide-mediated killing. *FEBS Lett.*, **321** : 59-62.

Riens, B. and Heldt, H.W. (1992). Decrease of nitrate reductase activity in spinach leaves during a light dark transition. *Plant Physiol.*, **98** : 573-577.

Rockel, P., Rockel, A., Wildt, J. and Segschneider, H.J. (1996). Nitric oxide (NO) emission by higher plants. In: *Progress in Nitrogen Cycling Studies* (Eds. Van Cleemput, O., Hofman, G. and Vermoesen, A.) Kluwer, Dordrecht, pp 603-606.

Rockel, P., Strube, F., Rockel, A., Wildt, J. and Kaiser, W.M. (2002). Regulation of nitric oxide (NO) production by plant nitrate reductase in vivo and in vitro. *J. Exp. Bot.*, **366** : 103-110.

Ruoff, P. and Lillo, C. (1990). Molecular oxygen as electron acceptor in the NADH-nitrate reductase system. *Biochem. Biophys. Res. Comm.*, **172** : 1000-1005.

Saint-Ges, V., Roby, C., Bligny, R., Pradet, A. and Douce, R. (1991). Kinetic studies of variations of cytosolic pH, nucleotide triphosphates (31P-NMR) and lactate during normoxic and anoxic transitions in maize root tips. *Eur. J. Biochem.*, **200** : 477-482.

Stöhr, C., Strube, F., Marx, G., Ulrich, W.R. and Rockel, P. (2001). A plasma menbrane-bound enzyme of tobacco roots catalyses the formation of nitric oxide from nitrite. *Planta*, **212** : 835-841.

Yamasaki, H., Sakihama, Y. and Takahashi, S. (1999). An alternative pathway for nitric oxide production in plants: new features for an old enzyme. *Trends in Plant Sci.*, **4** : 128-129.

Yamasaki, H. and Sakihama, Y. (2000). Simultaneous production of nitric oxide and peroxynitrite by plant nitrate reductase: in vitro evidence for the NR-dependent formation of active nitrogen species. *FEBS Lett.*, **468** : 89-92.

Chapter 10

NITRIC OXIDE SIGNALING THROUGH THE HYPERSENSITIVE DISEASE RESISTANCE RESPONSE

María C. Romero-Puertas and Massimo Delledonne*

Dipartimento Scientifico e Tecnologico, Università degli Studi di Verona, Verona, Italy

*Corresponding author : E-mail : massimo.delledonne@univr.it
Fax : +39-045-8027929

Summary

The mechanisms involved in plant defense show several similar characteristics with the innate immune systems of vertebrates and invertebrates. In the mammalian immune system, NO cooperates with ROS to induce apoptosis of tumor cells and macrophage killing of bacteria. In plants, a similar mechanism has evolved to prevent tissue invasion by pathogens. Avirulent pathogens induce NO synthesis in several plants and the biosynthetic origin of this NO has been reported recently. NO functions together with ROS in triggering hypersensitive cell death, an active process in which cysteine proteases appears to play an important role. Also, NO activates the expression of several defense genes (i.e. pathogenesis-related genes, phenylalanine ammonialyase, chalcone synthase) and could play a role, dependent on the function of SA, in pathways leading to systemic acquired resistance. In this chapter, we will focus on the signaling functions of NO during plant-pathogen interactions, when it is channeled through the cell death pathway by ROS.

In : Nitric Oxide Signaling in Higher Plants, 2004
(Eds Jose R. Magalhaes, Rana P. Singh and Leonidas P. Passos)
Studium Press, LLC, Houston, USA, pp 199-218

Keywords : Cell death, hypersensitive response, nitric oxide, reactive oxygen species.

1. INTRODUCTION

A widespread feature of plant disease resistance is the hypersensitive response (HR), which is characterized by the formation of necrotic lesions at the infection site and by the restriction of pathogen growth and spread (Heath, 1998). One of the earliest events in the HR is the rapid accumulation of reactive oxygen species (ROS) and nitric oxide (NO) through the activation of enzyme systems similar to neutrophil NADPH oxidase (Keller *et al.*, 1998; Lamb and Dixon, 1997) and nitric oxide synthase (NOS) (Durner *et al.*, 1998; Delledonne *et al.*, 1998; Chandok *et al.*, 2003). Activation of the oxidative burst is a central component of a highly amplified and integrated system that also involves salicylic acid, activation of ion fluxes, changes in protein phosphorylation patterns, extracellular pH, membrane potential, oxidative cross-linking of plant cell wall proteins, and perturbations in cytosolic Ca^{2+}. These events trigger the expression of disease and resistance mechanisms that mediate a systemic signal network involved in the establishment of plant immunity (Fig. 1). ROS are also required to trigger localized host cell death, but are not sufficient for an efficient response. It has been shown that NO cooperates with ROS in the induction of the hypersensitive cell death, and functions independently of such intermediates in the induction of defense related genes (Delledonne *et al.*, 1998; Durner *et al.*, 1998). In this chapter, we will discuss the involvement of NO in the hypersensitive disease resistance response.

1.1. Accumulation of NO during the Disease Resistance Response

In plant cells, it has been demonstrated that NO accumulates following infection with bacterial, viral and fungal pathogens. Thus, soybean and *Arabidopsis* culture cells generate a rapid augment of NO in response to challenge by avirulent bacteria strains (Delledonne *et al.*, 1998; Clarke *et al.*, 2000). In addition, elicitation of epidermal tobacco cells induces overproduction of NO within minutes (Foissner *et al.*, 2000). Likewise, high levels of NO have been detected in tobacco plants 24 h after infection with tobacco mosaic virus (TMV) (Chandok *et al.*, 2003). Furthermore, NO accumulation has been

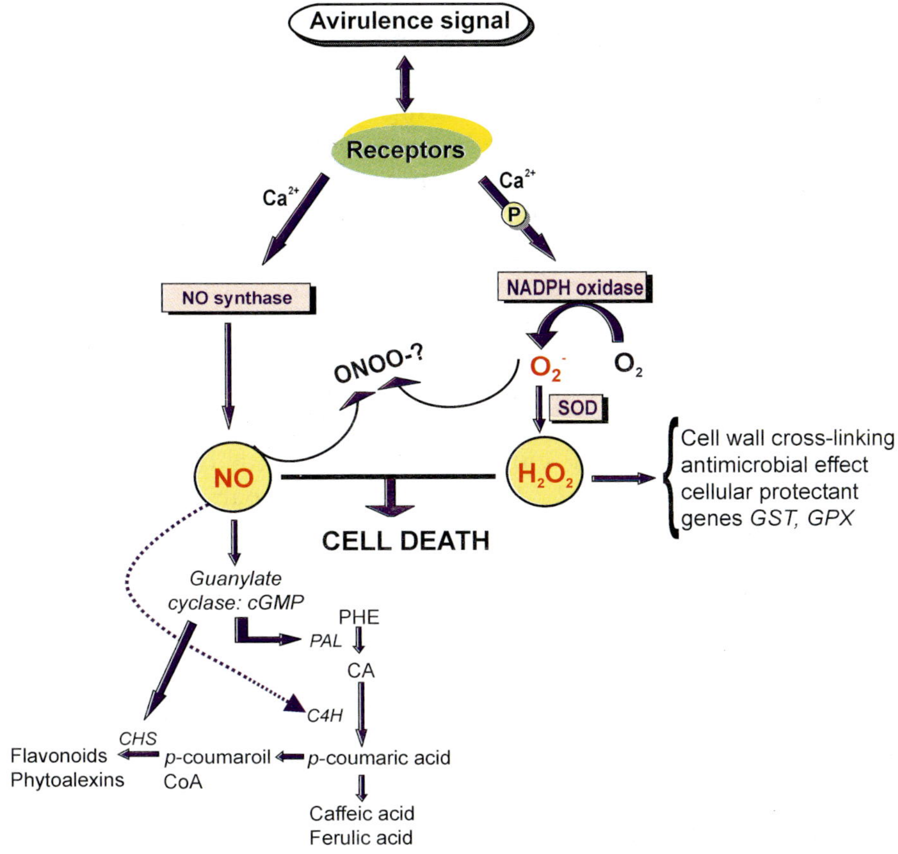

Figure 1 : Model of action of NO in the hypersensitive response. Abbreviations are as follows. CA, cinnamic acid; Ca^{2+}, calcium influx; cGMP, cyclic GMP; CHS, chalcone synthase; C4H, cinnamic acid 4-hydroxylase; GPX, glutathione peroxidases; GST, glutathione *S*-transferases; H_2O_2, hydrogen peroxide; NO, nitric oxide; $ONOO^-$, peroxynitrite; P, phosphorylation-dependent step; PAL, phenylalanine ammonia lyase; PHE, phenylalanine; SOD, superoxide dismutase.

shown under conditions in which the generation of ROS is also activated (Delledonne *et al.*, 1998; Clarke *et al.*, 2000; Foissner *et al.*, 2000; Allan and Fluhr, 1997) and this is concomitant with the avirulent gene-dependent oxidative burst that occurs immediately prior to the onset of hypersensitive cell death (Delledonne *et al.*, 1998; Durner *et al.*, 1998).

1.2. NO-generating Systems in the HR

In animal cells, the biosynthesis of NO is primarily catalyzed by

NOS, which oxidize L-arginine to form L-citrulline and NO. There are different isoforms of NOS, although each enzyme is a bi-domain consisting of a N-terminal oxygenase and a C-terminal reductase (Mayer and Hemmens, 1997). During the last few years, several lines of evidence have indicated that NOS is ubiquitous in plants. For example, TMV infection of resistant, but not susceptible tobacco, resulted in elevated NOS activity (Durner *et al.*, 1998). Blocking NOS with the competitive inhibitor *N*-nitro-L-arginine (L-NNA) prevented the induction of hypersensitive cell death by avirulent *Pseudomonas* in soybean suspension cells (Delledonne *et al.*, 1998) and co-infiltration of *Arabidopsis* leaves with L-NNA and with an avirulent *Pseudomonas,* inhibited the formation of the HR lesion triggered by recognition of the avirulent pathogen and promoted the development of spreading chlorosis observed with isogenic virulent bacteria (Delledonne *et al.*, 1998). In addition, partial inhibition of triazole fluorescence NO-specific dye DAF2-DA in tobacco leaves, treated with the elicitor cryptogein, has been demonstrated in the presence of L-NMMA, another inhibitor of animal NOS (Foissner *et al.*, 2000). A NOS-like activity has also been detected in soybean cotyledons in response to the *Diaporthe phaseolorum* f. sp *meridionalis* elicitor (Modolo *et al.*, 2002).

NO is involved in a number of physiological functions, and NOS-like activity is expected not to be detected only during plant-pathogen interactions. Indeed, NOS-like activity has been reported in extracts of the legume *Mucuna hasjoo* (Ninnemann and Maier, 1996), in *Lupinus albus* nodules (Cueto *et al.*, 1996), and in pea leaf peroxisomes (Barroso *et al.*, 1999). Moreover, inhibitors of mammalian NOS inhibited the generation of NO that precedes apoptosis in callus cells and foliar tissues of *Kalanchoe daigremontiana* and *Taxus brevifolia* (Pedroso *et al.*, 2000) and also inhibited the production of NO in *Arabidopsis*, parsley and tobacco cell cultures after treatment with cytokines (Tun *et al.*, 2001). NOS activity has also been detected in the soluble fractions of root tips and young leaves of maize seedlings (Ribeiro *et al.*, 1999).

Evidence for the existence of NOS in plants has also been obtained by immunoreactivity experiments. In recent years, a number of groups have shown that antibodies against mammalian NOS enzymes cross-react with plant proteins (Kuo *et al.*, 1995; Sen and Cheema, 1995; Cueto *et al.*, 1996; Huang and Knopp, 1998; Barroso *et al.*, 1999,

Ribeiro *et al.*, 1999). Despite these results, the existence of NOS in plants has been questioned based on the following evidence. First, the carboxyl-terminal domains of NOS have homology with many plant cytochrome P450 reductases (Beligni and Lamattina, 2001) and mammalian NOS antibodies also recognize many NOS-unrelated plant proteins (Butt *et al.*, 2003; Chandok *et al.*, 2003). Lastly, and more importantly, a sequence with high homology to the mammalian NOS is conspicuously absent in the genome of *Arabidopsis* (Neill *et al.*, 2002). Based on these observations, alternatives sources of NO have been proposed (Beligni and Lamattina, 2001; Yamasaki and Sakihama, 2000; Wojtaszek, 2000).

However, a recent report of pathogen-inducible NOS in plants (Chandok *et al.*, 2003) confirmed the suspicion that this enzyme exists. The major pathogen-induced NOS is a variant form of the P protein of the glycine decarboxylase complex (GDC) and is specifically induced during the resistance response to pathogen infection. The purified protein has a subunit with approximately 120 kDa and most likely functions as a dimer. Plant NOS share some biochemical characteristics with animal NOS such as sensitivity to inhibitors. It has been shown that the purified protein is sensitive to both L-NMMA and guanidine. Both plant and animal NOS require the cofactors H_4B, FAD, NADPH and O_2, and in addition require both Ca^{2+} and CaM (Chandok *et al.*, 2003). Moreover, the activity of plant NOS is suppressed by P protein inhibitors such as carboxymethoxylamine and aminoacetonitrile (Chandok *et al.*, 2003). The availability of a plant NOS will allow NO levels to be manipulated at a genetic level in plants. This will eliminate the side reactions and other unknown effects that are related to the pharmacological alteration of NO levels, allowing for a more precise characterization of NO and NOS function in plants. In fact, still there are numerous points to understand such as the localization of plant NOS, the role of the protein in disease resistance, and the eventual regulation of its activity. Furthermore, since in animals the production of NO is catalyzed by different isoforms, both constitutive and inducible, plants may possess such a division of NO synthesis as well. Thus, iNOS may not be the only source of NO in plants and the existence of a constitutive NOS would explain the NOS activity detected under diverse conditions (Ninnemann and Maier, 1996; Cueto *et al.*, 1996; Barroso *et al.*, 1999; Pedroso *et al.*, 2000; Ribeiro *et al.*, 1999).

In addition, there are other potential sources of NO in plants as it may be generated either non-enzymatically via the conversion of NO_2 by carotenoids, or enzymatically via nitrate reductase activities (Wendehenne *et al.*, 2001). The physiological implications of NO production by these sources are now the object of extensive investigation (Yamasaki, 2000) and are discussed in other chapters of this book.

2. NITRIC OXIDE AND CELL DEATH

NO is a near ubiquitous intra- and intercellular signaling molecule involved in the regulation of an impressive spectrum of diverse cellular functions. The effect of NO on animal cells depends on many complex conditions, such as the rate of production and diffusion, the concentration of ROS, and the level of enzymes involved in ROS scavenging, such as SOD and catalase (Tamir *et al.*, 1993). When unregulated production of NO occurs, cell death can occur through oxidative stress, disrupted energy metabolism, DNA damage, activation of poly (ADP-ribose) polymerase, and dysregulation of cytosolic calcium (Murphy, 1999).

It is well known that accumulation of NO and ROS during the HR is responsible for PCD in plants (Delledonne *et al.*, 1998; Clarke *et al.*, 2000). Cell death occurs in tobacco cv BY-2 cells grown in presence of NO and H_2O_2 generators as well (de Pinto *et al.*, 2003), presenting the hallmarks of PCD such as chromatin condensation and cytoplasmic shrinkage (Clarke *et al.*, 2000; de Pinto *et al.*, 2003).

2.1. What is the Role of Peroxynitrite?

In many biological systems, the cytotoxic effects of NO and ROS derive from the diffusion-limited reaction of NO with $O_2^{\cdot -}$ to form the peroxynitrite anion $ONOO^-$, which then interacts with many cellular components (Fang, 1997; Koppenol *et al.*, 1992) and may modulate the signaling functions of NO (Beckman *et al.*, 1990). Although the peroxynitrite anion is relatively unreactive, its conjugate acid, peroxynitrous acid, is a quite versatile species able to perform both one- and two-electron oxidation of a variety of molecules (amines, thiols, nitrogenous bases, ascorbate, etc.) thus damaging lipids and denaturing DNA and proteins (Pryor and Squadrito 1995; Stamler *et al.*, 1992). The mechanisms of these oxidative reactions are not completely understood, but at least some are mediated by formation of tyrosyl radicals (Arteel *et al.*, 1999).

Although excessive production of $ONOO^-$ can damage normal tissue, the reactive chemistry of $ONOO^-$ can be considered beneficial when the entire organism is considered, due to its cytotoxic effects on invading pathogens (Fang, 1997) and tumor cells (Lin *et al.*, 1995). In fact, peroxynitrite formed during the inflammatory response causes toxic effects like lipid peroxidation and cell death (Stamler *et al.*, 1992, 1994). Exposure of animal cells to $ONOO^-$ also produces concentration dependent cell death in the range from 1 to 1000 μM (Cookson *et al.*, 1998; Foresti *et al.*, 1999; Lin *et al.*, 1995). As incompatible plant-pathogen interactions induce production of both NO and ROS at high levels, peroxynitrite may contribute to host cell death and inhibition of pathogen growth similar to the mammalian inflammatory response. In fact, it has been shown that urate, a peroxynitrite scavenger that efficiently reduces the radicals centered on tyrosine residues (Pietraforte and Minetti, 1997), decreases lesion formation in *Arabidopsis* leaves treated with an abiotic peroxynitrite-generating system or with an $ONOO^-$ solution. However, urate did not exert any protective effects against damage originating from NO + H_2O_2 (Alamillo and Garcia-Olmedo, 2001). Nonetheless, this result should be interpreted with caution as urate oxidase activity in plants catalyzes the oxidation of uric acid in the presence of O_2 to form allantoin and H_2O_2 (Fraisse *et al.*, 2002). The addition of urate could cause a shift in the equilibrium of the ROS present in plants, affecting the antioxidant status of the cells and the reactivity of NO. Furthermore, it has been shown that when soybean suspensions are exposed to a wide range of concentrations of $ONOO^-$ (either commercial or in a continuous steady-state generation, SIN-1) cellular viability remains unaltered (Delledonne *et al.*, 2001). Nonetheless, SIN-1 induces accumulation of the transcript encoding PR-1 in tobacco leaves (Durner *et al.*, 1998) and $ONOO^-$ induces protein nitration in soybean and *Arabidopsis* leading to changes in the redox state of the cell (Delledonne *et al.*, 2001). Therefore, although $ONOO^-$ does not appear to be an essential intermediate of NO-induced cell death, it is expected to have important physiological and signaling functions in the HR.

2.2. SOD Function in the Induction of Hypersensitive Cell Death

Cu, Zn-SOD is one of the key cellular enzymes by which neurons and other cells protect themselves from NO-mediated damage (Troy *et al.*, 1996). In macrophages undergoing apoptotic cell death, Cu,Zn-

SOD is down-regulated at both the transcriptional and translational levels (Brockhaus and Brune, 1999) indicating that the cell death program can be switched on or off by modulating SOD activity. Indeed, downregulation of Cu,Zn-SOD in PC12 cells lead to death via the $ONOO^-$ pathway (Troy *et al.*, 1996), whereas its overexpression protects RAW 264.7 macrophages against NO cytotoxicity (Brockhaus and Brune, 1999).

In plant cells, $O_2^{\cdot-}$ is an important initial product of the pathogen-induced oxidative burst (Lamb and Dixon, 1997). H_2O_2 can then be formed non-enzymatically by dismutation of $O_2^{\cdot-}$ (Fridovich, 1978) or enzymatically by the action of SOD. The Cu,Zn-SOD inhibitor DDC has been shown to block H_2O_2 accumulation in *Arabidopsis* cells undergoing an oxidative burst following harpin treatment. This leads to an increased concentration of $O_2^{\cdot-}$ and abolishes hypersensitive cell death (Auh and Murphy, 1995; Desikan *et al.*, 1996). In addition, DDC blocks NO-induced, ROS-dependent cell death and H_2O_2 accumulation in soybean suspension cells challenged with avirulent *Pseudomonas*. This cell death can be rescued by the addition of sublethal amounts of H_2O_2, whereas the augmentation of the endogenous oxidative burst had a very limited effect, which is consistent with the ineffectiveness of $O_2^{\cdot-}$ in triggering NO induced cell death (Delledonne *et al.*, 2001). These results show that the accumulation of H_2O_2 detected during the pathogen-induced oxidative burst derives from the SOD-catalyzed dismutation of $O_2^{\cdot-}$ and further suggests that H_2O_2, and not $O_2^{\cdot-}$, is the key effector in NO/ROS-dependent hypersensitive cell death (Delledonne *et al.*, 2001).

2.3. NO/ROS Regulation of Cell Death

The role of NO in the activation of mechanisms leading to hypersensitive cell death has been characterized by studying the effect of treatment of cultured soybean cells with the NO donor sodium nitroprusside (SNP). In *Arabidopsis* cells in suspension, SNP and Roussin's black salt (RBS) induce cell death at concentrations that release NO in quantities similar to those generated by cells challenged by avirulent bacteria (Clarke *et al.*, 2000). In fact, SNP amplifies the induction of cell death by exogenous H_2O_2 in a synergistic manner and causes a dramatic increase in ROS induced cell death (Delledonne *et al.*, 1998). In the absence of ROS, neither high nor low concentrations of NO affect cell viability. The same result was shown in tobacco cultured cells where the addition of an NO donor and a H_2O_2 generator separately have no effect on the viability of the cells. However, when NO and H_2O_2 are supplied simultaneously cells die in a concentration dependent manner (de Pinto *et al.*, 2003).

From these results, it appears to be that the induction of the cell death program is dependent on an equilibrium between the levels of NO and ROS, and independent perturbations in either NO or H_2O_2 can be compensated by a change in the other component in order to re-establish an effective balance in this binary signal system. This may explain the fact that NO has been observed to exert both toxic and protective effects in different experimental systems/laboratories. Different agitation speeds for the suspension cultured cells or other dissimilar experimental conditions that are difficult to standardize (filtering, washing and manipulation of the cells), may all provoke oxidative stress (Yahraus *et al.*, 1995) and thus changes in the redox state of the cells. In turn, these changes would affect that NO/ROS ratio. For example, elicitors able to induce hypersensitive cell death upon infiltration into plant leaves will not activate the cell death program in cultured cells (He *et al.*, 1994; Piedras *et al.*, 1998).

The relatives rates of $O_2^{\cdot -}$ dismutation to H_2O_2 and reaction with NO to generate $ONOO^-$ are critical in the integration of the signal system to deliver NO and H_2O_2 that are required as co-activators of hypersensitive cell death. An equilibrium model for the interactions of NO and ROS has been proposed (Delledonne *et al.*, 2001). In this model, if the NO/ $O_2^{\cdot -}$ balance is in favor of $O_2^{\cdot -}$, NO is scavenged before it can react with H_2O_2. If the balance is in favor of NO, $O_2^{\cdot -}$ is scavenged before it is dismutated to H_2O_2. Scavenging NO with $O_2^{\cdot -}$ or scavenging $O_2^{\cdot -}$ with NO leads to the formation of $ONOO^-$, which is not an essential intermediate of NO-mediated cell death and thus functions as sink for potentially toxic signals. The observation that cell death during the HR is under control of the ratio of NO/ROS, and is not dictated by changes in the concentration of one of the two components has physiological consequences. The formation of ROS is an inevitable event in normal cell metabolism (Polle, 1996) and excesses of ROS accumulate during exposure to various stresses (Kliebenstein *et al.*, 1998; Mittler, 2002). The NO concentration is also subject to significant variation. The emission of NO from plants occurs under stress situations, such as herbicide treatment or pathogen attack, as well as under normal growth conditions. In most cases, the production of NO in plant tissues has been linked to the accumulation of NO_2 (Klepper, 1990). NO can be produced from NO_2 either non-enzymatically through light–mediated conversion by carotenoids or enzymatically by NADPH nitrate reductases (Klepper, 1990). Based on these observations, it may be hypothesized that $ONOO^-$ is continuously formed in healthy cells. Consequently, plant cells may have developed specific mechanisms to overcome the toxicity of ONOO, and may have adopted different NO/ROS signals for triggering cell death during the hypersensitive response.

2.4. Mechanism of NO-mediated Cell Death

NO has been shown to cooperate with H_2O_2 to induce DNA fragmentation and cell lysis in murine lymphoma cells, hepatoma cells, and endothelial cells (Farias-Eisner *et al.*, 1996; Filep *et al.*, 1997). The molecular mechanisms of this interaction are not clearly understood. *In vitro* studies have suggested that a reaction between gaseous NO and H_2O_2 produces singlet oxygen or hydroxyl radicals (Noronha-Dutra *et al.*, 1993). Alternatively, the toxicity of NO/H_2O_2 may be due to the production of a potent oxidant formed via a trace metal, H_2O_2, and an NO dependent process (Farias-Eisner *et al.*, 1996).

However a great deal of experimental evidence indicates that NO induces cell death by triggering an active process in which proteases appear to play a crucial role. Cystatin-sensitive proteases have been found to be critical regulators for HR in a soybean model system (Solomon *et al.*, 1999) and overexpression of AtCYS1, a cysteine protease inhibitor, was found to block cell death activated by either avirulent pathogens or by nitrosative stress in *A. thaliana* suspension cultured cells and in transgenic tobacco plants (Belenghi *et al.*, 2003). In addition, Ac-YVAD-CMK, an irreversible inhibitor of mammalian caspase-1, a class of cysteine proteases involved in apoptosis, was shown to block NO-induced cell death (Clark *et al.*, 2000).

Another clear evidence of the involvement of NO in the execution of a cell death program is the requirement of cGMP synthesis, a well-known second messenger of NO in animal systems (Denninger and Marletta, 1999). cGMP is produced by guanylate cyclase, which is activated either by NO binding to the heme-iron or through S-nitrosylation at critical cysteine residues (Stamler, 1994). By activation of different types of calcium channels, cGMP can both directly and indirectly regulate different physiological functions in animal cells (Wendehenne *et al.*, 2001). One such example in plants might be the requirement for cell death activation of a nucleotide-gated channel (AtCNGC2) identified in *Arabidopsis* affected in AtCNGC2 and fail to develop HR, although resistance is unaffected (Leng *et al.*, 1999; Clough *et al.*, 2000).

3. NO AND ACTIVATION OF GENES INVOLVED IN DEFENSE RESPONSES

NO functions together with ROS in triggering hypersensitive cell death, but it is also involved in other defense functions complementary to and independent of ROS. H_2O_2 drives oxidative cross-linking of tyrosine-rich structural proteins of the cell wall (Bradley *et al.*, 1992) and induces genes involved in cellular protection, including glutathione *S*-transferase (*gst*) and glutathione peroxidase (*gpx*) (Lamb and Dixon, 1997). ROS do not however appear to be the primary signal for rapid induction of defense genes, such as those encoding enzymes involved in the synthesis of phenylpropanoids, antibiotics, lignin, and salicylic acid (Levine *et al.*, 1994). On the contrary, there is evidence that signal transduction pathway inducing the production of phytoalexin stimulated by NO exists in higher plants cells (Noritake *et al.*, 1996). Moreover, inhibition of NOS activity has only a marginal effect on the initial induction of *gst* in response to avirulent *Pseudomonas*, whereas it markedly reduces the accumulation of transcripts encoding phenylalanine ammonia-lyase (PAL), the first enzyme of phenylpropanoid biosynthesis pathway, and chalcone synthase (CHS), the first enzyme of the branch specific for flavonoids and isoflavonoid-derived antibiotics (Delledonne *et al.*, 1998; Dixon and Paiva, 1995). It has been shown that the response of soybean cotyledons to elicitors from *Diaporthe phaseolorum* f. sp *meridionalis* imply that NO production via a NOS-like enzyme triggers the biosynthesis of antimicrobial flavonoids (Modolo *et al.*, 2002). Additionally, it has been found a transcriptional increase in cinnamate-4-hydroxylase (C4H) following infiltration of *Arabidopsis* leaves with a NO donor in cell death-inducing conditions (Polverari and Delledonne, submitted). C4H is considered a key enzyme in the synthesis of phenolic compounds related to disease resistance, and this result provides substantial support to the notion that defense responses related to the phenylpropanoid pathway are induced by NO in plants.

The role of NO in the induction of genes involved in cellular defense is further supported by experiments in tobacco (Durner *et al.*, 1998). Infiltration of tobacco leaves with commercially available NOS results in a significant accumulation of PAL and of pathogenesis-related protein 1 (PR-1), a well-defined marker of plant disease resistance, whose expression during the defense response generally lags behind that of PAL. The simultaneous increase of NO and ROS in tobacco bright-yellow 2 cells induces a rise in PAL activity as well (de Pinto *et al.*, 2002).

3.1. NO-mediated Signal Transduction

Accumulation of PAL and PR1 has been detected in tobacco cell suspensions treated with either NO donors or a membrane permeable analogue of cGMP. Furthermore, NO induction of PAL in tobacco suspension cells can be suppressed by several inhibitors of guanylate cyclase, suggesting the involvement of cGMP-dependent components in NO-dependent defense gene activation (Durner *et al.*, 1998). However, PAL expression is not fully blocked by inhibitors of cGMP production, implying that different modes of PAL induction operate downstream of NO. Although the involvement of cGMP in several plant signal transduction pathways has been demonstrated (Bowler *et al.*, 1994), it remains to be determined whether or not NO is the physiological activator of plant guanylate cyclase.

Cyclic ADP ribose (cADPR) has been implicated as another second messenger for NO signaling in animals, acting in a cGMP-dependent signaling cascade to mediate calcium mobilization (Denninger and Marletta, 1999). Treatment with cADPR induces PAL and PR-1 expression in tobacco (Durner *et al.*, 1998), which can be inhibited by ruthenium red, indicating its calcium dependence. Moreover, a cADPR antagonist suppressed NO induction of PR-1; however, this effect was incomplete, indicating that NO activation of defense responses may occur through more than one pathway (Klessig *et al.*, 2000). In addition to the activation of cGMP-dependent signaling cascades, nitrosylation of key components of cell metabolism is another mode of NO-signaling in animals (Stamler *et al.*, 2001) that may have parallels in plants. Strong evidence indicates that calcium release can also be regulated by channel S-nitrosylation or oxidation

(Anzai *et al.*, 2000; Xu *et al.*, 1998). The existence of multiple mechanisms of NO action makes the dissection of specific pathways rather difficult, but might explain the incomplete inhibition observed when individual metabolic steps are blocked (Klessig *et al.*, 2000).

Both positive and negative regulation of plant defense responses operating downstream of the generation of NO and ROS have been attributed to mitogen-activated protein kinase (MAPK) cascades. A MAPK was recently found to be activated by NO in *Arabidopsis* (Clarke *et al.*, 2000), but its role in the induction of genes involved in defense has yet to be investigated. However, research performed during the past few years has revealed that at least two MAPKs function as early positive regulators in plant defense signaling (Cardinale *et al.*, 2000). The tobacco SIPK (SA-induced protein kinase) and WIPK (wounding-induced protein kinase) are activated upon infection, treatment with elicitors, and in response to other types of abiotic stress; homologues have also been shown to function somewhat differently in diverse species (*Arabidopsis*, parsley, alfalfa). SIPK is typically induced by SA and by H_2O_2 (Yang *et al.*, 2001) and shows SA-mediated NO inducibility (Kumar and Klessig 2000), while WIPK is not induced by either SA or NO. The NtMEK2 kinase represents an upstream link between SIPK and WIPK as it can specifically activate both kinases (Yang *et al.*, 2001). A number of other kinases, and kinase kinases, are being identified that might constitute a complex signaling network leading to resistance, which may at least partially overlap other responses to a number of different stresses (Nurnberger and Scheel, 2001).

3.2. NO and Systemic Acquired Resistance

NO was also shown to induce accumulation of SA and its conjugates (Durner *et al.*, 1998), although the underlying mechanisms have not been fully elucidated. SA is required in incompatible interactions for the amplification of early signal(s) deriving from plant-pathogen recognition, resulting in the stimulation of an oxidative burst, defense gene expression, and hypersensitive cell death (Shirasu *et al.*, 1997). Besides the involvement in HR and defense activation, NO could play a role in pathways leading to systemic acquired resistance (SAR) (Durner *et al.*, 1999). It has been shown that in tobacco, NO-donors reduce the size of lesions caused by TMV on both treated and non-treated leaves (Song and Goodman, 2001). Furthermore, treatment of tobacco plant with inhibitors of NOS or a NO scavenger attenuated, but not abolished, SA-induced systemic acquired resistance (Song and Goodman, 2001). Experiments with transgenic tobacco plants *NahG*, that are unable to accumulate SA, show that NO had no effect on lesion size following TMV infection (Song and Goodman, 2001). From these results, it is likely that NO plays an important role in the induction of SAR signaling pathway(s) in tobacco, although its activity is fully dependent on the function of SA.

The establishment of SAR involves the existence of a putative systemic signal that migrates from infected to systemic, non-infected leaves. Compelling evidence indicates that SA, although necessary both for local resistance and for SAR induction, is not the long-distance signal molecule that triggers systemic resistance (Hunt and Ryals, 1996). In mammals, NO circulates in the blood as S-nitroso adducts of proteins, or as low molecular weight S-nitroso thiols, such as nitroso glutathione (GSNO). This molecule, believed to act as both an intra- and intercellular NO carrier, is a powerful inducer of plant defense genes (Durner *et al.*, 1998). Since glutathione is a major metabolite in the phloem, where the SAR signal is transmitted, it can be hypothesized that excess NO produced during the HR binds to glutathione; in this form, it may act as a long distance SAR signal (Durner *et al.*, 1999). Recently a GSNO-catabolizing enzyme and its encoding gene (GS-FDH) have been characterized, and a mutant yeast affected in gene functionality show enhanced susceptibility to nitrosative stress (Liu *et al.*, 2001). This gene has been identified also in pea and *Arabidopsis* (Shafqat *et al.*, 1996; Sakamoto *et al.*, 2002), suggesting that plants may be able to modulate the bioactivity and signaling function of this stabilized form of NO.

4. CONCLUSIONS AND FUTURE PROSPECTS

Currently, the protection of crops from pathogens and insects depends on the use of chemical products. The understanding of mechanisms underlying stress response, and especially the identification and characterization of the battery of genes that function in disease resistance will provide alternatives for protection of crops by using their own natural defenses. It should be considered, however, that plant cultivars without major resistance genes to their pathogens also carry the system for the hypersensitive cell death and expression of the defense genes and metabolism. From this viewpoint, there is no need to introduce resistance-encoding genes if the mechanisms for triggering accumulation of ROS and NO are known and can be controlled. In this context, the recent identification of a plant NOS will soon lead to the characterization and manipulation of signaling mechanisms leading to accumulation of NO.

ACKNOWLEDGEMENT

This work was supported by the Italian Ministry of University and by the EMBO Young Investigators Program.

REFERENCES

Alamillo, J.M. and Garcia-Olmedo, F. (2001). Effects of urate, a natural inhibitor of peroxynitrite-mediated toxicity, in the response of *Arabidopsis thaliana* to the bacterial pathogen *Pseudomonas syringae*. *Plant J.*, **25** : 529-540.

Allan, A.C. and Fluhr, R. (1997). Two distinct sources of elicited reactive oxygen species in tobacco epidermal cells. *Plant Cell*, **9** : 1559-1572.

Anzai, K., Ogawa, K., Ozawa, T. and Yamamoto, H. (2000). Oxidative modification of ion channel activity of ryanodine receptor. *Antioxid. Redox. Signal.*, **2** : 35-40.

Arteel, G.E., Briviba, K. and Sies, H. (1999). Protection against peroxynitrite, *FEBS Lett.*, **445** : 226-230.

Auh, C.K. and Murphy, T.M. (1995). Plasma membrane redox enzyme is involved in the synthesis of O_2^- and H_2O_2 by *Phytophthora* elicitor-stimulated rose cells, *Plant Physiol.*, **107** : 1241-1247.

Barroso, J.B., Corpas, F.J., Carreras, A., Sandalio, L.M., Valderrama, R., Palma, J.M., Lupianez, J.A. and del Rio, L.A. (1999). Localization of nitric-oxide synthase in plant peroxisomes. *J. Biol. Chem.*, **274** : 36729-36733.

Becker, K., Savvides, S.N., Keese, M., Schirmer, R.H. and Karplus, P.A. (1998). Enzyme inactivation through sulfhydryl oxidation by physiologic NO-carriers. *Nat. Struct. Biol.*, **5** : 267-71.

Beckman, J.S., Beckman, T.W., Chen, J., Marshall, P.A. and Freeman, B.A. (1990). Apparent hydroxyl radical production by peroxynitrite: implications for endothelial injury from nitric oxide and superoxide, *Proc Natl Acad Sci USA*, **87** : 1620-1624.

Belenghi, B., Acconcia, F., Trovato, M., Perazzolli, M., Bocedi, A., Polticelli, F., Ascensi, P. and Delledonne, M. (2003). AtCYS1, a cystatin from *Arabidopsis thaliana*, suppresses hypersensitive cell death. *Eur. J. Biochem.*, **270** : 1-12.

Beligni, M.V. and Lamattina, L. (2001). Nitric oxide in plants: the history is just beginning, *Plant Cell Environ.*, **24** : 267-278.

Bodgan, C. (2001). Nitric oxide and the regulation of gene expression. *Trends Cell Biol.*, **11** : 66-75.

Bowler, C., Van Camp, W., Van Montagu, M. and Inze, D. (1994). Superoxide dismutase in plants. *Crit. Rev. Plant Sci.*, **13** : 199-218.

Bradley, D.J., Kjellbom, P. and Lamb, C.J. (1992). Elicitor-induced and wound-induced oxidative cross-linking of a proline-rich plant-cell wall protein - a novel, rapid defense response. *Cell*, **70** : 21-30.

Brockhaus, F. and Brune, B. (1999). Overexpression of CuZn superoxide dismutase protects RAW 264.7 macrophages against nitric oxide cytotoxicity, *Biochem. J.*, **338** : 295-303.

Butt, Y.K., Lum, J.H. and Lo, S.C. (2003). Proteomic identification of plant proteins probed by mammalian nitric oxide synthase antibodies. *Planta*, **216** : 762-771.

Cardinale, F., Jonak, C., Ligterink, W., Niehaus, K., Boller, T. and Hirt, H. (2000). Differential activation of four specific MAPK pathways by distinct elicitors, *J. Biol. Chem.*, **275** : 36734-36740.

Chandok, M.R., Ytterberg, A.J., van Wijk, K.J. and Klessig, D.F. (2003). The pathogen-inducible nitric oxide synthase (iNOS) in plants is a variant of the P protein of the glycine decarboxilase complex. *Cell*, **113** : 469-482.

Clarke, A., Desikan, R., Hurst, R.D., Hancock, J.T. and Neill, S.J. (2000). NO way back: nitric oxide and programmed cell death in *Arabidopsis thaliana* suspension cultures. *Plant J.*, **24** : 667-677.

Clough, S.J., Fengler, K.A., Yu, I.C., Lippok, B., Smith, R.K. and Jr. Bent, A.F. (2000). The *Arabidopsis dnd1* "defense, no death" gene encodes a mutated cyclic nucleotide-gated ion channel. *Proc. Natl. Acad. Sci. USA,* **97** : 9323-9328.

Cookson, M.R., Ince, P.G. and Shaw, P.J. (1998). Peroxynitrite and hydrogen peroxide induced cell death in the NSC34 neuroblastoma x spinal cord cell line: role of poly (ADP-ribose) polymerase. *J. Neurochem.*, **70** : 501-508.

Cueto, M., Hernandez-Perea, O., Martin, R., Bentura, M.L., Rodrigo, J., Lamas, S. and Golvano, M.P. (1996). Presence of nitric oxide synthase activity in roots and nodules of *Lupinus albus. FEBS Lett.*, **398** : 159-164.

Darley-Usmar, V., Wiseman, H. and Halliwell, B. (1995). Nitric oxide and oxygen radicals: a question of balance, *FEBS Lett.*, **369** : 131-135.

Delledonne, M., Xia, Y., Dixon, R.A. and Lamb, C. (1998). Nitric oxide functions as a signal in plant disease resistance. *Nature*, **394** : 585-588.

Delledonne, M., Zeier, J., Marocco, A. and Lamb, C. (2001). Signal interactions between nitric oxide and reactive oxygen intermediates in the plant hypersensitive disease resistance response. *Proc. Natl. Acad. Sci. USA*, **98** : 13454-13459.

Denninger, J.W. and Marletta, M.A. (1999). Guanylate cyclase and the (NO)-N-./ cGMP signaling pathway. *Biochim. Biophys. Acta-Bioenerg.*, **1411** : 334-350.

Desikan, R., Hancock, J.T., Coffey, M.J. and Neill, S.J. (1996). Generation of active oxygen in elicited cells of *Arabidopsis thaliana* is mediated by a NADPH oxidase-like enzyme. *FEBS Lett.*, **382** : 213-217.

Dixon, R.A. and Paiva, N. (1995). Stress-induced phenylpropanoid metabolism. *Plant Cell,* **7** : 1085-1097.

Dolferus, R., Osterman, J.C., Peacock, W.J. and Dennis, E.S. (1997). Cloning of the Arabidopsis and rice formaldehyde dehydrogenase genes : implications for the origin of plant ADH enzymes. *Genetics*, **146** : 1131-1141.

Durner, J., Gow, A.J., Stamler, J.S. and Glazebrook, J. (1999). Ancient origins of nitric oxide signaling in biological systems, *Proc. Natl. Acad. Sci. USA,* **96** : 14206-14207.

Durner, J., Wendehenne, D. and Klessig, D.F. (1998). Defense gene induction in tobacco by nitric oxide, cyclic GMP, and cyclic ADP-ribose. *Proc. Natl. Acad. Sci. USA*, **95** : 10328-10333.

Durner, J. and Klessig, D.F. (1999). Nitric oxide as a signal in plants. *Curr. Opin. Plant Biol.*, **2** : 525.

Fang, F.C. (1997). Perspectives series: host/pathogen interactions. Mechanisms of nitric oxide-related antimicrobial activity. *J. Clin. Invest.*, **99** : 2818-2825.

Farias-Eisner, R., Chaudhuri, G., Aeberhard, E., Fukuto, J.M. (1996). The chemistry and tumoricidal activity of nitric oxide/hydrogen peroxide and the implications to cell resistance/susceptibility. *J. Biol. Chem.*, **271** : 6144-6151.

Filep, J.G., Lapierre, C., Lachance, S. and Chan, J.S. (1997). Nitric oxide co-operates with hydrogen peroxide in inducing DNA fragmentation and cell lysis in murine lymphoma cells. *Biochem. J.*, **321** : 897-901.

Fliegmann, J. and Sandermann, Jr. H. (1997). Maize glutathione-dependent formaldehyde dehydrogenase cDNA : a novel plant gene of detoxification. *Plant Mol. Biol.*, **34** : 843-854.

Foissner, I., Wendehenne, D., Langebartels, C. and Durner, J. (2000). *In vivo* imaging of an elicitor-induced nitric oxide burst in tobacco. *Plant J.*, **23** : 817-824.

Foresti, R., Sarathchandra, P., Clark, J.E., Green, C.J. and Motterlini, R. (1999). Peroxynitrite induces haem oxygenase-1 in vascular endothelial cells: a link to apoptosis. *Biochem. J.*, **339** : 729-736.

Fraisse, L., Bonnet, M.C., de Farcy, J.P., Agut, C., dersigny, D. and Bayol, A. (2002). A colorimetric 96-well microtiter plate assay for the determination of urate oxidase activity and its kinetic parameters. *Anal. Biochem.*, **309** : 173-179.

Fridovich, I. (1978). The biology of oxygen radicals. *Science*, **201** : 875-880.

He, S.Y., Bauer, D.W., Collmer, A. and Beer, S.V. (1994). Hypersensitive response elicited by *Erwinia amylovora* harpin requires active plant metabolism. *Mol. Plant-Microbe Interact.*, **7** : 289-292.

Heath, M.C. (1998). Apoptosis, programmed cell death and the hypersensitive response, *Eur. J. Plant Pathol.*, **104** : 117-124.

Huang, J.S. and Knopp, J.A. (1997). Involvement of nitric oxide in *Ralstonia solanacearum* induced hypersensitive reaction in tobacco. In *Proceedings of the Second International Wilt Symposium* (Eds. Prior, P., Elphinstione, J. and Allen, C.). INRA, Versailles, France.

Hunt, M.D. and Ryals, J.A. (1996). Systemic acquired resistance signal transduction. *Crit. Rev. Plant. Sci.*, **15** : 583-606.

Keller, T., Damude, H.G., Werner, D., Doerner, P., Dixon, R.A. and Lamb, C. (1998). A plant homolog of the neutrophil NADPH oxidase gp91^{phox} subunit gene encodes a plasma membrane protein with Ca^{2+} binding motifs. *Plant Cell,* **10** : 255-266.

Klepper, L. (1990). Comparison between Nox Evolution Mechanisms of Wild-type and nr1 Mutant soybean leaves. *Plant Physiol.*, **93** : 26-32.

Klessig, D.F., Durner, J., Noad, R., Navarre, D.A., Wendehenne, D., Kumar, D., Zhou, J.M., Shah, J., Zhang, S., Kachroo, P., Trifa, Y., Pontier, D., Lam, E. and Silva, H. (2000). Nitric oxide and salicylic acid signaling in plant defense, *Proc. Natl. Acad. Sci. USA,* **97** : 8849-8855.

Kliebenstein, D.J., Monde, R.A. and Last, R.L. (1998). Superoxide dismutase in *Arabidopsis*: an eclectic enzyme family with disparate regulation and protein localization. *Plant Physiol.*, **118** : 637-650.

Koppenol, W.H., Moreno, J.J., Pryor, W.A., Ischiropoulos, H. and Beckman, J.S. (1992). Peroxynitrite, a cloaked oxidant formed by nitric oxide and superoxide. *Chem. Res. Toxicol.,* **5** : 834-842.

Kumar, D. and Klessig, D.F. (2000). Differential induction of tobacco MAP kinases by the defense signals nitric oxide, salicylic acid, ethylene, and jasmonic acid. *Mol. Plant Microbe Interact.*, **13** : 347-351.

Kuo, W.N., Kuo, T.W., Jones, D.L. and Baptiste, J. Jr. (1995). Nitric oxide synthase immunoreactivity in baker's yeast, lobster and wheat germ. *Biochem. Arch.*, **11** : 238-244.

Lamb, C. and Dixon, R.A. (1997). The oxidative burst in plant disease resistance. *Ann. Rev. Plant Physiol. Plant. Mol. Biol.*, **48** : 251-275.

Leng, Q., Mercier, R.W., Yao, W. and Berkowitz, G.A. (1999). Cloning and first functional characterization of a plant cyclic nucleotide-gated cation channel. *Plant Physiol.,* **121** : 753-761.

Levine, A., Tenhaken, R., Dixon, R. and Lamb, C. (1994). H_2O_2 from the oxidative burst orchestrates the plant hypersensitive disease resistance response. *Cell,* **79** : 583-593.

Lin, K.T., Xue, J.Y., Nomen, M., Spur, B. and Wong, P.Y. (1995). Peroxynitrite-induced apoptosis in HL-60 cells. *J. Biol. Chem.,* **270** : 16487-16490.

Liu, L., Hausladen, A., Zeng, M., Que, L., Heitman, J. and Stamler, J.S. (2001). A metabolic enzyme for S-nitrosothiol conserved from bacteria to humans. *Nature*, **410** : 490-4.

Martinez, M.C., Achkor, H., Persson, B., Fernandez, M.R., Shafqat, J., Farrés, J., Jornvall, H. and Parés, X. (1996). *Arabidopsis* formaldehyde dehydrogenase. Molecular properties of plant class III alcohol dehydrogenase provide further insights into the origins, structure and function of plant class p and liver class I alcohol dehydrogenases. *Eur. J. Biochem.*, **241** : 849-857.

Mayer, B. and Hemmens, B. (1997). Biosynthesis and action of nitric oxide in mammalian cells. *Trends Biochem. Sci.,* **22** : 477-481.

Mittler, R. (2002). Oxidative stress, antioxidants and stress tolerance. *Trends Plant Sci.,* **7** : 405-410.

Modolo, L.V., Cunha, F.Q., Braga, M.R. and Salgado, L. (2002). Nitric oxide synthase-mediated phytoalexin accumulation in soybean cotyledons in response to the *Diaporthe phaseolorum* f. sp *meridionalis* elicitor. *Plant Physiol.*, **130** : 1288-1297.

Murphy, M.P. (1999). Nitric oxide and cell death. *Biochim. Biophys. Acta*, **1411** : 401-414.

Navarre, D.A., Wendehenne, D., Durner, J., Noad, R. and Klessig, D.F. (2000). Nitric oxide modulates the activity of tobacco aconitase. *Plant Physiol.,* **122** : 573-582.

Neill, S.J., Desikan, R., Clarke, A., Hurst, R.D. and Hancock, J.T. (2002). Hydrogen peroxide and nitric oxide as signalling molecules in plants. *J. Exp. Bot.*, **53** : 1237-1247.

Ninnemann, H. and Maier, J. (1996). Indications for the occurrence of nitric oxide synthases in fungi and plants and the involvement in photoconidiation of *Neurospora crassa. Photochem. Photobiol.,* **64** : 393-398.

Noritake, T., Kawakita, K. and Doke, N. (1996). Nitric oxide induces phytoalexin accumulation in potato tuber tissues. *Plant and Cell Physiol.*, **37** : 113-116.

Noronha-Dutra, A.A., Epperlein, M.M. and Woolf, N. (1993). Reaction of nitric oxide with hydrogen peroxide to produce potentially cytotoxic singlet oxygen as a model for nitric oxide-mediated killing. *FEBS Lett.*, **321** : 59-62.

Nurnberger, T. and Scheel, D. (2001). Signal transmission in the plant immune response. *Trends Plant. Sci.,* **6** : 372-379.

Pedroso, M.C., Magalhaes, J.R. and Durzan, D. (2000). A nitric oxide burst precedes apoptosis in angiosperm and gymnosperm callus cells and foliar tissues. *J. Exp. Bot.*, **51** : 1027-1036.

Piedras, P., Hammond Kosack, K.E., Harrison, K. and Jones, J.D.G. (1998). Rapid, Cf-9- and Avr9-dependent production of active oxygen species in tobacco suspension cultures. *Mol. Plant-Microbe Interact.*, **11** : 1155-1166.

Pietraforte, D. and Minetti, M. (1997). Direct ESR detection or peroxynitrite-induced tyrosine-centred protein radicals in human blood plasma. *Biochem. J.,* **325** : 675-684.

de Pinto, M.C., Tommasi, F. and De Gara, L. (2002). Changes in the antioxidant systems as part of the signaling pathway responsible for the programmed cell death activated by nitric oxide and reactive oxygen species in tobacco bright-yellow 2 cells. *Plant Physiol.*, **130** : 698-708.

Polle, A. (1996). Mehler reaction: friend or foe in photosynthesis?. *Bot. Acta,* **109** : 84-89.

Pryor, W.A. and Squadrito, G.L. (1995). The chemistry of peroxynitrite: a product from the reaction of nitric oxide with superoxide. *American J. Physiol.,* **268** : 699-722.

Ribeiro, E.A., Cunha, F.Q., Tamashiro, W.M.S.C. and Martin, I.S. (1999). Growth phase-dependent subcellular localization of nitric oxide synthase in maize cells. *FEBS Lett.*, **445** : 283-286.

Sakamoto, A., Ueda, M. and Morikawa, H. (2002). *Arabidopsis* glutathione-dependent formaldehyde dehydrogenase is an S-nitrosoglutathione reductase. *FEBS Lett.*, **515** : 20-24.

Sen, S. and Cheema, I.R. (1995). Nitric oxide synthase and calmodulin immunoreactivity in plant embryonic tissue. *Biochem. Arch.*, **11** : 221-227.

Shafqat, J., El-Ahmad, M., Danielsson, O., Martinez., M.C., Persson, B., Parés, X. and Jornvall, H. (1996). Pea formaldehyde-active class III alcohol dehydrogenase : common derivation of the plant and animal forms but not of the corresponding ethanol-active forms (classes I and P). *Proc. Natl. Acad. Sci. USA*, **93** : 5595-5599.

Shirasu, K., Nakajima, H., Rajasekhar, V.K., Dixon, R.A. and Lamb, C. (1997). Salicylic acid potentiates an agonist-dependent gain control that amplifies pathogen signals in the activation of defense mechanisms. *Plant Cell,* **9** : 261-270.

Solomon, M., Belenghi, B., Delledonne, M., Menachem, E. and Levine, A. (1999). The involvement of cysteine proteases and protease inhibitor genes in the regulation of programmed cell death in plants. *Plant Cell*, **11** : 431-444.

Song, F. and Goodman, R.M. (2001). Activity of nitric oxide is dependent on, but is partially required for function of, salicylic acid in the signaling pathway in tobacco systemic acquired resistance. *Mol. Plant-Microbes Interact*, **14** : 1458-1462.

Stamler, J.S., Singel, D.J. and Loscalzo, J. (1992). Biochemistry of nitric oxide and its redox-activated forms. *Science,* **258** : 1898-1902.

Stamler, J.S. (1994). Redox signaling: nitrosylation and related target interactions of nitric oxide. *Cell,* **78** : 931-936.

Stamler, J.S., Lamas, S. and Fang, F.C. (2001). Nitrosylation: The prototypic redox-based signaling mechanism. *Cell,* **106** : 675-683.

Tamir, S., Lewis, R.S., de Rojas, W.T., Deen, W.M., Wishnok, J.S. and Tannenbaum, S.R. (1993). The influence of delivery rate on the chemistry and biological effects of nitric oxide. *Chem. Res. Toxicol.*, **6** : 895-899.

Troy, C.M., Derossi, D., Prochiantz, A., Greene, L.A. and Shelanski, M.L. (1996). Downregulation of Cu/Zn superoxide dismutase leads to cell death via the nitric oxide-peroxynitrite pathway. *J. Neurosci.*, **16** : 253-261.

Tun, N.N., Holk, A. and Scherer, G.F.E. (2001). Rapid increase of NO release in plant cell cultures induced by cytokinin. *FEBS Lett.*, **509** : 174-176.

Uotila, L. and Koivusalo, M. (1979). Purification of formaldehyde and formate dehydrogenases from pea seeds by affinity chromatography and S-formylglutathione as the intermediate of formaldehyde metabolism. *Arch. Biochem. Biophys.*, **196** : 33-45.

Wendehenne, D., Pugin, A., Klessig, D.F. and Durner, J. (2001). Nitric oxide: comparative synthesis and signaling in animal and plant cells. *Trends Plant Sci.,* **6** : 177-183.

Wojtaszek, P. (2000). Nitric oxide in plants : to NO or not to NO. *Phytochem.*, **54** : 1-4.

Xu, L., Eu, J.P., Meissner, G. and Stamler, J.S. (1998). Activation of the cardiac calcium release channel (ryanodine receptor) by poly-S-nitrosylation. *Science,* **279** : 234-237.

Yahraus, T., Chandra, S., Legendre, L. and Low, P.S. (1995). Evidence For a Mechanically Induced Oxidative Burst. *Plant Physiol.*, **109** : 1259-1266.

Yamasaki, H. (2000). Nitrite-dependent nitric oxide production pathway: implications for involvement of active nitrogen species in photoinhibition *in vivo. Philos. Trans. R. Soc. Lond. B. Biol. Sci.,* **355** : 1477-1488.

Yamasaki, H. and Sakihama, Y. (2000). Simultaneous production of nitric oxide and peroxynitrite by plant nitrate reductase: in vitro evidence for the NR-dependent formation of active nitrogen species. *FEBS Lett.*, **468** : 89-92.

Yang, K.Y., Liu, Y. and Zhang, S. (2001). Activation of a mitogen-activated protein kinase pathway is involved in disease resistance in tobacco. *Proc. Natl. Acad. Sci. USA,* **98** : 741-746.

Chapter 11

NITRIC OXIDE AND PLANT MITOCHONDRIA

Michela Zottini

Dipartimento di Biologia Università degli Studi di Padova - Via U. Bassi, 58/B - I-35131 Padova - ITALY
Tel. : +39-049-8276247; Fax : +39-049-8276300
E-mail : mzottini@civ.bio.unipd.it

Summary

Study of the interactions between nitric oxide (NO) and mitochondria are important for two major reasons. First, mitochondrial respiration is responsible for energy production in many metabolic functions in plant cells. Second, both apoptosis and necrosis, the two types of cell death, are closely linked to mitochondrial functions. NO can act at the mitochondrial level by binding to cytochrome oxidase, the terminal enzyme of the respiratory chain, thus inhibiting respiration. This interaction is thought to be responsible for the pro-apoptotic action of NO, since in NO treated cells various processes are observed, including the release of cytochrome c, production of reactive oxygen species, a decrease in the concentration of ATP, and opening of the permeability transition pore. However, NO-induced inhibition of respiration may also play a protective role. In reality, under limiting concentrations of O_2, NO plays a major role in reducing plant cell metabolism in order to decrease the consumption of O_2 and avoid anoxia. It has been suggested that NO may provide a sensing mechanism for O_2, allowing the cell to perceive variations in O_2 tension at O_2 concentrations well above limiting concentrations for mitochondrial electron transport.

In : Nitric Oxide Signaling in Higher Plants, 2004
(Eds Jose R. Magalhaes, Rana P. Singh and Leonidas P. Passos)
Studium Press, LLC, Houston, USA, pp 219-237

Keywords : Nitric oxide; mitochondria; programmed cell death; nitric oxide synthase; anoxic stress; O_2 sensing mechanisms; cytochrome oxidase; alternative oxidase.

1. INTRODUCTION

Mitochondria play a central role in energy and carbon metabolism of eukaryotic cells, being the site of both the tricarboxylic acid cycle (TCA) and oxidative phosphorylation (Siedow and Day, 2000). In the form of ATP, energy is required by plants for growth, development and maintenance, and is mainly produced by two mechanisms, namely photophosphorylation in chloroplasts and oxidative phosphorylation in mitochondria. The contribution of each organelle to the production of ATP is subject to spatial-temporal regulation. In mature green leaves, for example, mitochondria account for nearly all ATP production in the dark, whereas chloroplasts are the main source of ATP in the light. However, under illumination in leaf tissues, mitochondria also are believed to contribute, at least partially, to the total ATP required for sucrose synthesis, CO_2 fixation, metabolite transport, and protein synthesis (Affourtit *et al.*, 2001).

In addition to playing a catabolic role in plant metabolism, mitochondria also fulfill an anabolic function, particularly in the presence of light. The carbon skeleton required for amino acid synthesis is derived from intermediates of the TCA cycle (Hoefnagel *et al.*, 1998). This cycle must work independently of the energy charge of the cell. For this reason, the plant mitochondrial respiratory chain possesses several non-protonmotive protein complexes that allow electron transfer along the respiratory chain without concomitant energy production. In this way, the TCA cycle can remain active if the ATP/ADP ratio in the cell is high. Tight regulation of mitochondrial respiratory activity is of crucial importance so that the different metabolic demands of the cell are satisfied (Vanlerberghe and McIntosh, 1997).

Recent data suggest that plant mitochondria also play a central role in control of programmed cell death pathways, representing both a stress sensor and dispatcher (Jones 2000). Programmed cell death is an essential physiological process that is required for the normal development and maintenance of tissue homeostasis (Kuriyama and Fukuda, 2002), but also occurs in response to various stresses (Heath, 2000; Beers and McDowell, 2001). Cytochrome c and other factors released from mitochondria are considered to be important signals for the triggering of programmed cell death, in animals as well as in plants (Lam *et al.*, 2001). It has been shown that when induced to cell death by the addition of D-mannose, maize cells release cytochrome c from the mitochondria (Sun *et al.*, 1999). Moreover, translocation of cytochrome c from the mitochondria to the cytosol has also been observed in cucumber plants during heat shock-induced programmed cell death (Balk *et al.*, 1999). The involvement of mitochondria in pathogen-induced plant defense responses and programmed cell death has been recently demonstrated. Xie and Chen (2000) reported that harpin-induced hypersensitive cell death is associated with ATP depletion of mitochondrial dysfunction. More recently, it has been demonstrated that the fungal toxin victorin induced apoptotic cell death in oats by altering mitochondrial functionality (Curtis and Wolpert, 2002; Yao *et al.*, 2002). Taken together, this data indicate that mitochondria are important regulators in the control of programmed cell death in plants.

In animal systems, mitochondria have been recognized for some time as a major target of NO, an important signal molecule in several physiological and pathological pathways (Durner and Klessig, 1999; Delledonne *et al.*, 1998). During the last five years, several studies have been performed concerning NO action in plant cells (Neill *et al.*, 2003), although little is still known about the signal transduction pathways by which NO interacts with plants, giving rise to specific cellular responses.

2. CHARACTERISTIC FEATURES OF THE RESPIRATORY CHAIN OF PLANT MITOCHONDRIA

The respiratory chain of plant mitochondria is different in many aspects from the animal counterpart. In plant mitochondria, apart from complex I that utilizes matrix NADH as a substrate, an additional four NADH dehydrogenases (insensitive to rotenone, the classical complex I inhibitor; Møller, 2001) also exist that allow oxidation of both the matrix and the cytosolic NAD(P)H. Moreover, plant mitochondria possess an alternative pathway to the classic cytochrome pathway, through which electrons are transferred to molecular oxygen. This pathway branches from the ubiquinone pool and consists of a single enzyme, namely alternative oxidase. Cytochrome oxidase and alternative oxidase compete for electrons in plant mitochondria, and the inhibition of one pathway redirects flow to the other when alternative oxidase is activated by pyruvate (Hoefnagel *et al.*, 1995). The two pathways can be differentiated by inhibitors such as cyanide, antimycin A, myxothiazol or carbon monoxide (acting on the cytochrome pathway), and n-propyl gallate (nPG) or salicylhydroxamic acid (SHAM, acting on alternative oxidase). Alternative oxidase, which also enables plants to respire in the presence of toxic compounds such as cyanide and carbon monoxide, is non-protonmotive (Vanlerberghe and McIntosh, 1997). Similarly, the external, rotenone-insensitive NAD(P)H dehydrogenases are able to translocate protons through the inner mitochondrial membrane (Hoefnagel *et al.*, 1998). Due to this structural characteristic, plant mitochondria can theoretically respire in such a way that is not energy conserving and, therefore, beyond the control of the protonmotive force.

3. NO INHIBITS THE CYTOCHROME PATHWAY IN MITOCHONDRIA RESPIRATION

It has been known for some time that NO is able to inhibit mitochondrial respiration (Brown, 1999) and is accomplished by two different means. First, NO itself can inhibit cytochrome oxidase in a rapid, selective and potent, but reversible manner (Giuffrè *et al.*, 1996). Second, reactive species of nitrogen (RSN) can inhibit many mitochondrial components in a slow, non selective and weak, but irreversible manner (Brown and Borutaite, 2002; Cooper, 2002). The potential biological relevance of NO action became evident after the discovery, in the 1980s, that NO is a biological mediator and the demonstration, in the 1990s, that it inhibits respiration in mammalian cells. The finding that nanomolar concentrations of NO reversibly inhibit cytochrome oxidase and compete with molecular oxygen indicated that this interaction may have a potential physiological role in the control of cell respiration. Moreover, this inhibitory effect might be involved in certain pathologies (Moncada and Erusalimsky, 2002).

Cytochrome oxidase is situated on the inner mitochondrial membrane and is the terminal complex of the mitochondrial respiratory chain in animal systems. The enzyme catalyzes the oxidation of cytochrome c with the reduction of O_2 to water in a process linked to proton pumping out of the mitochondrial matrix. Cytochrome oxidase contains two heme centers (a and a3) and two copper centers (CuA and CuB) from which the heme iron of cytochrome a3, together with CuB, both in the reduced foıms, constitute the O_2 binding site. Like carbon monoxide, NO binds to the same proteic sites as O_2 (Cooper, 2002). The inhibition due to the interaction of NO with heme a3 is competitive with O_2, reversible by light, and favored by high levels of cytochrome reduction and low O_2 concentrations. Inhibition due to interaction with CuB is favored by a low level of cytochrome reduction and high levels of O_2. Both forms of inhibition are rapidly reversible and may occur *in vitro*, but *in vivo* the inhibition is largely competitive and reversible by light. This suggests that the inhibition due to reversible binding to heme a3 may be more important when in a cellular context (Sarti *et al.*, 2000).

In 1996, Millar and Day demonstrated that NO can also inhibit mitochondrial respiration in plant systems and in fact were the first to report that the activity of the cytochrome pathway in isolated soybean mitochondria is highly sensitive to NO. The degree of inhibition of cytochrome oxidase depends on the concentration of dissolved oxygen and is fully reversed upon decay of NO, as in animal mitochondria. The specificity of NO is analogous to the well-known effect of carbon monoxide as a competitive inhibitor of cytochrome oxidase, but not of the alternative oxidase in plant mitochondria. The inhibition of the cytochrome pathway in the presence of ADP at low NO concentrations is probably due to a control coefficient for cytochrome oxidase that is greater during oxidative phosphorylation than when proton leak rate is a controlling factor in the absence of ADP (Millar and Day, 1996).

4. PLANT PCD AS A RESPONSE TO MITOCHONDRIAL RESPIRATORY STATUS

It has been recently demonstrated that NO-induced inhibition of respiration in plant cells can be associated with apoptotic cell death (Zottini *et al.*, 2002; Saviani *et al.*, 2002). These experiments assigned a central role to mitochondria in triggering/regulating programmed cell death and contributed to the identification of a possible mechanism through which NO may induce death in plants.

It has been reported that NO has specific effects at the mitochondrial level, also *in vivo*, on carrot cell suspensions (Zottini *et al.*, 2002); this is also true for *Arabidopsis* (Zottini, unpublished data). NO decreases the total respiration by about 50% in carrot cell suspensions treated with the NO-donor sodium nitroprusside (SNP). By using specific inhibitors for the cytochrome pathway (antimycin A) or the alternative pathway (SHAM), it has been demonstrated that NO specifically inhibits the former, while activity is induced through the latter pathway (Fig. 1). The induction of alternative oxidase activity is not merely due to a diversion of electrons, since cytochrome oxidase is inhibited, but is attributable to an induction of protein expression (Zottini *et al.*, 2002). This result has been confirmed by cDNA microarray analysis in *Arabidopsis*, where NO treatment induces the transcription of *Aox1a* within three hours (Huang *et al.*, 2002).

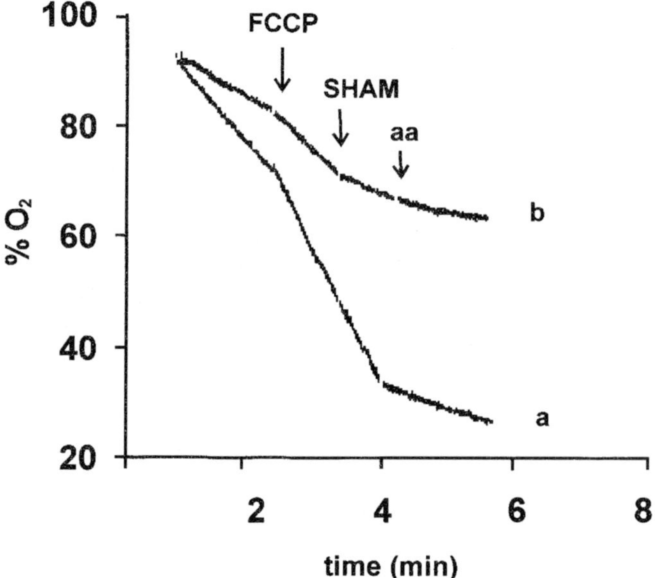

Figure 1 : Effect of NO exposure on respiration of carrot cells. a) oxygen consumption recordered in control untreated cells; b) oxygen consumption recordered in cells treated for 24h with 1mM SNP. FCCP (carbonyl cyanide p-(trifluoromethoxy) phenylhydrazone (1μM), SHAM (0.1mM) and aa (antimycin a 5μM) were added as indicated by the arrows. Cells were analyzed for O2 consumption in a glass chamber, maintained at 25°C, equipped with a Clark-type oxygen electrode.

An immediate effect of the inhibition of respiration is the reduction of the mitochondrial membrane potential (Beltrán *et al.*, 2000), which is known to be an important mechanism associated with apoptosis in animal systems (Zamzami *et al.*, 1995). Variations in the extent of the mitochondrial membrane potential have been evaluated in both carrot (Zottini *et al.*, 2002) and citrus cells (Saviani *et al.*, 2002) by staining with the cationic fluorescent dyes tetramethylrhodamine (TMRM), and chloromethyl X-rhosamine (CMXros) (Sarti *et al.*, 1999), respectively, upon NO treatment. Carrot cells, incubated with TMRM, display a clear mitochondrial fluorescence pattern. The accumulation of the dye concerns only the mitochondria and depends on their energization state. SNP treatment, by collapsing the electrical component of the membrane potential ($\Delta\Psi m$), causes the fluorescence signal to decrease (Fig. 2; Zottini *et al.*, 2002). Moreover, in citrus cells it has been observed that NO, causing a depolarization of mitochondrial membrane, induces opening of the permeability transition pore (PTP) (Saviani *et al.*, 2002). Reduction of the $\Delta\Psi m$, together with an increase in the matrix calcium concentration, is one of the factors that activate the opening of the PTP in animal mitochondria, where PTP opening is associated with the triggering

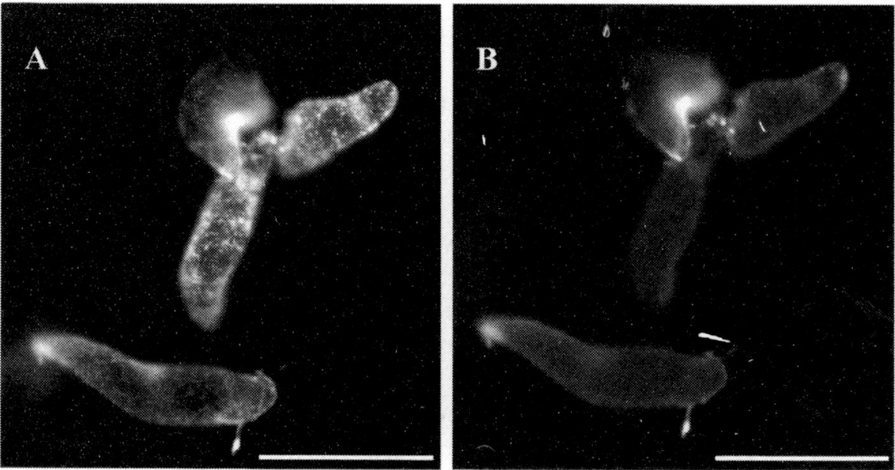

Figure 2 : Mitochondrial fluorescence patterns of carrot cells loaded with tetramethylrhodamine (TMRM) and analyzed by epifluorescence microscope. A) Mitochondrial fluorescence pattern of untreated cells. B) The same cells as in A after the 30 minute treatment with 1 mM SNP. Aliquots of cells were incubated in culture medium containing 500 nM TMRM, at 25°C for 15 minutes, washed three times with culture medium and layered on coverslips for analysis. Bar = 50 μm.

of apoptosis (Petronilli *et al.*, 2001). Very little is known about PTP in plants (Arpagaus *et al.*, 2002), although there are some indications that it might play an important role in the regulation of programmed cell death. Actually, in citrus cells, cyclosporin A, by blocking the opening of pore, prevents the depletion in potential of the mitochondrial membrane and the subsequent DNA degradation observed after exposure to NO. Thus, it can be concluded that cyclosporin A prevents NO-induced apoptosis (Saviani *et al.*, 2002).

By preventing electron flux to cytochrome oxidase, prolonged exposure of mitochondria to NO causes the over-reduction of the respiratory chain components and, consequently, an increase in reactive oxygen species (ROS; Boveris and Chance, 1973). Furthermore, there is a reduction in the levels of antioxidant enzymes (de Pinto *et al.*, 2002) that also make mitochondria more vulnerable to the effects of ROS (Miyamoto *et al.*, 2003).

On the other hand, NO treatment induces expression of alternative oxidase (Zottini *et al.*, 2002). It has been observed that a sharp decline in the capacity of the cytochrome pathway (induced e.g. by antimycin A) is accompanied by a strong induction in the capacity of alternative oxidase, which maintains respiration and cell viability (Elthon *et al.*, 1989). Vanlerberghe *et al.* (2002) showed that in transgenic cells lacking alternative oxidase, a loss of the capacity of the cytochrome pathway results in a complete loss of respiratory capacity (since these cells cannot induce alternative oxidase) and results in massive apoptotic cell death. Inhibition of the cytochrome pathway, in plant as in animals, triggers a death program, but this activation can be strongly attenuated, in plants, by the presence and/or induction of alternative oxidase (Vanlerberghe *et al.*, 2002). Under many different physiological and pathological conditions, regulation of the expression of alternative oxidase may represent an important mechanism by which cells can actively modulate entry into a programmed cell death pathway. We hypothesize that the ability of alternative oxidase to attenuate cell death may be related to its ability to generate ATP or may even be related to its ability to dampen mitochondrial generation of ROS by preventing over-reduction of respiratory chain components (Maxwell *et al.*, 1999). This function might be of crucial importance, since ROS are greatly involved in several programmed cell death process in both animals and in plants (Neill *et al.*, 2002a). Since inhibition of cytochrome oxidase by NO induces alternative oxidase, it is possible to hypothesize that this latter enzyme may play an important role in reducing the generation of ROS when NO is produced in excess.

NO production is stimulated during various types of stress. Alternative oxidase is also induced under similar conditions and has the ability to counteract these stresses. These two events permit the suggestion that at the mitochondrial level, NO plays a major role in activating a defense system. If the altered situation is prolonged, the perturbed cell will die. Both the rate and extent of a range of events consequent to NO treatment determine the final destiny of the cell. If the stress is limited in time, the cell will survive. If, on the other hand, the stress is prolonged, the cell will die. If the cell can maintain a sufficient level of ATP, the cell will eventually enter programmed cell death, otherwise it will die necrotically.

5. NO AND CONTROL OF MITOCHONDRIAL RESPIRATION

The modulation of mitochondrial respiration by NO may help explain experimental observations relating to homeostasis and O_2 consumption. NO, regulating and/or reducing mitochondrial respiratory activity, acts as both a messenger and a regulator of O_2 consumption when O_2 concentration are low, and the metabolic activities of the cell must continue nonetheless.

The supply of sufficient O_2 to internal tissues is a fundamental physiological challenge to multicellular organisms. While animals evolved specialized respiratory organs and a circulation system, plants lack specific systems for O_2 delivery. Transient flooding, water logging and microbial activity in the soil can rapidly lead to O_2 deprivation within the roots, affecting the growth and distribution of terrestrial plants and leading to major reductions in crop yield (Kozlowski 1984). Anoxic conditions are not only due to reduction of the external O_2 supply, but plants often face endogenous anoxia problems. Moreover, in a well-oxygenated surrounding, plant tissue having a high metabolic activity can become hypoxic. One example is given by tissues that lack large intracellular air space, are poorly vacuolated, or are located at the center of organs that are remote from the sites where O_2 enters the plant (Geigenberger, 2003). This is most likely because a large diffusion gradient is required to drive O_2 movement into the tissue at a rate that is sufficient to match the rate of O_2 consumption. During anoxia, when cytochrome oxidase activity is limited by low O_2 concentrations, synthesis of ATP by oxidative phosphorylation is inhibited and ATP is made through fermentation (Drew, 1997). In this situation, both cell metabolism and functionality are damaged because the efficiency of ATP synthesis is drastically reduced. Moreover, fermentation results in an acidification of the cytosol, induction of glycolysis, and the accumulation of lactate and ethanol. Reoxygenation of anoxic tissues induces generation of dangerous O_2 radicals that lead to peroxidation (Drew, 1997).

Under limiting O_2 concentrations, plant cells reduce their metabolism in order to decrease O_2 consumption thereby avoiding anoxia. It has been reported that at low O_2 concentrations plants respond by producing NO (Dordas *et al.*, 2003), most likely by activating nitrate reductase (NR; Rockel *et al.*, 2002) and induction of alternative oxidase, which contribute to the reduction of reoxygenation-induced injuries (Amor *et al.*, 2000). Hypoxic pretreatment greatly extends viability during a subsequent period of anoxia. Acclimatization is in part due to induction of so-called anaerobic genes that encode glycolytic enzymes as well as those involved in alcoholic fermentation. This implies a reduction in respiration with a subsequent decrease in the consumption of ATP, resulting in widespread inhibition of biosynthetic processes (Hoefnagel *et al.*, 1998).

In animal systems, it is reported that the Km for O_2 of purified cytochrome oxidase is about 10-20 times higher than that measured *in vivo* (Tamura *et al.*, 1989). This difference could be due to the presence of NO within cells. Furthermore, changes in gene expression will undoubtedly occur in cells that are exposed to varying degrees of hypoxia. One interesting aspect is that these responses are activated at O_2 concentrations that are well above those that are limiting for mitochondrial electron transport (Jiang *et al.*, 1996). This could be explained, at least in part, by the action of endogenous NO, which, by increasing the Km for O_2 of cytochrome oxidase, causes the cell to sense hypoxia at O_2 concentrations that are relatively high (Moncada and Erusalimsky, 2002). This could be part of an adaptive mechanism in which changes in the Km for O_2, caused by NO, could be significant. Thus, it has been suggested that the NO/O_2 ratio at the level of cytochrome oxidase could provide a sensing mechanism for O_2 that will regulate O_2 consumption to the precise point at which it is used (Clementi *et al.*, 1999).

Traditional studies indicated that the control of O_2 consumption by mitochondria is exerted by ATP-turnover reactions by around 50% (Brand and Murphy 1987). Control by cytochrome oxidase occurs only when O_2 is nearly absent. The interaction between NO and O_2 would lead to a situation in which, when in the presence of an adequate supply of O_2, cytochrome oxidase is able to exert considerable control on the respiratory rate by changing its Km for O_2. In conclusion, there should be a regulatory mechanism, lying proximal to ATP-turnover, which operates at the level of O_2 supply rather than demand of energy (Moncada and Erusalimsky, 2002). If this hypothesis is confirmed, it would be interesting to investigate how these two regulatory mechanisms are coordinated.

One immediate consequence of inhibiting cytochrome oxidase is the increased O_2 availability for the surrounding tissues of plants. Low O_2 concentrations induce a drastic drop in cellular metabolic activity and the induction of metabolic pathways and strategies that conserve energy, and hence reduce O_2 consumption and leave ATP available for other metabolic activities (Geigenberger *et al.*, 2000). Moreover, the plant undergoes adaptive long-term modifications (aerenchyma formation) that both restricts metabolic activity and improves the supply of O_2 by reducing the respiration rate per unit volume of tissue (Drew *et al.*, 2000). The activation of programmed cell death process for aerenchyma formation is linked to the induction of synthesis and signaling of ethylene (Drew *et al.*, 1979). Actually, there are suggestions that ethylene generation may be regulated by NO (Haramaty and Leshem, 1997) and that NO may also be involved in a signal amplification loop involving ethylene in the plant immune system (McDowell and Dangl, 2000). It would be of great interest to investigate the direct role of NO on aerenchyma formation in the light of recent reports on the role of NO in triggering apoptosis in plants (Clarke *et al.*, 2000).

These various adaptive responses, occurring in the plant to counteract anoxia, are activated at O_2 concentrations well above of those that are limiting for electron transport at the mitochondrial level; this suggests the possibility that an O_2 sensing mechanism is involved. Oxygen-sensing systems are elaborated in bacteria and yeast

(Bunn and Poyton, 1996), although the molecular mechanisms of O_2 sensing in multicellular organisms are yet to be understood (Lopez-Barneo *et al.*, 2001). In plants, non-symbiotic hemoglobins have been suggested to be involved in O_2 sensing. It has been reported that over-expression of a non-symbiotic hemoglobin in *Arabidopsis* gives an improvement in growth in normoxic conditions and an increase of survival in strong hypoxic conditions (Hunt *et al.*, 2002). Very recently it has been reported that alfalfa root cultures induce NO emission under hypoxic stress and that NO levels might be regulated by endogenous hemoglobins (Dordas *et al.*, 2003). Such a result would lead to hypothesize a role of hemoglobins in modulating NO levels in plant cells, and, on the other hand, would demonstrate the tight relationship between NO and low O_2 stress conditions.

It is possible to propose a model similar to that proposed in animals in which there is a NO-dependent autoregulation and an adaptive mechanism to hypoxia (Goligorsky, 2000). A drop in O_2 tension results in the release of NO. This could occur from the preexisting pool of heme protein-bound NO (e.g. non-symbiotic hemoglobins) as a result of a decreased of affinity for NO (similar to the mechanism described for hemoglobin by Stamler *et al.*, 1997) or through the activation of a nitric oxide synthase (NOS) located at the mitochondrial level (possibly that just described by Chandok *et al.*, 2003). An increased NO concentration would result in the inhibition of cytochrome oxidase, with a consequent reduction in O_2 consumption, a decrease in the proton-motive force for the mitochondrial calcium-uniporter, and a drop in ATP production. In this way, NO would also account for avoiding an overload of calcium in mitochondria where it functions as trigger for PTP opening and thus as a programmed cell death elicitor. Collapse of $\Delta\Psi m$ would be counteracted by inversion of ATPase, which by hydrolyzing glycolytic ATP, allows proton pumping out of the mitochondrial matrix with the final effect of sustaining the $\Delta\Psi m$. Thus, the mitochondrial membrane potential will be gradually restored and, in the meantime, the inhibition of oxidative phosphorylation induces an increase of glycolysis, assuring ATP as a substrate for ATPase. The contemporary production of ROS, which could react with NO, might be the feedback mechanism by which the cell tries to restore mitochondrial respiration. This altered situation would not last for a sustained period; if physiological conditions are not restored, the cell enters into a death process.

6. DOES A PLANT mtNOS EXIST?

While NOS activity has been described in several organisms ranging from bacteria to animals, in plants a NOS enzyme with homology to the mammalian enzyme has not been isolated to date. However, NOS activity has been described in many plant species (Barroso *et al.*, 1999; Ribeiro *et al.*, 1999; Modolo *et al.*, 2002). Different studies carried out under various experimental conditions resembling both physiological and pathological situations suggest the existence of at least three enzymes responsible for NO production (in addition to non-enzymatic NO generation; Neill *et al.*, 2003): i) a NR responsible for basal, physiological and non-elicited production of NO, ii) an inducible NOS responsible for the dramatic and prolonged increase of NO upon pathogen attack in resistant plants and iii) a constitutive NOS that can be activated rapidly and transiently by pathogen or pathogenic-derived elicitors. This division in NO production resembles that found in mammals in which distinct NOS isoforms fulfill different roles in dissimilar physiological conditions. Very recently an enzyme has been isolated and characterized with NOS activity that is induced by pathogens (Chandok *et al.*, 2003). Upon purification, a protein band of 120kD has been identified in tobacco: it shows a high sequence homology with a subunit (P protein) of glycine decarboxylase, a mitochondrial multienzyme system catalyzing the catabolism of glycine molecules flooding from peroxisomes during photorespiration (Douce *et al.*, 2001). Chandok *et al.* (2003) demonstrated that this iNOS is a variant of the P protein and is likely to be located at the mitochondrial level as it shows a N-terminal pre-sequence for mitochondrial localization. The lack of sequence homology with animal NOS enzymes justifies the previous failure to isolate protein using antibodies specific for the animal NOS (Butt *et al.*, 2003). Despite the lack of homology, the biochemical properties of the plant enzyme appear to be very similar to those of the animal counterpart. Experimental data demonstrate that the plant iNOS requires various cofactors, such as NADPH, FAD, FMN, and calcium calmodulin, as well as calcium ions. It has also been observed that the NOS activity of the purified protein is sensitive to specific inhibitors of animal NOS (l-NMMA and aminoguanidine). Moreover the plant iNOS, like the animal enzyme, is specifically induced during the resistance to pathogen infection (Chandok *et al.*, 2003).

Additional studies on the localization and functional regulation have yet to be performed in order to investigate if this recently characterized iNOS plays a major role in other conditions in which NO burst are observed, such as mechanical stress and hormonal induction (Neill *et al.*, 2003,: Carimi *et al.*, 2002).

7. FUTURE PROSPECTS

Mitochondria are a main target of NO. Since mitochondria are a pivotal organelle in plant metabolism and life, it is of critical importance to study the interactions between mitochondria and NO, identifying other possible targets other than those already discovered (i.e., alternative oxidase and cytochrome oxidase). Plant mitochondria posses enzymes unique to the plant kingdom (e.g. rotenone-insensitive NAD(P)H dehydrogenases) and it would be interesting to understand if they are affected by NO and, moreover, what the physiological significances of this interaction are.

As mitochondria are a critical site in modulating many stress responses in plants, it would be of great interest to determine if NO release localized at the mitochondrial level occurs and if there is a spatio-temporal regulation of NO production. It can be hypothesized that the effects of NO may be confined to specific microdomains. This concept is already accepted in the case of calcium signaling and, as recently suggested, for hydrogen peroxide and cyclic nucleotides as well (Neill *et al.*, 2002b)

It is of extreme interest to fully characterize the recently identified iNOS (Chandok *et al.*, 2003), probably situated at the mitochondrial level, in order to evaluate if its enzymatic activity is responsible for the modulation of mitochondrial function. This would demonstrate that the previously described modulation of respiration occurs due to a specific and localized production of NO. Furthermore, this would help to elucidate if and how the NO generating enzyme takes part in an O_2-sensing mechanism, permitting the definition of the role and mechanism of action of NO during hypoxia. Dordas *et al.*, (2003) exclude that the NO produced under low O_2 conditions could originate from a NOS enzyme, since it requires O_2 for its function. At any rate, it must be taken into account that the cell perceives O_2 reduction at concentrations well above those that are limiting for mitochondrial electron transport. However, this finding is open to several interpretations.

Lastly, the role of NO as signal molecule in the interaction between mitochondria and chloroplasts must be defined as well. It is known that respiration and photosynthesis are interdependent processes. Respiration depends on photosynthesis for its substrates and photosynthesis depends on respiration for various compounds. In particular, mitochondria play a major role in oxidizing excess reducing equivalents, even if the demand for ATP is low. Experimental evidence indicates that mitochondrial activity in light conditions can reduce photoinhibition by removing excess photosynthetic reducing equivalents (Hoefnagel *et al.*, 1998). This is also possible because of the existence of NAD(P)H dehydrogenases situated on the external face of the internal mitochondrial membrane that are able to oxidize cytosolic NAD(P)H. This process can also occur without the synthesis of ATP through the alternative oxidase pathway. It would be interesting to understand the role of NO in modulating the interactions between mitochondria and chloroplasts, since it is known that NR (that is supposedly the NO generating enzyme in non-elicited conditions), not only plays a central role during photosynthesis, but its transcription is also light-dependent and follows a diurnal pattern (Hoff *et al.*, 1994; Rockel *et al.*, 2002).

ACKNOWLEDGMENTS

I wish to thank Mario Terzi, Fiorella Lo Schiavo and Francesco Carimi for helpful comments on the manuscript. This work was supported by "Progetto giovani ricercatori" of the University of Padova.

REFERENCES

Affourtit, C., Krab, K. and Moore, A. (2001). Control of plant mitochondrial respiration. *Biochim. Biophys. Acta,* **1504** : 58-69

Amor, Y., Chevion, M. and Levine, A. (2000). Anoxia pretreatment protects soybean cells against H2O2-induced cell death: possible involvement of peroxidases and of alternative oxidase. *FEBS Lett.,* **447** : 175-180

Arpagaus, S., Rawyler, A., Braendle, R. (2002) Occurrence and characteristics of the mitochondrial permeability transition in plants. *J. Biol. Chem.,* **277** : 1780-1787.

Balk, J., Leaver, C.J., McCabe, P.F. (1999) Translocation of cytochrome c from the mitochondria to the cytosol occurs during heat-induced programmed cell death in cucumber plants. *FEBS Lett.*, **463** : 151-154

Barroso, J.B., Corpus, F.J., Carreras, A., Sandalio, L.M., Valderrama, R., Palma, J.M., Lupianez, J.A. and del Rio, L.A. (1999). Localization of nitric-oxide synthase in plant peroxisomes. *J. Biol. Chem.*, **274** : 36729-36733.

Beers E.P and McDowell J.M. (2001). Regulation and execution of programmed cell death in response to pathogens, stress and developmental cues. *Curr. Opin. Plant Biol.*, **4** : 561-567

Beltrán, B., Mathur, A., Duchen, M.R., Erusalimsky, J.D. and Moncada, S. (2000). The effect of nitric oxide on cell respiration: A key to understanding its role in cell survival or death. *Proc. Natl. Acad. Sci. USA,* **97** : 14602-14607

Boveris, A. and Chance, B. (1973). The mitochondrial generation of hydrogen peroxide. *Biochem. J.*, **134** : 707-716

Brand, M.D. and Murphy, M.P. (1987). Control of electron flux through the respiratory chain in mitochondria and cells. *Biol. Rev. Camb. Philos. Soc.*, **62** : 141-193

Brown, G.C. (1999). Nitric oxide and mitochondrial respiration. *Biochim. Biophys. Acta,* **1411** : 351-369

Brown, G.C. and Borutaite, V. (2002). Nitric oxide inhibition of mitochondrial respiration and its role in cell death. *Free Radic. Biol. Med.*, **33** : 1440-1450

Bunn, H.F. and Poyton, R.O. (1996). Oxygen sensing and molecular adaptations to hypoxia. *Physiol. Rev.*, **76** : 839-885

Butt, Y.K., Lum, J.H. and Lo S. (2003). Proteomic identification of plant proteins probed by mammalian nitric oxide synthase antibodies. *Planta,* **216** : 762-771

Carimi, F., Zottini, M., Formentin, E., Terzi, M., Lo Schiavo, F. (2002). Cytokinins, new apoptotic inducers in plants. *Planta,* **216** : 413-421.

Chandok, M.R., Ytterberg, A.J., van Wijk, K.J. and Klessig, D.F. (2003). The pathogen-inducible nitric oxide synthase (iNOS) in plants is a variant of the P protein of the glycine decarboxylase complex. *Cell,* **113** : 469-482.

Clarke, A., Desikan, R., Hurst, R.D., Hancock, J.T. and Neill, S.J. (2000). NO way back: nitric oxide and programmed cell death in Arabidopsis thaliana suspension cultures. *Plant J.*, **24** : 667-677.

Clementi, E., Brown, G.C., Foxwell, N. and Moncada, S. (1999). On the mechanism by which vascular endothelial cells regulate their oxygen consumption. *Proc. Natl. Acad. Sci. USA,* **96** : 1559-1562.

Cooper, C.E. (2002). Nitric oxide and cytochrome oxidase: substrate, inhibitor or effector? *Trends Biochem. Sci.*, **27** : 33-39

Curtis, M.J. and Wolpert, T.J. (2002). The oat mitochondrial permeability transition and its implication in victorin binding and induced cell death. *Plant J.*, **29** : 295-312

de Pinto, M.C., Tommasi, F. and De Gara, L. (2002). Changes in the antioxidant systems as part of the signaling pathway responsible for the programmed cell death activated by nitric oxide and reactive oxygen species in tobacco Bright-Yellow 2 cells. *Plant Physiol.*, **130** : 698-708.

Delledonne, M., Xia, Y., Dixon, R.A. and Lamb, C. (1998). Nitric oxide functions as a signal in plant disease resistance. *Nature,* **394** : 585-588.

Dordas, C.,. Hasinoff, B.B.,. Igamberdiev, A.U, Manac'h, N., Rivoal, J. and Hill, R.D. (2003). Expression of a stress-induced hemoglobin affects NO levels produced by alfalfa root cultures under hypoxic stress. *Plant J.,* **351** : 763-770

Douce, R., Bourguignon, J., Neuburger, M. and Rébeillé, F. (2001) The glycine decarboxylase system: a fascinating complex. *Trends Plant Sci.,* **6** : 167-176

Drew, M.C. (1997). Oxygen deficiency and root metabolism: Injury and Acclimation Under Hypoxia and Anoxia. *Ann. Rev. Plant Physiol. Plant Mol. Biol.,* **48** : 223-250.

Drew, M.C., He, C.J. and Morgan, P.W. (2000). Programmed cell death and aerenchyma formation in roots. *Trends Plant Sci.,* **5** : 123-127

Drew, M.C., Jackson, M.B. and Giffard, S. (1979). Ethylene-promoted adventitious rooting and development of cortical air spaces (aerenchyma) in roots may be adaptive responses to flooding in Zea mays L. *Planta,* **147** : 83-88.

Durner, J. and Klessig, D.F. (1999). Nitric oxide as a signal in plants. *Curr. Opin. Plant Biol.,* **2** : 369-374.

Elthon, T.E., Nickels, R.L. and McIntosh, L. (1989). Mitochondrial events during development of termogenesis in Sauratum guttatum. *Planta,* **180** : 82-89

Geigenberger, P., Fernie, A.R., Gibon, Y., Christ, M. and Stitt, M. (2000). Metabolic activity decreases as an adaptive response to low internal oxygen in growing potato tubers. *Biol. Chem.,* **381** : 723-740.

Geigenberger, P. (2003). Response of plant metabolism to too little oxygen. *Curr. Opin. Plant Biol.,* **6** : 247-256

Giuffré, A., Sarti, P., D'Itri, E., Buse, G., Soulimane, T. and Brunori, M. (1996). On the mechanism of inhibition of cytochrome c oxidase by nitric oxide. *J. Biol. Chem.,* **271** : 33404-33408

Goligorsky, M.S. (2000). Making Sense out of Oxygen Sensor. *Circ. Res.,* **86** : 824-826.

Haramaty, E. and Leshem, Y.Y. (1997). Ethylene regulation by the nitric oxide (NO) free radical: a possible mode of action of endogenous NO. In: Biology and Biotechnology of the Plant Hormone Ethylene (Kannelis, A.K., Chang, C., Kende, H. and Grierson, D., eds). Dordrecht: Kluwer Academic Publishers, pp. 253-258.

Heath, M.C. (2000). Hypersensitive response-related death. *Plant Mol. Biol.,* **44** : 321-334.

Hoefnagel, M.H.N., Atkin O.K. and Wiskich, J.T. (1998). Interdependence between chloroplasts and mitochondria in the light and the dark, *Biochim. Biophys. Acta,* **1366** : 235-255

Hoefnagel, M.H.N., Millar, A.H., Wiskich, J.T. and Day, D.A. (1995). Cytochrome and alternative respiratory pathways compete for electrons in the presence of pyruvate in soybean mitochondria. *Arch. Biochem. Biophys.,* **318** : 394-400

Hoff, T,, Truongm H-N. and Caboche M. (1994). The use of mutants and transgenic plants to study nitrate metabolism. *Plant Cell Env.,* **17** : 486-506.

Huang, X., von Rad, U. and Durner, J. (2002). Nitric oxide induces transcriptional activation of the nitric oxide-tolerant alternative oxidase in Arabidopsis suspension cells. *Planta,* **215** : 914-923.

Hunt, P.W., Klok, E.J., Trevaskis, B., Watts, R.A., Ellis, M.H., Peacock, W.J. and Dennis, E.S. (2002). Increased level of hemoglobin 1 enhances survival of hypoxic stress and promotes early growth in Arabidopsis thaliana. *Proc. Natl. Acad. Sci. USA,* **99** : 17197-17202.

Jiang, B.H., Semenza, G.L., Bauer, C. and Marti, H.H. (1996). Hypoxia-inducible factor 1 levels vary exponentially over a physiologically relevant range of O2 tension. *Am. J. Physiol.,* **271** : C1172-C1180.

Jones, A. (2000). Does the plant mitochondrion integrate cellular stress and regulate programmed cell death? *Trends Plant Sci.,* **5** : 225-230.

Kozlowski, T.T. (1984). Extent, causes and impacts of floodin. In : Flooding and plant growth (Ed Kozlowski TT). New York Academic Press pp: 1-7

Kuriyama, H. and Fukuda, H. (2002). Developmental programmed cell death in plants *Curr. Opin. Plant Biol.,* **5** : 568-573

Lam, E., Kato N, Lawton M (2001). Programmed cell death, mitochondria and the plant hypersensitive response. *Nature,* **411** : 848-853.

López-Barneo, J., Pardal, R. and Ortega-Sáenz, P. (2001). Cellular mechanism of oxygen sensing *Ann. Rev. Physiol.,* **63** : 259-287

Maxwell, D.P., Wang, Y., McIntosh, L. (1999). The alternative oxidase lowers mitochondrial reactive oxygen production in plant cells. *Proc. Natl. Acad. Sci. USA,* **96** : 8271-8276

McDowell, J.M. and Dangl, J.L. (2000). Signal transduction in the plant immune response. *Trends Biochem. Sci.,* **25** : 79-82.

Millar, A.H. and Day, D.A. (1996). Nitric oxide inhibits the cytochrome c oxidase but not the alternative oxidase of plant mitochondria. *FEBS Lett.,* **398** : 155-158

Miyamoto, Y., Koh, Y.H., Park, Y.S., Fujiwara, N., Sakiyama, H., Misonou, Y., Ookawara, T., Suzuki, K., Honke, K. and Taniguchi, N. (2003). Oxidative stress caused by inactivation of glutathione peroxidase and adaptive responses. *Biol. Chem.,* **384** : 567-574.

Modolo, L.V., Cunha, F.Q., Braga, M.R. and Salgano, I. (2002). Nitric oxide synthase-mediated phytoalexin accumulation in soybean cotyledons in response to the Diaporthe phaseolorum f. sp. meridionalis elicitor. *Plant Physiol.,* **130** : 1288-1297.

Møller, I.M. (2001). Plant mitochondria and oxidative stress: electron transport, NADPH turnover, and metabolism of reactive oxygen species. Ann. Rev. Plant Physiol. *Plant Mol. Biol.,* **52** : 561-591.

Moncada, S., Erusalimsky, J.D. (2002). Does nitric oxide modulate mitochondrial energy generation and apoptosis? *Nat. Rev. Mol. Cell. Biol.,* **3** : 214-220.

Neill, S.J., Desikan, R., Clarke, A., Hurst, R.D. and Hancock, J.T. (2002a). Hydrogen peroxide and nitric oxide as signalling molecules in plants. *J. Exp. Bot.,* **53** : 1237-1247

Neill, S.J., Desikan, R. and Hancock, J.T. (2002b). Hydrogen peroxide signalling, *Curr. Opin. Plant Biol.,* **5** : 388-395.

Neill, S.J., Desikan, R. and Hancock, J.T. (2003). Nitric oxide signaling in plants *New Phytologist,* **159** : 11-35

Petronilli, V., Penzo, D., Scorrano, L., Bernardi, P. and Di Lisa, F. (2001). The mitochondrial permeability transition, release of cytochrome c and cell death. Correlation with the duration of pore openings in situ. *J. Biol. Chem.,* **276** : 12030-12034.

Ribeiro, E.A., Cunha, F.Q., Tamashiro, W.M.S.C.and Martins, I.S. (1999). Growth phase-dependent subcellular localization of nitric oxide synthase in maize cells. *FEBS Lett.,* **445** : 283-286.

Rockel, P., Strube, F., Rockel, A., Wildt, J. and Kaiser, W.M. (2002). Regulation of nitric oxide (NO) production by plant nitrate reductase in vivo and in vitro. *J. Exp. Bot.,* **53** : 103-110.

Sarti, P., Lendaro, E., Ippoliti, R., Bellelli, A., Benedetti, P.A. and Brunori, M. (1999). Modulation of mitochondrial respiration by nitric oxide: investigation by single cell fluorescence microscopy. *FASEB J.,* **13** : 191-197.

Sarti, P., Giuffré, A., Forte, E., Mastronicola, D., Barone, M.C. and Brunori M. (2000). Nitric Oxide and Cytochrome c Oxidase: Mechanisms of Inhibition and NO Degradation. *Biochem Biophys Res Commun,* **274** : 183-187.

Saviani, E.E, Orsi, C.H., Oliveira, J.F.P., Pinto-Maglio, C.A.F.and Salgado I. (2002). Participation of the mitochondrial permeability transition pore in nitric oxide-induced plant cell death. *FEBS Lett.,* **510** : 136-140

Siedow, J.N. and Day, D.A. (2000). Respiration and photorespiration. In: B Buchanan, W. Gruissem, R. Jones, eds, Biochemistry and Molecular Biology of Plants. American Society of Plant Physiologists, Rockville, MD, pp 676-728

Stamler, J.S, Jia, L., Eu, J.P., McMahon, T., Demchenko, I., Bonaventura, J., Gernert, K. and Piantadosi, C.A. (1997). Blood flow regulation by S-nitrosohemoglobin in the physiological oxygen gradient. *Science,* **276** : 2034-2037

Sun, Y-L., Zhao, Y., Zhao, Y., Hong, X., Zhai, Z-H. (1999). Cytochrome c release and caspase activation during menadione-induced apoptosis in plants. *FEBS Lett.,* **462** : 317-321

Tamura, M., Hazeki, O., Nioka, S. and Chance, B. (1989). In vivo study of tissue oxygen metabolism using optical and nuclear magnetic resonance spectroscopies. *Ann. Rev. Physiol.,* **51** : 813-834

Vanlerberghe, G.C. and McIntosh, L. (1997). Alternative oxidase: from gene to function. *Ann. Rev. Plant Physiol. Plant Mol. Biol.,* **48** : 703-734

Vanlerberghe, G.C., Robson, C.A. and Yip, J.Y.H. (2002). Induction of mitochondrial alternative oxidase in response to a cell signal pathway down-regulating the cytochrome pathway prevents programmed cell death. *Plant Physiol.,* **129** : 1829-1842

Xie, Z. and Chen, Z. (2000). Harpin-induced hypersensitive cell death is associated with altered mitochondrial functions in tobacco cells. *Mol. Plant Microbe Interact.,* **13** : 183-190

Yao, N., Tada, Y., Sakamoto, M., Nakayashiki, H., Park, P, Tosa, Y. and Mayama, S. (2002). Mitochondrial oxidative burst involved in apoptotic response in oats. *Plant J.,* **30** : 567-579.

Zamzami, N., Marchetti, P., Castedo, M., Zanin, C., Vayssiere, J.L., Petit, P.X. and Kroemer, G. (1995). Reduction in mitochondrial potential constitutes an early irreversible step of programmed lymphocyte death in vivo. *J. Exp. Med.,* **181** : 1661-1672

Zottini, M., Formentin, E., Scattolin, M., Carimi, F., Lo Schiavo, F. and Terzi, M. (2002). Nitric oxide affects plant mitochondrial functionality in vivo. *FEBS Lett.,* **515** : 75-78.

Chapter 12

NITRIC OXIDE AND SEED GERMINATION

Zlatko Giba[1], Dragoljub Grubišic[2] and Radomir Konjevic[1]★

[1]Institute of Botany, Faculty of Biology, University of Belgrade, and [2]Institute for Biological Research "Siniša Stankovic", 29. novembra 142., 11060 Belgrade, Yugoslavia

★Corresponding author : E-mail : konjevic@bfbot.bg.ac.yu
Fax : 381-11-769903

Summary

The first discovered phenomenon promoted by nitric oxide in plants was the stimulation of light-induced seed germination. Although the effectiveness of nitrogenous compounds, such as nitrates and nitrites, are known from the beginning of the 20th century, their activity in stimulation of seed germination is still far from being completely understood. Recent data connect nitrate effectiveness with the activity of free radical gaseous nitrogen oxides such as NO and NO_2. The sensitivity of dormant seeds to external nitrogen oxides could explain promotive effect of smoke evolved during and after wild fires. This phenomenon can be considered as a special case of a general mechanism by which ecosystems regulate timing and sequence of germination of different species during annual seasonal changes. Nitrogen oxides, present in the soil as soil-trace gases, could be considered as potent information carriers. They would provide seeds with information, not only on nitrate level in the soil, but also on microbial activity and therefore soil quality. Because of annual variation of soil trace gas fluxes, seeds could be provided with information on season and climate changes in their surroundings.

In : Nitric Oxide Signaling in Higher Plants, 2004
(Eds Jose R. Magalhaes, Rana P. Singh and Leonidas P. Passos)
Studium Press, LLC, Houston, USA, pp 239-275

Thus, nitrogen oxides would be the outer information carriers providing the seeds with integral information on all of these important factors for plant growth and development.

Keywords : nitrogen oxides, seed germination, dormancy, light, soil trace gases

1. INTRODUCTION

During the evolution, plants developed a special organ, the seed, which ensures their spatio-temporal distribution, and perpetuation of the species. Besides the protective covering, the whole structure containing embryo, is generously provided with nutrient reserves to support the emerging seedling until the moment when it becomes a self-sufficient autotrophic organism.

A particular property of seeds, in comparison with other plant organs and tissues, is that they are resting organs, generally with a low moisture content ranging from 7 – 15%. In this state, with very low (almost at the standstill) but measurable metabolic activity, seeds are able to survive for long time periods, often for many years. Dry (unimbibed) seeds, in which none of the germination processes occur, are assigned as quiescent. To germinate, under conditions of suitable temperature and aeration, quiescent seeds need only to be hydrated. However, seeds of majority of species do not germinate even under optimum external conditions. These seeds are assigned as dormant and additionally they must undergo certain special situations before they can germinate. To terminate the state of dormancy, different seeds must be exposed to various factors such as low, high or alternating temperatures, light of certain duration and quality, different chemical stimuli, mechanical factors *etc.*, or to specific combinations of these factors. Because the main function of the seed is to establish a new plant, it is peculiar that dormancy, being an intrinsic block of germination, exists at all (Bewley, 1997). However, there is no unambiguous definition of this phenomenon. This is probably due to the fact that dormancy is manifested, and broken, in various ways in different species (Bewley and Black, 1982; Nikolaeva *et al.*, 1985). Commonly, seed dormancy is regarded as an incapability of intact, viable seeds to complete germination under otherwise favorable conditions. Dispersion of the seeds, mediated by animals or wind, obviously could be considered as an adaptive trait that ensures "distribution in space" of seed germination. Similarly, dormancy could be considered as an adaptive trait that ensures and optimizes the "distribution in time" of seed germination. To understand the phenomenon of dormancy one must bear in mind that in ecological sense, timing is everything. Plants are sessile organisms and the fate of the future plant is almost completely determined when seed "decides " to germinate. A number of systems and mechanisms have appeared during the evolution serving to detect surrounding conditions and define appropriate point in time for germination. All of them are collectively known as dormancy mechanisms. To ensure survival of the future seedling, environmental conditions have to be detected, integrated and translated through the appropriate endogenous but also exogenous signaling molecules, very early at the seed level, even before the onset of germination (Giba *at al.*, 2003).

From physiological point of view, germination process could be considered as triphasic. It consists of: imbibition (a rapid initial water uptake), a physical process characteristic also for dead seeds; the plateau phase (a small change of water content but high metabolic activity), and subsequent increase of water content coincident with radicle protrusion and growth. In terms of regulation of germination, the plateau phase is of a principal interest. Length of this germination phase is extended by dormancy mechanisms, while factors promoting germination do so by shortening it. In general, dormancy-breaking factors which include exogenously applied dormancy-breaking substances need to be in contact with seeds prior to or only during this phase.

2. PERCEPTION AND ACTION OF LIGHT

Light-mediated dormancy represents one of the most delicately regulated types of dormancy. As an environmental stimulus, light usually acts as a dormancy-breaking agent, but it can also inhibit the germination. Thus, light-sensitive seeds are divided into positive or negative photoblastic seeds. Interestingly, even non-dormant, non-photoblastic seeds, such as common agricultural species which readily germinate when sufficient moisture and oxygen are available, can be, under certain conditions, light-sensitive (Thanos and Mitrakos,

1979). As a rule, positively photoblastic seeds are not equipped with large amounts of nutrient reserves (Dedoner *et al.*, 1988). Germination of these seeds under unfavorable conditions for photosynthesis could lead to demise of the future vegetative plant. Light signals from surroundings must be detected and translated early and very precisely at the seed level. These seeds are therefore equipped with appropriate photoreceptors.

Obviously, light is not only the source of energy for plants. It is a dominant environmental factor for adaptation to changes in the surroundings and the regulation of plant growth and development. Signal transduction induced by light as an information carrier, starts with the perception of light by specialized photoreceptors. Apart from photosynthetic pigments, to detect light quality, quantity and direction, plants possess specialized red/far-red photoreversible phytochrome(s), blue light - absorbing cryptochromes and phototropins, and still unidentified UV/B photoreceptors (Briggs and Olney, 2001). They are cytoplasmic or nuclear proteins present in various types of plant organs and tissues. Different physiological processes, such as stem elongation, flowering or seed germination, are regulated by light, *i.e.* they are under control of these photoreceptors.

Among plant photoreceptors only phytochromes have been shown to directly mediate induction of germination (Bortwick *et al.*, 1954). Using *Arabidopsis thaliana* seeds as a model system, it was shown that light regulation of seed germination is primarily mediated by PHY B stored in the seeds. Prolonged imbibition in darkness leads to higher accumulation levels of PHY A which also promotes germination after being converted into the active form (Shinomura *et al.*, 1994; 1996). In general, red light (λ=660nm) induces germination of light-requiring seeds, while far-red light (λ=730nm) inhibits germination. Phytochrome B and phytochrome A regulate red light-induced germination, whereas induction of germination by far-red light, when it occurs, is mediated only by phytochrome A (Shinomura *et al.*, 1996). Recently, Henning *et al.* 2002 showed that phytochrome E might be involved in light-induced seed germination, as well. When triggered, these photoreceptors initiate a signal transduction cascade, which probably involves G-proteins, cGMP, Ca^{2+}/Cam complex (Bowler *et al.*, 1994; Mustilli and Bowler, 1997).

Finally, light-stimulation of gibberellin synthesis leads to the completion of germination (Toyomasu *et al.*, 1998).

3. INORGANIC NITRATES STIMULATE SEED GERMINATION

Nitrogenous compounds, such as potassium nitrate, promote germination in many seeds, especially in light-sensitive, *i.e.* photoblastic seeds. Nitrates significantly increase the light sensitivity and decrease the light requirement of seeds (Toole *et al.*, 1955). Although this effect of nitrates and nitrites has been known for a long time (Lehmann, 1909), and several hypotheses on their interaction with phytochrome have been formulated (Hilton, 1985, Grubišic and Konjevic, 1990.), almost nothing was known about the mechanism of action. The proposed connection between nitrogenous compounds, the pentose-phosphate pathway and seed dormancy (Roberts, 1973) has never been experimentally confirmed (Adkins *et al.*, 1984; Cohn, 1989). During the course of investigations in the 20th century various ideas and suggestions have appeared. Nitrate activation of postulated phytochrome receptor has been proposed to explain the germination of *Sysimbrium officinale* seeds (Hilhorst and Karssen, 1989). Inorganic nitrates also stimulate the phytochrome-controlled germination of *Spirodela polyrhiza* turions (Appenroth *et al.*, 1992), as well as phytochrome-controlled germination of fern spores, where their effectiveness was explained on the basis of electron-accepting properties of nitrates (Haas and Scheuerlein, 1991).

Similar to nitrates, inorganic nitrites stimulate seed germination of different species (Bewley and Black, 1982). In majority of cases, stimulative effect is pH- dependent. Maximal activity of inorganic nitrites could be obtained at pH values around pH = 3 (Cohn *et al.*, 1983). Interestingly, it was shown that stimulative effect of nitrites could be achieved even without direct contact of seeds with nitrite solution. The stimulation of germination could be attained by placing the grains in a closed container in the presence of a vessel containing acidified nitrite solution. Based on this observation, assuming dormancy-breaking activity of some gaseous compound evolved from acidified nitrite solution, Cohn and coworkers have shown that nitrogen dioxide (NO_2) breaks dormancy of *Oryza sativa* when applied to either imbibed or unimbibed caryopses (Cohn and Caste, 1984). These results were largely overlooked until they came into the focus again at the moment when it was shown that different NO-releasing compounds, such as organic nitrates, are very potent stimulators of seed germination (Grubišic *et al.*, 1991, 1992).

4. ORGANIC NITRATES STIMULATE SEED GERMINATION

Chemically, organic nitrates are esters of polyhydroxylic alcohols and nitric acid (Fig. 1).

Although used as antianginal drugs in human medicine for more than 100 years (see for commentary: Ignarro, L.J., 2002), until the beginning of the last decade of the 20th century, no data about physiological effects of organic nitrates (*e.g.* nitroglycerin) in plants have been published. The physiological effectiveness of organic nitrates in plants has been shown, for the first time, in stimulation of *Stellaria media* seed germination (Grubišic *et al.*, 1991). The effect of organic nitrates was further demonstrated and examined using light-induced *Paulownia tomentosa* seed germination as a model system (Grubišic *et al.*, 1992).

The germination of *Paulownia tomentosa* seeds is phytochrome-controlled. The light requirement for maximum germination of these seeds may vary from very brief exposure to several hours of red light. However, in seeds that otherwise require long periods of light (in some cases up to 18 hours), it can be reduced to a single pulse of 5 min red light by the exogenous application of potassium nitrate (Grubišic and Konjevic, 1990). It was found that similar effects in

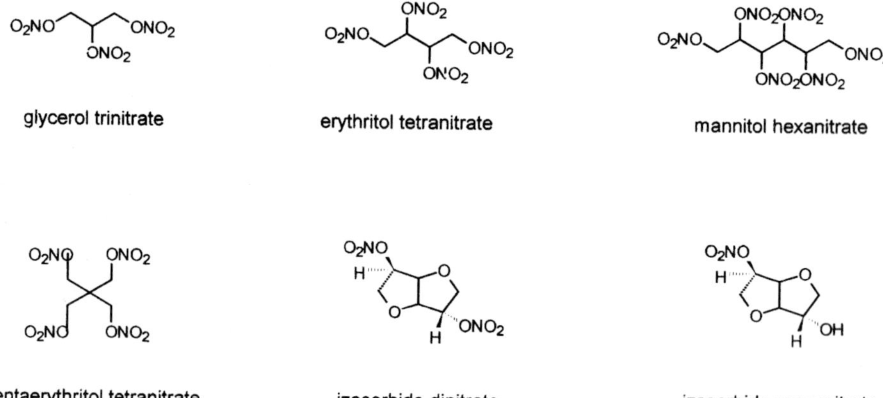

Figure 1 : The chemical structures of some organic nitrates frequently used in biologycal systems.

reducing light requirement for maximal germination of *P. tomentosa* seeds, as well as in promoting germination of *S. media* seeds in darkness, could be obtained by organic nitrates such as nitroglycerin, isosorbide mono- and di-nitrate or pentaerythritol tetranitrate (Grubišic *et al.*, 1991, 1992). They were more effective than inorganic nitrates. In both cases, *i.e.* in *P. tomentosa* and *S. media* seed germination, nitroglycerine for example, produced the same stimulative effect as potassium nitrate but in approximately hundred times lower concentrations (Fig. 2). Primarily, these results were discussed in terms of so-called 'anesthetic theory' which states that the changes in physical properties of the membrane, after administration of some short chain alcohols and carboxylic acids, enables attachment of different signaling proteins to plasmalemma

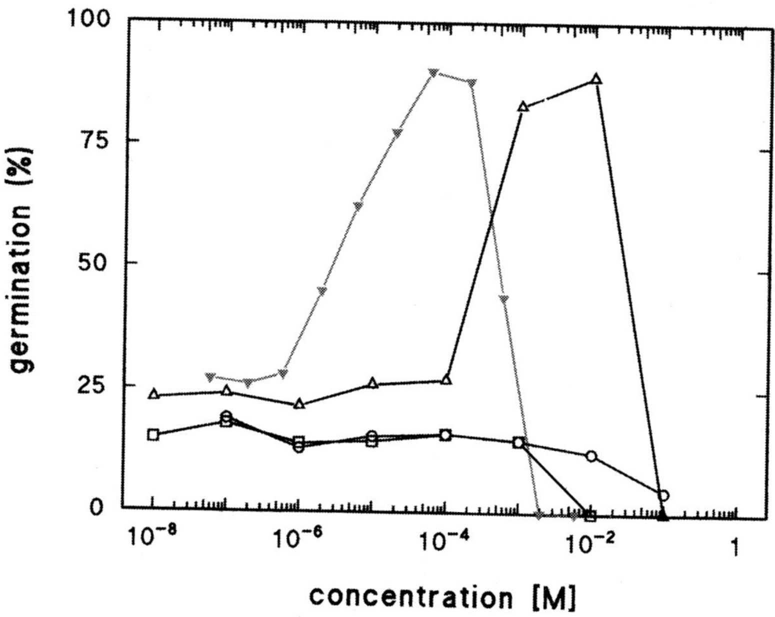

Figure 2 : The effect of nitroglycerin and potassium nitrate (KNO_3) on the germination of *Paulownia tomentosa* seeds induced by a pulse of red light. Seeds were imbibed for 3 days in different concentrations of potassium nitrate (Δ), nitroglycerin (▼), glycerol or glycerol triacetate (□,○) in darkness, irradiated with 5 min red light, and left to germinate. The germination of control samples (seeds imbibed in water) was 27%. Germination was scored 7 days after the red light pulse. Standard errors are not shown since they never exceeded 3%. (N=400 seeds per point) (after Grubišic *et al.*, 1992).

and this would be a clue for understanding dormancy-breaking activity of these and similar substances (for a review see Hallet and Bewley, 2002). However, chemical analogues of nitroglycerin, such as triacetin (glycerol triacetate) or glycerol alone, are completely ineffective in *P. tomentosa* seed germination (Fig. 2) undoubtedly suggesting the importance of the presence of nitro groups in the molecule for its physiological activity.

5. THE EFFECT OF SUBSTANCES WITH ELECTRON-ACCEPTING PROPERTIES

Inorganic and organic nitrates differ greatly with respect to their solubility in water and in organic solvents. However, a high effectiveness of organic nitrates in seed germination could not be explained only on the basis of this difference. In animal systems organic nitrates exert their function by undergoing biological reduction and releasing nitric oxide. Biological reduction by itself could be critical signaling event in breaking seed dormancy. During investigations on the effects of different substances with electron accepting properties in *P. tomentosa* seed germination, it was found that only high concentrations (around 10^{-2} M) of compounds with standard redox potential equal to or higher than E_0'= + 360 mV, like inorganic nitrates, hexacyanoferrate (III), or hexachloroirridate (IV) potentiate the germination of *P. tomentosa* seeds (Giba *et al.*, 1994). Substances with lower redox potential than that, such as dichlorophenolindophenol, phenazine methosulphate and some others, are completely ineffective. Similar results were observed in seeds of different species and were reported by the others (Plummer, *et al.*, 2001). Some authors reported the critical redox potential for dormancy-breaking activity to be at slightly lower levels of about E_0'= + 300 mV (Estabroock and Yoder, 1998). Although the effectiveness of strong oxidants in seed germination is still not well understood, a remarkable exception from all of these findings was stimulative effectiveness of sodium nitroprusside in seed germination (Giba *et al*, 1994). In light-induced *P. tomentosa* seed germination, this low potential electron acceptor, was effective in concentrations similar to organic nitrates but with even a broader range of promotive concentrations (Fig. 3). Its structural analogue without nitrosonium ion (NO^+) as the sixth ligand, potassium hexacyanoferrate (II), which is therefore unable to release nitric oxide, expressed a negligible effect in a high concentrations around 10^{-2} M.

5.1. Sodium Nitroprusside Stimulates Seed Germination due to its NO-releasing Properties

Sodium nitroprusside (Na_2 $[Fe(CN)_5NO]$) is an official clinical drug used as direct-acting vasodilator in animal systems. It is a complex of ferrous ion (Fe^{2+}), with five cyanide anions (CN^-) and nitrosonium ion (NO^+). Because of high water solubility, nitroprusside is employed as an exogenous source of NO, thus serving as an excellent experimental tool in animal physiology.

The possibility that nitroprusside exerts its effectiveness in seed germination due to the action of released nitric oxide was tested by the application of its third analogue, carbonyl prussiate ($K_3[Fe(CN)_5CO]$) (Giba 1997a, Giba *et al.*, 1998a). This molecule contains, in the inner ligand sphere, as the sixth ligand, isoelectronic carbon monoxide (CO), instead of nitrosonium ion (NO^+). Similar to hexacyanoferrate (II), carbonyl prussiate expressed minor effect on seed germination at concentration of 10^{-2} M (Fig 3), It should be outlined that NO^+, CO, and CN^-, are isoelectronic ligands providing corresponding ferrous complexes (Fig. 4) practically the same electronic configuration. However, only sodium nitroprusside is able to release nitric oxide into the solution. Possible byproducts of nitroprusside decomposition such as ferrous iron or cyanide ions are ineffective in *P. tomentosa* seed germination (Giba *et al.*, 1998). It was obvious that effectiveness of sodium nitroprusside in these processes depends on the presence of NO^+ in the molecule, *i.e.* it is active in seed germination because of its NO releasing properties. Later, stimulative effect of sodium nitroprusside was demonstrated and confirmed in another, well known model system for investigations of light-induced seed germination. The seeds of *Lactuca sativa* c.v. Grand Rapids germinate even in darkness in the presence of 10^{-4} M sodium nitroprusside (Beligni and Lamattina, 2000).

Furthermore, in light-induced *P. tomentosa* seed germination, used as a model system, it was shown that other, very different compounds with NO-releasing properties, express the same promotive effect. Again, appropriate inactive compounds, chemical and structural analogues unable to release NO, expressed either a negligible or no effect. For example, S-nitroso acetylpenicillamine (SNAP) also potentiated germination of these seeds. The maximum of activity was reached at the concentrations between 10^{-4} M and 10^{-3} M. Control

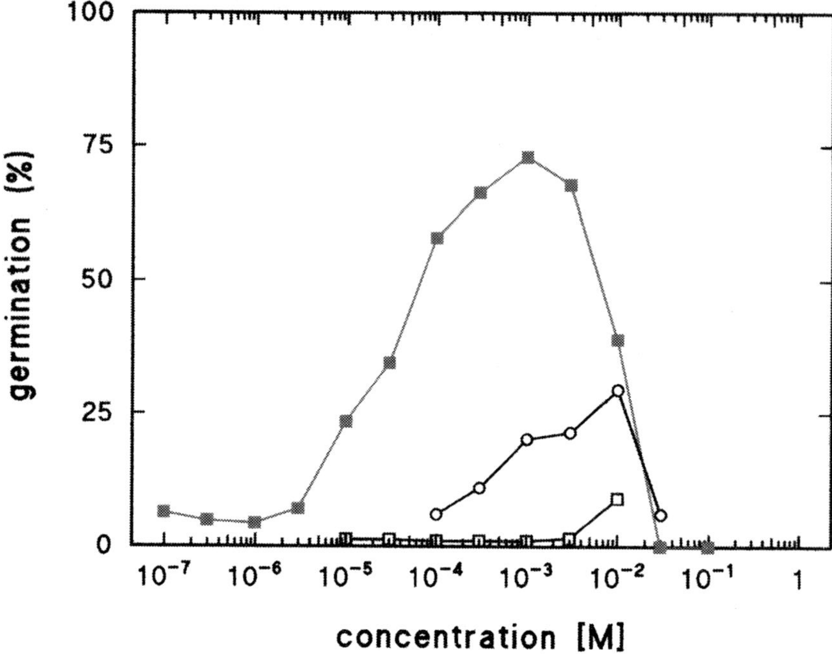

Figure 3 : The effect of sodium nitroprusside, haxacyanoferrate (II) and carbonyl prussiate on the germination of *Paulownia tomentosa* seeds induced by a pulse of red light.
Seeds were imbibed for 3 days in different concentrations of sodium nitroprusside (■), potassium haxacyanoferrate (II) (○), and potassium carbonyl prussiate (□) in darkness, irradiated with 5 min red light, and left to germinate. The germination of control samples (seeds imbibed in water) was up to 10%. Germination was scored 7 days after the red light pulse. Standard errors are not shown since they never exceeded 3%. (N=400 seeds per point) (after Giba *et al.*, 1994 and Giba *et al.*, 1988a).

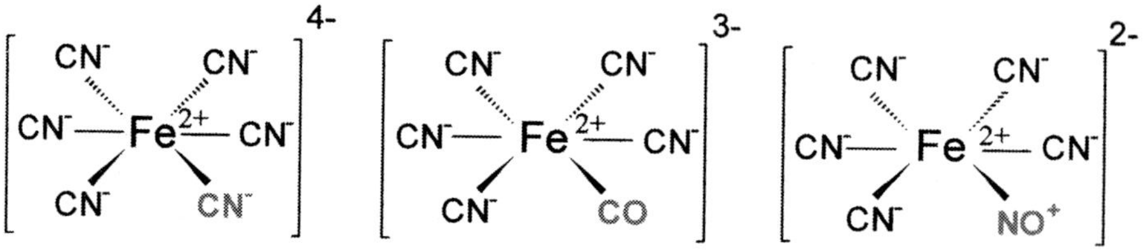

$[Fe^{2+}(CN^-)_6]^{4-}$ $[Fe^{2+}(CN^-)_5(CO)]^{3-}$ $[Fe^{2+}(CN^-)_5(NO^+)]^{2-}$

Figure 4 : The structures of ferrous complexes used.
From left to right: the anionic parts of three analogues complex ferrous compounds potassium hexacyanoferrate (II), potassium carbonyl prussiate and sodium nitroprusside.

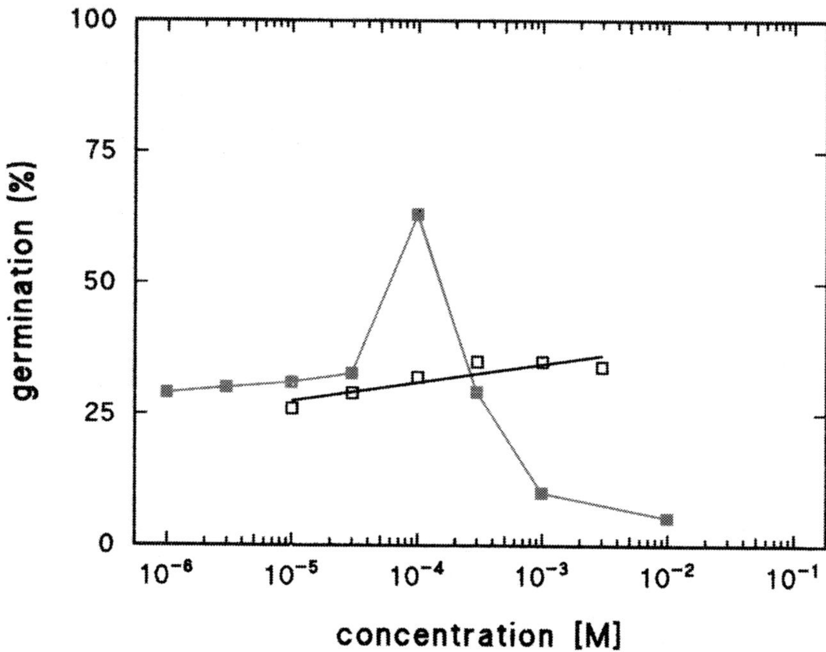

Figure 5 : The effect of SIN-1 on the germination of *Paulownia tomentosa* seeds induced by a pulse of red light. Seeds were imbibed for 3 days in different concentrations of SIN-1 (■), and molsidomine (□) in darkness, irradiated with 5 min red light, and left to germinate. The germination of control samples (seeds imbibed in water) was up to 28%. Germination was scored 7 days after the red light pulse. Standard errors are not shown since they never exceeded 3%. (N=400 seeds per point) (after Giba *et al.*, 1988a).

compound, without nitric oxide in the S-position, acetylpenicillamine, exhibited a modest effect but in almost hundred times higher concentration (not shown). 3-morpholinosydnonimine (SIN-1) was effective at concentrations around 10^{-4} M (Fig. 5). The control experiment was run with etoxycarbonyl 3-morpholinosydnonimine (molsidomine). Molsidomine (acylated sydnonimine) is stable in water solution and exerts its effect in animal systems due to the action of liver esterases transforming it to SIN-1. This compound exhibited no effect.

Obviously, chemically completely different compounds with a common property, to be NO donors under certain circumstances, act as stimulators of seed germination. Additionally, all of them differ with regard to the mechanism of NO release. For example, NO release from sodium nitroprusside occurs only after nitroprusside has undergone reduction and released a part, or all of the bound cyanides. In the case of SIN-1, sydnonimine ring opens, in pH-dependent, base-catalyzed reaction, to yield N-morpholinoiminoacetonitrile which then reacts with molecular oxygen to form unstable intermediate that spontaneously releases NO. Similar to organic nitrates, all of NO donors tested exert dose-dependent stimulation showing maximum activity at concentrations around 10^{-4} M. Taking presented results as a strong indication of NO involvement in phytochrome-controlled germination of *P. tomentosa* seeds, nitric oxide (NO) was detected directly.

6. DIRECT DETECTION OF NITRIC OXIDE IN SEED GERMINATION

Nitric oxide is a poisonous, unstable, simple (30 Da), diatomic, free radical gas with numerous functions in biological systems. Although several detection methods for its direct (chemiluminescence, electrochemical methods, photoacustic spectroscopy) or indirect (colorimetric, fluorimetric methods) were proposed and performed (for a review see Lamattina *et al.*, 2003) electron paramagnetic resonance (EPR) spectroscopy is practically the only suitable method for its direct detection in seeds.

Although it is so-called open shell structure (free radical) and possesses an unpaired electron, nitric oxide is, by itself, EPR silent. To overcome this, the EPR spin-trapping method is developed enabling stabilization of the free radical, and its detection in different biological tissues (Yoshimura *et al.*, 1996, Zweier *et al.*, 1995, Giba *et al.*, 1998a, Caro and Puntarulo, 1999). Iron complexes with dithiocarbamate derivatives such as dianion *N*-(dithiocarboxy) sarcosine, (Fe-DTCS), have a high affinity for NO; they can trap, stabilize and accumulate it. The formed nitrosyl iron adduct, NO-Fe-DTCS, is relatively stable and detectable by EPR spectrometry. Following slightly modified procedure of Tsuchiya *et al.*, (1996) it was found that well known NO donor, sodium nitroprusside, releases nitric oxide under reductive conditions. It was shown that the obtained spectrum with three line hyperfine structure (Fig 6A) and two characteristic g-values, is actually attributable to formed $[Fe^{II}(DTCS)_2(NO)]^{2-}$ complex, a product of released NO and spin

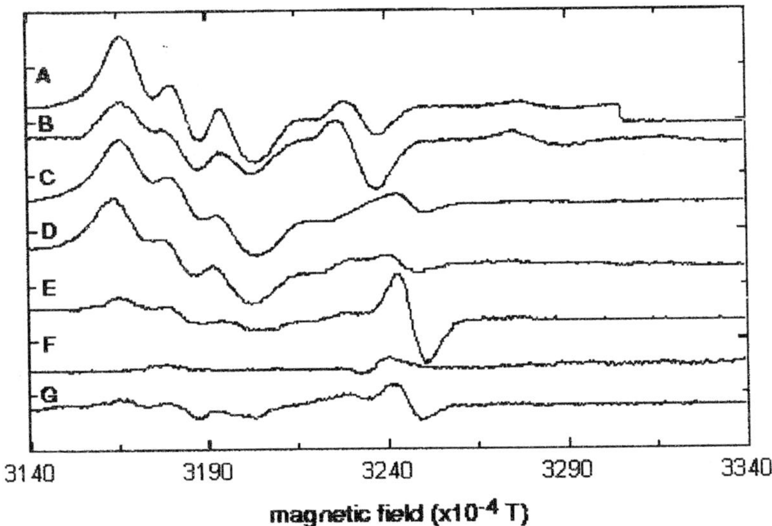

Figure 6 : EPR spectra obtained under different experimental conditions:
(A) Egg albumin + Fe(DTCS)2 + sodium nitroprusside + reductant
(B) Egg albumin + Fe(DTCS)2 + KNO_2 + reductant
(C) Seed wings + Fe(DTCS)2 + KNO_2 without reductant, acidified to pH=3
(D) Seed wings + Fe(DTCS)2 + KNO_2 without reductant, acidified to pH=2.5
(E) Seed wings + Fe(DTCS)2 + KNO_2 + reductant at pH=6
(F) Seed wings + Fe(DTCS)2 + KNO_2 without reductant at pH=6
Concentrations of reactants: [Fe^{2+}] = [DTCS] = 3.3 mM; [egg albumin] = 33mg/l; reductant [$Na_2S_2O_4$] = 2M; concentrations of substancies used as NO source: [sodium nitroprusside] = 0.3 mM; [KNO_2] = 0.1 mM
(G) Whole seeds were imbibed in darkness in 33 mM solution of DTCS for 24h. Solution was decanted and replaced by 1mM nitroglycerine solution supplemented with 33 mM $FeSO_4$. Three hours later the seeds were introduced in EPR tube and frozen.
The EPR spectra in (A), (B), (C), (D), (E) and (G) show the signal with triplet hyperfine structure (Giba *et al.*, 1988a).

trap used (for the details see Giba *et al.*, 1998a). A similar spectrum was obtained after chemical reduction of nitrite under the same experimental conditions (Fig. 6B). Furthermore, using the same reductant ($Na_2S_2O_4$), it was observed that inorganic nitrates and nitroglycerine do not show characteristic EPR spectrum, under the same experimental conditions, indicating that they could not be reduced by this reductant. This fact was important because it implicates necessity for biological reduction of both substances. Consequently, it was shown that NO could be evolved, in plant tissues, from nitrate during the enzymatic reduction by nitrate reductase (Yamasaki *et al.*, 1999). Similarly, it has been demonstrated that plant tissues could easily metabolize nitroglycerine into dinitro- and mononitro- derivatives (Goel *et al.*, 1996). Its conversion to NO is mediated by the activity of aldehide dehydrogenase in animal systems (Zhiqiang *et al.*, 2002); the fact which is not yet confirmed in plants. However, during EPR measurements it was found that nitrite could release NO spontaneously under certain experimental conditions. This reaction is strongly pH-dependent and at pH values around pH = 3, nitrite liberates NO without addition of a reductant (Fig 6C, D). The reaction could be described as dismutation of formed nitrous acid under acidic conditions (for the discussion see Van Cleemput and Samater, 1996). Unless reduced, nitrite could not liberate NO spontaneously at higher pH values (Fig. 6F). In some of the experimental procedures during EPR measurements, to avoid precipitation of trapping-complex, the dust of *P. tomentosa* seed wings was used because of its excellent dispersing agent properties, and. additionally, because of mimicking the presence of the seeds in the EPR tube. The EPR spectrum in Fig. 6G shows the result of *in vivo* detection of NO after the treatment of imbibed *P. tomentosa* seeds with nitroglycerine. A weak signal was obtained 3 hours after nitroglycerine administration to the seeds. Interestingly, this is a same (minimal) time period of incubation, sufficient for stimulative effect of nitroglycerin in *P. tomentosa* seed germination (Grubišic *et al.*, 1992).

In addition to earlier findings, subsequently conducted physiological experiments confirmed NO activity in seed germination and agreed well with EPR measurements. For example, the presence of nitrite in the incubation medium stimulates light-induced *P. tomentosa* seed germination. Maximal effect could be achieved at concentrations similar to inorganic nitrates, around 10^{-2} M. Acidification of nitrite solution, down to pH=3.5 does not contribute greatly to its effectiveness. Although the extent of germination was slightly higher, maximum of nitrite activity remains at the same concentration (Fig. 7). However, at pH equivalent to its pK (3.3), or lower, *i.e.* under conditions when nitrite releases NO spontaneously and behaves as NO donor (Fig. 5C, D), its effectiveness dramatically increases. Under these conditions, nitrite stimulated the germination in almost two orders of magnitude lower concentrations. This is

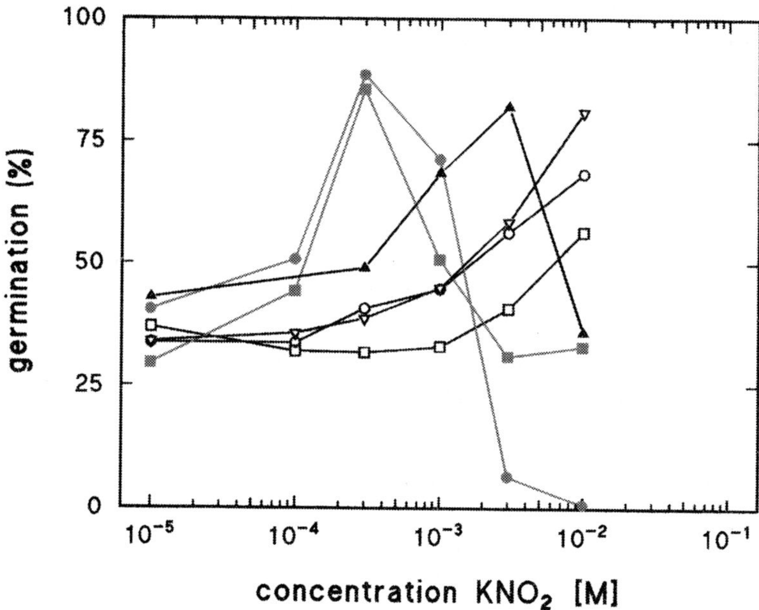

Figure 7 : The effect of potassium nitrite (KNO_2) on light-induced seed germination of *Paulownia tomentosa* seeds under different experimental conditions.
After one day of imbibition in darkness seeds were transfered to solution of potassium nitrite adjusted to pH values 6 (□), 4 (○), 3.5 (▽), 3 (■), and 2.5 (●). After two days the seeds were washed, solutions replaced by water, irradiated by 5min red light pulse, and left to germinate for an additional 7 days.
After one day of imbibition in darkness seeds were transfered to solution of potassium nitrite and immediately a few drops of $Na_2S_2O_4$ were added to give a final concentration of 50 mM (▲). After two days of incubation, the seeds were washed, solution replaced by water, irradiated by 5min of red light, and left to germinate for an additional 7 days.
The experiment was performed by using a seed sample which germinate up to 35% with 5min red light pulse. Standard errors are not shown sinve they never exceeded 3% (N=400 seeds per point) (Giba *et al.*, 1988a).

similar to the activity of other NO donors tested in the same experimental system. Slightly smaller shift in effectiveness could be obtained if the seeds are treated with potassium nitrite and reductant mixture which produces NO at pH close to neutral (Fig. 7).

Assuming NO activity in seed germination, one could explain different effects of NO_2 activity in dry seeds, and/or effects of NO_2-treated filter paper on germination (Cohn and Castle, 1984; Keeley and Fotheringham, 1997). In these cases NO is physiologically active compound. After moistening, treatments providing NO evolvement (from formed nitrous acid), such as acidity or reductive conditions led to the stimulation of germination, while treatment inhibiting NO release, such as alkalinization of the medium, produced an opposite effect. Therefore, the dormancy-breaking activity of nitric oxide undoubtedly demonstrated by Giba *et al.*, (1998a) may be the basis for the stimulative effects of NO_2 in dry and imbibed seeds (Cohn and Castle, 1984; Keeley and Fotheringham, 1997), for the effect of acidified nitrite solutions (Cohn *et al.*, 1983; Giba *et al.*, 1998a), and for the effect of different types of substances with NO-releasing properties (Grubišic *et al.*, 1991, 1992, Giba *et al.*, 1997a, 1998a).

7. SOME PHYSIOLOGICAL REFLECTIONS ON NO EFFECTIVENESS IN SEED GERMINATION

Possible effectiveness of NO as a dormancy breaking agent in seed germination was for the first time anticipated by Hendricks and Taylorson (1974). Using different types of respiratory inhibitors, such as sodium azide and potassium cyanide, they contemplatively came to the conclusion that NO might be physiologically active, acting as dormancy breaking agent in seed germination. More than one decade later, studying effects of azide in animal systems, Ferid Murad came to a similar conclusion additionally connecting NO effectives with cGMP formation. Several years later, Ignarro identified postulated EDRF (endothelium-derived relaxing factor) to be NO. Together with Robert Furchgot, because of biological significance of newly discovered physiologically active molecule, they were awarded the Nobel Prize in Physiology and Medicine for the year 1998.

The majority of NO effects in biological systems appear to be mediated through three main possibilities (Stamler, 1992). First, NO could react with molecular oxygen in aqueous solution producing nitrate and nitrite in varying proportions dependent on the presence of additional oxidizing species. Second, NO could react with superoxide radicals forming peroxynitrite anion radical, hydroperoxynitrite, and in the last instance NO_2 and hydroxyl radicals. Third, NO could react with transition metals from active centers of various metaloenzymes. In the first two cases, NO would produce its effect through formed nitrates, nitrites and peroxynitrite as a strong oxidant. Calculated on the basis of concentrations of supplied NO donors, around 10^{-4} M, the quantity of the mentioned products, would be insufficient to induce any significant effect in this experimental system. Namely, electron acceptors, as well as nitrates, are effective in seed germination in the concentrations two orders of magnitude higher than NO-releasing compounds, around 10^{-2} M (Giba *.et al*, 1994)). The third possibility is the most conceivable. A haem-containing enzyme, guanylate cyclase, is primary target for NO where NO serves as an activator of this enzyme transforming GTP to cGMP. In *P. tomentosa* seed germination, methylene blue, one of the inhibitors of soluble guanylate cyclase known from animal systems (Feelish and Noack, 1987), inhibits light-induced and NO donors-potentiated germination (Fig. 8). This inhibition could be achieved when methylene blue was applied before NO donors. If NO donors were added before methylene blue, the latter inhibited the germination at a slightly higher concentration. Similar results were obtained when SIN-1 or nitroglycerine were used instead of sodium nitroprusside. However, gibberellin-induced germination was not affected by methylene blue applied in concentrations that inhibit light-induced, and NO donors-potentiated germination (Giba *et al.*, 1998a). The fact that gibberellin completely overcomes inhibitory effect of methylene blue excludes possible side effects of this compound and leads to the conclusion that the site of methylene blue action involves earlier events of the germination process, before gibberellin action takes place. Based on this observation one can assume site of action of NO in germination process. Because of the inhibitory effect of methylene blue in NO donors-stimulated germination, the site of NO action could be only in the very early events of germination process, where, in the case of *P. tomentosa* seed germination, phytochrome pigment system is operative. Nitric oxide may affect steps in the phytochrome signal transduction pathway involving cGMP biosynthesis (Bowler *et al.*, 1994). It was shown that, upon the activation and before migrating to the nucleus (for a review see Nagy and Schäfer, 2000), phytochrome(s) interact(s) with some cytoplasmic constituents. The results of microinjection experiments suggest that phytochrome-controlled signaling starts in the cytoplasm and is mediated by heterotrimeric G-proteins, and by second messengers, Ca^{2+} and cGMP. (Bowler *et al.*,1994). Stimulation of cGMP formation by NO in plants was first reported in spruce needles, where the concentration of cGMP rapidly increased by almost four

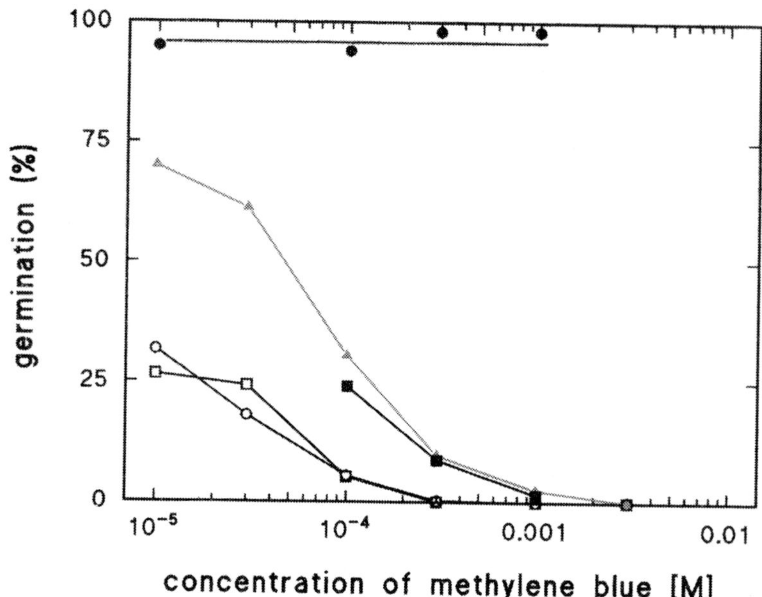

Figure 8 : The effect of methylene blue on sodium nitroprusside - potentiated and gibberellin - induced seed germination of Paulownia tomentosa.
Seeds were either irradiated with 5 min red light 3 days after the onset of imbibition or incubated in gibberellic acid from the onset of imbibition. Different concentrations of methylene blue alone were present from the beginning of imbibition until the assesment of germination (□). After 3 days of imbibition, methylene blue was replaced by 10^{-3} M sodium nitroprusside solution (○). After 3 days of imbibition 10^{-3} M sodium nitroprusside solution was replaced by different methylene blue concentrations (▲). Mixtures of different methylene blue concentrations (10^{-4} - 10^{-3} M) and and 10^{-3} M sodium nitroprusside were present throghout experimental period (■). Mixtures of different methylene blue concentrations and 10-3 M gibberellic acid were present throught experimental period without irradiation (●).
The experiment was performed by using a seed sample which germinates up to 25% with 5min red light pulse. Seeds incubated only in 10^{-3} M sodium nitroprusside for 3 days prior to irradiation germinated up to 75%. Standard errors are not shown since they never exceeded 3% (N=400 seeds per point) (Giba *et al.*, 1988a).

orders of magnitude after exposure to gaseous NO (Pfeiffer *et al.*, 1994). Rapid and transient effects of exogenously applied NO on the concentration of cGMP have been observed also in tobacco leaves and isolated cells (Durner *et al.*, 1998). Other stimuli, such as gibberellins in barley aleurone layer (Penson *et al.*, 1996), or light stimulation of bean cells (Brown and Newton, 1992), could cause transient increase in cGMP concentrations (for a review see Lamattina *et al.*, 2003). Interestingly, in spite of these findings, neither guanylate cyclase and the corresponding phosphodiesterase responsible for negative regulation, *i.e.* for cGMP degradation, nor cGMP-dependent protein kinase have been isolated and cloned in plants till present.

8. SOURCES OF NO IN SEEDS

Possible sources of NO in plant kingdom were discussed, for the first time, at European Symposium on Plant Photomorphogenesis held at Sitges (Spain) back in 1995. There was a general agreement that if there is any enzymatic endogenous NO production in plants it would not be likely mediated by animal-like NO-synthases (Giba *et al.*, 1995b, Pil-Soon Song, Mahackova - personal communication). Rather, apart from non-enzymatic routes, with regard to its enzymatic apparatus, plants would have several different opportunities to produce NO using available nitrates and nitrites already present in soil.

Although spontaneous endogenous production of nitric oxide during seed germination has been reported (Caro and Puntarulo, 1999), there is an apparent lack of evidence for the presence of NO-synthases (NOS) in plants, especially in seeds. While enzymes that transform L-arginine into citrulline and nitric oxide (Palmer *et al.*, 1988) are found in practically all kingdoms of living world (for a review, see Torreilles, 2001), they are not found in plants. Although some authors suggested such a possibility for plants (Leshem, 1996, Cueto *et al.*, 1996, Corpas *et al.*, 2001) neither the gene nor cDNA, nor any protein with a higher sequence similarity to known NOS were cloned or isolated. Additionally, there is no evidence for the presence of gene encoding NOS in the *Arabidopsis* genome. This suggests that in this species (at least) the presence of NOS is not likely. However, in plants, especially under stress conditions, nitric oxide may be produced nonenzymatically through light-mediated conversion of NO_2 by carotenoids (Cooney *et al.*, 1994), enzymatically by nitrate and nitrite reductases (Dean and Harper, 1988; Yamasaki *et al.*, 1999, Stöhr *et al.*, 2001) and possibly by other nitric oxide synthase-independent mechanisms (Kozlov *et al.*, 1999; Harrison, 2002). Examining stimulative effects of NO donors in barley seed germination Bethke *et al.*, (2003) have shown that aleurone cells have a powerful proton pumps allowing external pH to be under 3. Additionally, the reduced ascorbate is synthesized and exported by the aleurone into the apoplastic space. Because of the presence of reductant (ascorbate) and low pH, microenvironment around barley aleurone layer is ideally suited for nonenzymatic production of NO from nitrite. In fact, addition of nitrite was shown to result in an immediate increase of NO production as measured by mass spectrometry (Bethke *et al.*, 2003). These findings suggest the possibility that NO could be considered as an intercellular signaling molecule. The existence of enzymes such as plasma membrane nitrite: NO-reductase (Ni-NOR) is very important if NO is acting as intercellular messenger in plants. This enzyme is well placed to produce and release NO from the cell (Stöhr *et al.*, 2001). Interestingly, there is a clear evidence that irradiated cells in one tissue can communicate with other non-irradiated cells which than show a typical phytochrome-controlled effect (Schütz and Furuya, 2001). Moreover, irradiation even of a single cell of a *Spirodela polyrhiza* turion is sufficient to trigger germination of the whole structure, and this stimulus is chemically transduced (Tirlapur *et al.*, 1999).

On the other hand, because of the presence of nitrogen oxides in soil and/or lower troposphere, it would be reasonable to expect that plants have developed specific mechanisms to protect against elevated concentrations of nitrogen oxides rather than additional mechanisms for their production. Some authors speculated that the presence of the alternative oxidase (AOX) in plants, which decreases the probability of superoxide formation, serves additionally for such purposes (Millar and Day, 1996). When the cytochrome pathway is restricted or inhibited by nitric oxide (or cyanide ions) during terminal oxidation electron flow, operation of the alternative oxidase (which is resistant against both cyanide ions and nitric oxide) would additionally decrease the formation of the extremely strong oxidant, peroxynitrite, a product of the superoxide and nitric oxide reaction (Beckman *et al.*, 1990). Recent studies, using microarray techniques, support this statement. Huang *et al.*, (2002).demonstrated transcriptional activation of alternative oxidase (*AOX*) gene family as one of the earliest events after treatment with NO donor NOR-3 in *Arabidopsis* cell suspension cultures.

As mentioned previously, in majority of cases nitrogenous compounds stimulate germination of light-sensitive seeds. Seeds of

Sisymbrium officinale, are even not photoblastic if they do not germinate in the presence of potassium nitrate, or if endogenous concentrations of nitrates in seeds are too low (Karssen and Hilhorst, 1992). Moreover, even seeds of plant species (some members of the family *Gentianaceae, e.g. Gentiana cruciata*) that are usually not in contact with nitrate-rich soils, respond to nitrate treatments. Although these seeds germinate if potassium nitrate is added before or during the irradiation period, they are completely insensitive if nitrates are added later. The sensitivity of *G. cruciata* seeds to exogenous gibberellins is not altered (Grubišic *et al.*, 1995). Very rarely, and we are aware of only one such example, nitrogenous compounds inhibit the germination: V*accinium myrtillus* seeds, although light-sensitive, are strongly inhibited if they germinate in the presence of potassium nitrate, organic nitrates or sodium nitroprusside (Giba, 1992; Giba *et al.*, 1995a). Seeds of some other *Vaccinium* species are just nitrate-insensitive (Stushnoff and Hough, 1968).

Even if we accept that stimulative effect of inorganic nitrates in light-induced germination is NO-mediated phenomena, the nature of the underlying interaction with the phytochrome pigment system and/or some component of phytochrome transduction chain is still unclear. It was found by *in vitro* measurements that NO-generating capacity of nitrate reductase (NR) could account only for approximately 1% of the total activity (Rockel *et al.*, 2002). Based on effective concentrations of NO donors, around 10^{-4} M, one can assume that most frequently used concentrations of nitrates in seed germination, around 10^{-2} M, would be sufficient for its physiological activity *via* NO pathway. Recently, based on elegant experimentation of Shinomura *et al.* (1994, 1996), who physiologically "dissected" PHY A - and PHY B-specific induced germination of *Arabidopsis thaliana* seeds, we have obtained a slightly better insight into these mechanisms. One can show that the presence of potassium nitrate or different NO donors in incubation medium during light-induced germination of *A. thaliana* seeds predominantly affected PHY A-specific induced germination, while PHY B-specific induced germination was affected to a far lesser extent (Batak *et al.*, 2002). It is important to emphasize that at least two downstream components of NO-signaling found in animals, cGMP or ADP-ribose, are functional in plants (Durner *et al.*, 1998, see for review: Wendehenne *et al.*, 2001). Because cGMP is only known member of PHY A transduction chain (Bowler *et al.*, 1994) that is also sensitive to nitric oxide in plants (Pfeiffer *et al.*, 1994), these data again establishes a potential interrelationship of NO effectiveness, cGMP formation, and PHY A activity in seed germination.

It is believed that PHY A is the most recent phytochrome species that appeared in evolution. The appearance of new photoreceptor which is activated by light of practically all parts of visible light spectrum, several order of magnitude higher photosensitivity if compared to PHY B - controlled reactions, and some other unique features (Shinomura *et al.*, 1996), are undoubtedly evolutionary step forward. Under thin layer of soil, seeds became able to detect surface light conditions (Furuya and Schäfer, 1996). It is not surprising that additional seed-sensitivity to the presence of nitrogenous compounds in surroundings was coupled with phytochrome activity, the basic pigment system by which the seeds detect environmental changes. This gave plants, already at the seed level, a new advantage from ecophysiological point of view.

9. SOME ECOLOGICAL REFLECTIONS ON NO IN SEED GERMINATION

Gaseous nitric oxide was shown to break dormancy of *Emenanthe penduliflora* seeds (Keeley and Fotheringham, 1998). It was also reported to be active in solution in seed germination of light-induced *Paulownia tomentosa, Pinus mugo,* (Giba *et al.*, 1997a, b), *Arabidopsis thaliana* (Batak *et al.*, 2002) and *Lactuca sativa* (Beligni and Lamattina, 2000), as well in *Stellaria media* seed germination in darkness (Grubišic *et al.*, 1991: Giba *et al.*, 1992). An interesting question is why would plants develop such a system in which NO would be involved, during transduction of light signal *via* phytochrome? The answer has arisen during investigations of plant species with so-called 'fire-triggered' germination.

Inorganic nitrates have long been known to promote seed germination in many species. As already mentioned, various suggestions have been made with regard to the role of nitrates. One of the hypotheses suggested a possible ecological role for nitrate requirement in seed germination. The nitrate requirement would provide seeds with information on the nitrogen status in the soil and the proximity of already established plants (Pons, 1989). Assuming that nitrate dependence of dormant seeds allows them to detect the nitrogen status of the ecosystem before the onset of germination, this capacity was considered to be a part of survival strategy for plants (Goudey *et al.*, 1988; Hsiao and Quick, 1996; Bell *et al.*, 1999). Additionally, Thanos and Rundel (1995) suggested that increased soil nitrate content after wild-fires (Polglase *et al.*, 1986) could stimulate germination from the soil seed bank, thus triggering repair of a fire-damaged ecosystem.

However, for some species with fire-triggered germination, a simple brief heat-shock between 80 and 120 °C is sufficient to induce germination. This heat shock may enhance the germination by allowing water uptake in localized regions, such as the hilum, chalazal cap and strophiolar plug (Keeley and Fotheringham, 1998), or by causing destruction of chemical inhibitors present in the seeds (Bell *et al.*, 1993). Besides the fact that seed coat may be ruptured by influence of high temperatures, some authors suggested that removal of seed dormancy by dry heat is not due to changes in seed coat but in the embryo (Ruge, 1955). In some species, heat-shock must be combined with additional stimuli such as stratification or light (Keeley *et al.*, 1985).

Heat-shock-stimulated germination is widespread in different families, such as Fabaceae, Rhamnaceae, Convolvulaceae, Malvaceae, Cistaceae, or Sterculiaceae. It was found in different ecosystems (Ballard, 1973; Bewley and Black, 1982; Thanos *et al.*, 1992; Bell *et al.*, 1993, Pierce *et al.*, 1995). For substantial number of species, heat-shock does not affect germination. Rather, dormancy is broken by combustion products or their derivatives (Keeley, 1991). Charred wood was first shown to stimulate seed germination in the case of *Emmenanthe penduliflora* (Hydrophylaceae), a post-fire annual species (Wicklow, 1977), and it is also effective for many other species (Keeley *et al.*, 1985). On the other hand, smoke evolved during wild-fires may be the most important chemical stimulus for germination of “fire-type” species (Brown and van Staden, 1997). De Lange and Boucher (1990) were the first to report that plant-derived smoke stimulates seed germination. They showed that smoke acts as a stimulus for breaking dormancy of *Audonia capitata* (Bruniaceae). After this discovery, smoke-stimulated germination has been reported for many fire- and non-fire-dependent species of different families such as Brassicaceae, Caryophylaceae, Papaveraceae, Lamiaceae, Apiaceae, Solanaceae, Ericaceae or

Asteraceae (Brown, 1993; Baldwin *et al.*, 1994; Baxter and van Staden, 1994; Dixon *et al.*, 1995; Drewes *et al.*, 1995; Thomas and van Staden, 1995; Brown and van Staden, 1997). It was unclear whether the chemicals in charred wood, which are active in triggering the germination, are the same as those responsible for smoke-induced germination. Several studies failed to identify the active component in both cases (Keeley and Pizzorno, 1986, van Staden *et al.*, 1995). It was shown that the promotive effect of smoke is independent on seed size or shape. The promotion of seed germination in response to smoke is widespread and occurs in annual, perennial, herbaceous species (Brown *et al.*, 1995; Brown and van Staden, 1997). A smoke response was observed in species from all plant life forms (trees to herbs) and reproductive strategies (Dixon *et al.*, 1995). Smoke appears to be an almost universal signal for germination. However, in majority of cases smoke stimulates light-sensitive seeds, improving germination of both positively and negatively photoblastic seeds (Brown and van Staden, 1997).

Smoke or aqueous smoke extracts (Doherty and Cohn, 2000), alter aspects of plant growth other than seed germination. The stimulation of flowering in some *Watsonia* species and in fire-lily (*Cyrthanthus ventricosus*) (Bean, 1962; Keeley, 1993), as well as increased root growth and development in mung bean (Taylor and van Staden, 1996), have been reported. Interestingly, both effects could be affected by exogenous application of nitric oxide (NO), as well (Gouvea *et al.*, 1997; Leshem *et al*, 1998). In addition, smoke can induce seed protection against microbial attack (Parmmeter and Uhrenholdt, 1975). Generally, nitric oxide is associated with hypersensitive response in plants (Noritake *et al.*, 1996; Durner *et al.*, 1998; Delledone *et al.*, 2001).

By 1997, after extensive investigations, it was concluded that nitrogen oxides, especially NO_2, are active components of smoke responsible for smoke-stimulated germination using *Emmenanthe penduliflora* seed germination as a model system (Keeley and Fotheringham, 1997). The physiological basis of stimulative nitrogen dioxide effect could be explained by the spontaneous generation of nitric oxide (NO) by analogy to animal systems where exogenously applied NO_2 mimics typical NO-mediated effects (Zweier *et al.*, 1995; Davidson *et al.*, 1996; see also Giba *et al.*, 1998b).

However, active components of smoke are generally available in ecosystems. Nitrogen oxides, which are physiologically active in germination processes, are also members of the group of so called "soil traces gases" (Conrad, 1996). Some gases of this group, such as CO_2 and CO, could be active in seed germination (see for example: Corbineau *et al.*, 1990). N_2O or CH_4 have no significant effect on germination in contrast to NO_2 and NO which are highly effective in this process. Nitrogen oxides are important trace gases in the troposphere and soil chemistry (Gallbaly and Roy, 1978; Crutzen, 1979; O'Neil 1985). The main source of these emissions into the atmosphere is biological activity related to the soil nitrification and de-nitrification (Remde *et al.*, 1989; Conrad, 1996; Van Clemput and Samater, 1996). Nitrogen oxide emissions depend upon ammonium and nitrate content of the soil (Tornton and Valente, 1996). Concentrations and/or deposition of NO_2, for example, greatly differ between tilled soil and soil with green vegetation, and clearly display annual changes, *i.e.* seasonal variation (Nielsen *et al.*, 1996). Based on the sensitivity of different seeds to the presence of nitrogen oxides, these compounds could well be considered as signals that provide seeds with information on nitrogen status in the soil, as well as seasonal and climate changes in the surroundings (Skiba *et al.*, 1982; Christianson and Cho, 1983; Powlson *et al.*, 1988). Since the quantity of nitrogen oxides in the soil is a direct function of soil microbial activity (Tortoso and Hutchinson, 1990; Ye *et al.*, 1994; Conrad, 1996), the seeds also would be provided with additional information concerning soil quality. In addition, abiotic factors such as soil moisture, carbon content, soil aeration and granulation, the presence of different metal ions such as iron (Guerinot and Yi, 1994), or soil pH, affect nitrogen oxide emission (Blackmer and Cerrato, 1986; Slemr and Seiler 1991; Davidson, 1992; Van Cleemput and Samater, 1996; Yamulki *et al.*, 1997). Thus, the presence of nitrogen oxides in the soil, or in lower troposphere, would provide the seeds with integral information on many important factors for successful plant growth and development, including disturbance events in an ecosystem, such as wild-fires. For example, post-fire soil fluxes of nitrogen oxides could be several fold higher comparing to unburned soils (Keeley and Fotheringham, 1997). Moreover, seeds may "memorize" treatments with smoke components, *i.e.* NO_2 (Cohn and

Castle, 1984; Brown and van Staden, 1997) for several months. Nitrogen dioxide may adsorb onto the outer seed coat (or soil particles) and later, after moistening, be hydrated to nitrous and nitric acid (Cohn and Castle, 1984; Cohn, 1996). In this mixture, dismutation of nitrous acid (Blackmer and Cerrato, 1986; Van Cleemput and Samater, 1996) would lead to a rise of physiologically active nitric oxide (Giba *et al.*, 1998a). This could be a part of explanation for mechanisms of annual (Baskin and Baskin, 1985), and/or seasonal (Karssen, 1982) changes in light sensitivity, *i.e.* dormancy of buried seeds (Derkx and Karssen, 1994). In this sense, NO_2 could be considered as a stable form of NO. Although radical by itself NO_2 is in fact additionally stabilized and exists as a dimer N_2O_4. It must be outlined that the reaction of nitric oxide with atmospheric triplet oxygen is the main reason why this molecule is termed “unstable”. Chemical oxidation of nitric oxide to nitrogen dioxide is the third order reaction and its rate depends on the square of nitric oxide concentration: k_3 $[NO]^2$ $[O_2]$. Hence, oxidation of nitric oxide is significant only at a relatively high nitric oxide concentration. At low concentrations, nitric oxide could be considered as “stable”. Half-life of nitric oxide in micromolar concentrations is of the order of hours, and in nanomolar concentrations several orders of magnitude higher (Conrad, 1996).

The effectiveness of nitrogen oxide treatments in seed germination significantly depends upon duration of treatments and species (Figure 9). In majority of cases, NOx exposure must be coupled with additional stimuli, particularly light (Grubišic *et al.*, 1992; van Staden *et al.*, 1995; Giba *et al.*, 1997a, 1998a; Keeley and Fotheringham, 1998). Variations in species responses, apparently related to differential sensitivity to active compound in a number of non-fire-dependent species, would be important for a sequence of their germination, thus ensuring community structure in the ecosystems during annual changes of seasons; for example after winter in temperate climate zone, or during winter rainfall regime in a Mediterranean-type climate. From this point of view, stimulative effect of nitrogenous compounds present in the smoke could be considered as a special case, *i.e.* as a part of a general mechanism by which ecosystems regulate the timing and germination sequence of various species.

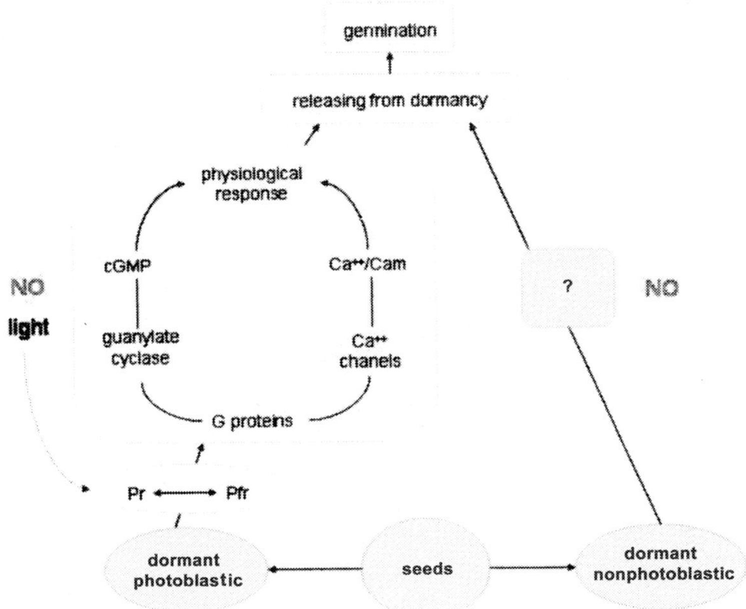

Figure 9 : The illustrative model of nitrogen oxide(s) activity in seed germination. Among the species with seeds that display the nitrogen oxide-stimulated germination, a large number are in the same time photoblastic, but others are not. This implies more than one mechanism, or more than one site of action for nitrogen oxide(s).

10. INTERACTION WITH OTHER CHEMICAL FACTORS AND SOME CONSEQUENCES

Although the role of nitrogen oxides in the germination sequence in ecosystems might be considered to be the most important (Giba *et al.*, 2003), the influence and interaction with other chemical factors should not be overlooked. For example, dormancy of red rice can be broken by certain concentrations of acidified nitrite solution or by nitrogen dioxide gas applied to imbibed or unimbibed grains. Interestingly, in the same experimental system one can show that there are another active dormancy-breaking components as shown for some commercially available liquid smoke preparations which stimulate germination (Doherty and Cohn, 2000). This is consistent with the dormancy-breaking activity of smoke obtained from pure cellulose (see for example: Baldwin *et al.*, 1994). Ethylene gas is considered by some authors as one of the primary hormones that regulates plant growth and development, including seed germination (Smith, 1999). However, some authors neglect its role from the ecological point of view (Keeley and Fotheringham, 1998). One can assume CO and CO_2 to be better candidates. The effects of phenyl propanoid species, such as coumarins, on germination may have ecological consequences in affecting the succession of plant population in a given growing area, since the coumarin derivatives exuded by one plant may inhibit the germination of another (Geissman, 1958), a phenomenon assigned as allelopathy. On the other hand, some fungal products and ferric chelates naturally present in the soil, with standard redox potential exceeding +300 mV could stimulate germination (Estabroock and Yoder, 1998). Based on their redox potential and physiological activity, this group can be extended by different true xenobiotics such as hexacyanoferrate (III) and hexachloroirridate (IV) which can stimulate light-induced germination (Giba *et al.*, 1994). However, this opens other questions concerning signaling in plants at the seed level and interaction with other environmental stimuli

Finally, it should be pointed out that regulation of germination processes in ecosystems could be easily disturbed. Namely, the effect of nitrogenous compounds as air pollutants (Brimblecombe and Stedman, 1982, Barnes and Wellburn, 1998) should be taken into account. Over 80% of nitric oxide emissions worldwide are generated by human activities. Higher levels of pollutants in lower troposphere may stimulate seed germination when other external conditions are inappropriate for seedlings growth. This process would be initiated by the activity of primary nitrogen pollutants, NO_2 and NO, collectively known as NOx, and/or by the activity of secondary generated atmospheric nitrogenous compounds such as organic nitrates, which include alkyl nitrates, aromatic nitrates and bifunctional alkyl nitrates (Schneider *et al.*, 1998). Organic nitrates are very potent stimulators of seed germination (Grubišiæ *et al.*, 1992). Pollutants could trigger the seed germination and consequently deplete the soil bank of stored seeds, with irreversible consequences for the future of populations and community structure. On the other hand, some researchers have tested the direct use of smoke and/or liquid smoke extracts applied to the soil and found them to be successful as germination triggers (*e.g.* Brown, 1994; reviewed by Brown and van Staden, 1997). The "smoke techniques" may at least in part substitute for starting localized and controlled fires, commonly performed in some countries in order to renew ecosystems.

11. CONCLUSIONS AND FUTURE PROSPECTS

It is well established that NO represents a signaling molecule in plants. Seeds are special plant organs with very specific requirements. Here we demonstrated that with regard to seeds, NO acts as an external signaling molecule which carries out an integrative information on conditions in the seeds surroundings. Clearly, additional studies are necessary to evaluate the offered hypotheses. Atmospheric chemical processes, as well as chemical and biochemical processes in the soil are far from being adequately and completely understood (see for example: Cattânio *et al.*, 2002). Although much work has already been done, our understanding of the signaling function of NO in plants, and especially in seeds, remains very limited. Deeper insights into the mechanisms underlying physiological effectiveness of reactive nitrogen and reactive oxygen species in biological systems (see for example Wentworth *et al.*, 2003) are still required.

REFERENCES

Adkins, S.W., Simpson, G.M. and Naylor, J.M. (1984) The physiological basis of seed dormancy in *Avena fatua*. IV. Alternative respiration and nitrogenous compounds. *Physiol. Plant.*, **60** : 234-238.

Appenroth, K.J., Augsten, H. and Mohr, H. (1992) Photophysiology of turion germination in *Spirodela polyrhiza* (L.) Schleiden. X. Role of nitrate in the phytochrome-mediated response. *Plant Cell Environ.*, **15** : 743-748.

Ballard, L.A.T. (1973) Physical barriers to germination. *Seed Sci. Technol.*, **1** : 285-303.

Baldwin, I.T., Suszak-Kozinski, L., and Davidson, R. (1994) Up in smoke. I. Smoke-derived germination cues for post-fire annual *Nicotiana attenuata* Torr. ex. Watson. *J. Chem. Ecol.*, **20** : 2345-2371.

Barnes, J.G. and Wellburn, A.R. (1998) Air pollutant combinations. In: *Responses of Plant Methabolism to Air Pollution and Global Change*. (Eds. De Kok, L.J. and Stulen, I.) Leiden, The Netherlands, Backhuys Publishers. pp. 147-164.

Batak, I., Devic, M., Giba, Z., Grubišic, D., Poff, K.L. and Konjevic, R. (2002) The effects of potassium nitrate and NO donors on phytochrome A- and phytochrome B-specific induced germination of *Arabidopsis thaliana* seeds. *Seed Sci. Res.*, **12** : 253-259.

Baxter, B.J.M. and Van Staden, J. (1994) Plant-derived smoke: an effective seed pretreatment. *Plant Growth Regul.*, **14** : 279-282.

Baskin, J.M. and Baskin, C.C (1985) The annual dormancy cycle in burried weed seeds: A continuum. *BioScience*, **35** : 492-498.

Bean, P.A. (1962) An enquiry into the effect of wild fires on certain geophytes. *M.Sc. Thesis*. University of Cape Town.

Beckman, J.S., Beckman, T.W., Chen, J., Marshall, P.A. and Freeman, B.A. (1990) Apparent hydroxyl radical production by peroxynitrite: Implication for endothelial injury from nitric oxide and superoxide. *Proc. Nat. Aca. Sci. USA,* **87** : 1620-1624.

Bell, D.T., Plummer, J.A. and Taylor, S.K. (1993) Seed germination ecology in southwestern Western Australia. *Bot. Rev.,* **59** : 24-73.

Bell, D.T., King, L.A. and Plummer, J.A. (1999) Ecophysiological effects of light quality and nitrate on seed germination in species from Western Australia. *Australian J. Ecol.,* **24** : 2-10.

Beligni, M.V. and Lamattina, L. (2000) Nitric oxide stimulates seed germination and de-etiolation, and inhibits hypocotyl elongation, three light-inducible responses in plants. *Planta,* **210** : 215-221.

Bethke, P., Beligni, V., Badger, M., Jacobsen, J. and Jones R. (2003) Nitric oxide as an endogenous regulator of seed physiology. *Plant Biology,* 25-30 July, Honolulu, (Hawai USA), Abs. 25004.

Bewley, J.D. (1997) Seed germination and dormancy. *Plant Cell,* **9** : 1055-1066.

Bewley, J.D. and Black, M. (1982) *Physiology and Biochemistry of Seeds in Relation to Germination* Springer-Verlag, New York, USA.

Blackmer, A.M. and Cerrato, M.E. (1986) Soil properties affecting formation of nitric oxide by chemical reaction of nitrite. *Soil Sci. Soc. American J.,* **50** : 1215-1218.

Bortwick, H.A., Hendricks, S.B., Toole, E.H., and Toole, V.K. (1954) Action of light on lettuce-seed germination. *Botanical Gazette,* **115** : 205-225.

Bowler, C., Neuhaus, G., Yamagata, H. and Chua, NH. (1994) Cyclic GMP and calcium mediate phytochrome phototransduction. *Cell,* **77** : 73-81.

Briggs, W.R. and Olney, M.A. (2001). Photoreceptors in plant photomorphogenesis to date. Five phytochromes, two cryptochromes, one phototropin, and one superchrome. *Plant Physiol.,* **125** : 85-88.

Brown, E.G. and Newton, R.P. (1992) Analytical procedures for cyclic nucleotides and their associated anzymes in plant tissue. *Phytochem. Anal.,* **3** : 1-13.

Brown, N.A.C. (1993) Promotion of germination of fynbos seeds by plant-derived smoke. *New Phytol.,* **123** : 575-584.

Brown, N.A.C. (1994) First the gas, now instant dehydrated smoke. *Veld & Flora,* **80** : 72-73.

Brown, N.A.C., Botha, P.A., and Prosh, D. (1995) Where there's smoke. *The Garden J. Royal Hort. Soc. London,* **120** : 402-405.

Brown, N.A.C., and Van Staden, J. (1997) Smoke as a germination cue: a review. *Plant Growth Regul.,* **22** : 115-124.

Brimblecombe, P. and Stedman, D.H. (1982) Historical evidence for a dramatic increase of nitrate component of acid rain. *Nature,* **298** : 460-462.

Caro, A. and Puntarulo, S. (1999) Nitric oxide generation by soybean embryonic axes. Possible effect on mitochondrial function. *Free Radical Res.,* **31** : S 205-212.

Cattanio, J.H., Davidson, E., Nepstad, D., Verchor, L. and Ackerman, I. (2002) Unexpected results of a pilot throughfall exclusion experiment on soil emissions of CO_2, CH_4, N_2O and NO in eastern Amazonia. *Biology and Fertility of Soils,* **36** : 102-108.

Cohn, M.A. (1989) Factors influencing the efficacy of dormancy-breaking chemicals. In : *Recent Advances in the Development and Germination of Seeds.* (Ed. Taylorson, R.B.) New York, Plenum. pp 261-267.

Cohn, M.A., Butera, D.L. and Hughes, L.A. (1983) Seed dormancy in red rice. III. Response to nitrite, nitrate and ammonium ions. *Plant Physiol.,* **73** : 381-384.

Cohn, M.A., and Castle, L. (1984) Dormancy in red rice. IV. Response of unimbibed and imbibing seeds to nitrogen dioxide. *Physiol. Plant.,* **60** : 552-556.

Cohn, M.A. and Hughes, J.A. (1986) Seed dormancy in red rice. V. Response to azide, hydroxylamine and cyanide. *Plant Physiol.,* **80** : 531-533.

Cohn, M.A. (1996): Chemical mechanisms of breaking seed dormancy. *Seed Sci. Res.,* **6** : 95-99.

Conrad, R. (1996) Soil microorganisms as controllers of atmospheric trace gases. (H_2, CO, CH_4, OCS, N_2O, and NO). *Microbiol. Rev.,* **60** : 609-640.

Corbineau, F., Bagniol, S., and Come D. (1990). Sunflower (*Helianthus annus* L.) seed dormancy and its regulation by ethylene. *Israel J. Bot.,* **39** : 313-325.

Corpas, F., Barosso, J. and Del Rio, A. (2001) Peroxisomes as a source of reactive oxygen species and nitric oxide signal molecules in plant cells. *Trends in Plant Sci.,* **6** : 145-150.

Cooney, R.V., Harwood, P.J., Custer, L.J. and Franke, A.A. (1994) Light-mediated conversion of nitrogen dioxide to nitric oxide by carotenoids. *Environ. Health Persp.,* **102** : 460-462.

Christianson, C.B. and Cho, C.M. (1983) Chemical denitrification of nitrite in frozen soils. *Soil Sci. Soc. America J.,* **7** : 38-42.

Crutzen, P.J. (1979) The role of NO and NO_2 in the chemistry of the troposphere and stratosphere. *Annual Review of Earth Planet Science,* **7** : 443-472.

Cueto, M., Hernandez-Perera, O., Martin, R., Bentura, M.L., Rodrigo, J., Lamas, S. and Golvano, M.P. (1996) Presence of nitric oxide synthase activity in roots and nodules of *Lupinus albus. FEBS Lett.,* **398** : 159-164.

Davidson, E.A. (1992) Sources of nitric and nitrous oxide following wetting of dry soil. *Soil Sci. Soc. America J.,* **56** : 95-102.

Davidson, C.A., Kaminski, P.M., Migdan, W., and Wollin, M.S. (1996) Nitrogen dioxide causes pulmonary arterial relaxation via thiol nitrosation and NO formation. *American J. Physiol.,* **270** : H1038-H1043.

Dean, J.V. and Harper, J.E. (1988) The conversion of nitrite to nitrogen oxide(s) by the constitutive NAD(P)H-nitrate reductase enzyme from soybean. *Plant Physiol.,* **88** : 389-395.

Dedoner, A., Rethy, R., De Petter, H., Fredericq, H., and De Greef, J. (1988). Preliminary screening experiments on the effects of light and GA_3 in the germination of different seed species. In : *Photoreceptors and Plant Development.* (Ed. De Greef, J.) Antwerpen, Univ. Press.

De Lange, J.H. and Boucher, C. (1990) Autoecological studies on *Audonia capitata* (Bruniaceae). I. Plant-derived smoke as a seed germination cue. *South African J. Bot.,* **56** : 700-703.

Delledonne, M., Zeier, J., Marocco, A. and Lamb, C. (2001) Signal interactions between nitric oxide and reactive oxygen intermediates in the plant hypersensitive disease resistance response. *Proc. Nat. Acad. Sci. USA*, **98** : 13454-13459.

Derkx, M.P.M. and Karssen, C.M. (1994) Are seasonal dormancy patterns in *Arabidopsis thaliana* regulated by changes in seed sensitivity to light, nitrate and gibberellin. *Ann. Bot.,* **73** : 129-136.

Dixon, K,W., Roche, S. and Pate, J.S. (1995) The promotive effect of smoke-derived from burnt native vegetation on seed germination of Western Australian plants. *Oecologia,* **101** : 185-192.

Doherty, L.C. and Cohn, M.A. (2000) Seed dormancy in red rice. XI. Commercial liquid smoke elicits germination. *Seed Sci. Res.,* **10** : 415-420.

Drewes, F.E., Smith, M.T. and Van Staden, J. (1995) The effect of plant-derived smoke extract on the germination of light-sensitive lettuce seed. *Plant Growth Regul.,* **16** : 205-209.

Durner, J., Wendehenne, D. and Klessig, D.F. (1998) Defense gene induction in tobacco by nitric oxide, cyclic GMP, and cyclic ADP-ribose. *Proc. Nat. Acad. Sci. USA,* **95** : 10328-10353.

Estabroock, E.M. and Yoder, J.I. (1998): Plant-plant communications: rhizosphere signalling between parasitic angiosperms and their hosts. *Plant Physiol.,* **116** : 1-7.

Feelish, M and Noack, E.A. (1987) Correlation between nitric oxide during degradation of organic nitrates and activation of guanylate cyclase. *Eur. J. Pharmacol.,* **139** : 19-30.

Furuya, M. and Schäfer, E. (1996) Photoperception and signalling of induction reactions by different phytochromes. *Trends in Plant Sciences,* **1** : 301-307.

Galbally, I.E. and Roy, C.R. (1978) Loss of fixed nitrogen from soils by nitric oxide exhalation. *Nature,* **275** : 734-735.

Geissman, T.A. (1958) The metabolism of phenylpropane derivatives in plants. In : *Encyclopedia of Plant Physiology, 10,* (Ed. W. Ruhland) Springer-Verlag. Berlin. pp 543-559.

Giba, Z. Grubišic, D. and Konjevic, R. (1992) Sodium nitroprusside - stimulated germination of common chick weed (*Stellaria media* L.) seeds. *Arch. Biol. Sci.,* **44** : 17P-18P.

Giba, Z. (1992) The effect of white light, growth regulators and temperature on the germination of blueberry (*Vaccinium myrtillus* L.) seeds. (In Serbian) *M.Sc. Thesis,* University of Belgrade.

Giba, Z., Grubišic, D., Konjevic, R. (1994) The effect of electron acceptors on the phytochrome-controlled germination of *Paulownia tomentosa* seeds. *Physiol. Plant.,* **91** : 290-294.

Giba, Z., Grubišic, D., Konjevic, R. (1995a) The involvement of phytochrome in light-induced germination of blueberry (*Vaccinium myrtillus* L.) seeds. *Seed Sci. Technol.,* **3** : 11-19.

Giba, Z., Grubišic, D., Konjevic, R. (1995b) The effect of nitric oxide- releasing compounds on the germination of *Paulownia tomentosa* seeds. *European Symposium on Photomorphogenesis in Plants,* 9th-13th July, p. 23, Sitges, (Barcelona, Spain).

Giba, Z., Grubišic, D., Sajc, L., Stojakovic, Đ, and Konjevic, R. (1997a) Effect of some nitric oxide donors and methylene blue on light-induced germination of *Paulownia tomentosa* seeds. *Arch. Biol. Sci.,* **49** : 15P-16P.

Giba, Z., Scordilis, A., Grubišiæ, D., Thanos, C.A. and Konjeviæ, R. (1997b) Control of seed germination in mountain pine. *First Balkan Botanical Congress.* 95, Thessaloniki, Greece.

Giba, Z., Grubišic, D., Todorovic, S., Sajc, L., Stojakovic, Đ, and Konjevic, R. (1998a) Effect of nitric oxide - releasing compounds on phytochrome - controlled germination of Empress tree seeds. *Plant Growth Regul.,* **26** : 175-181.

Giba, Z., Grubišic, D., Todorovic, S. and Konjevic, R. (1998b) Occurrence and regulatory roles of superoxide anion radical and nitric oxide in plants. *Iugoslavica Physiologica et Pharmacologica Acta* 34, 447-461. (also available on the web site : http://www.ippa.bg.ac.yu)

Giba, Z. (1998c) Modified phytochrome action in *Paulownia tomentosa* seed germination. *Ph.D. Thesis,* University of Belgrade.

Giba, Z., Grubišic, D., and Konjevic, R. (2003) Nitrogen oxides as environmental sensors for seeds. *Seed Sci. Res.,* **13** : 187-196.

Goudey, J.S., Saini, H.S., and Spencer, M.S. (1998) Role of nitrate in regulating germination of *Synapsis arvensis* L (wild mustard). *Plant Cell and Environ.,* **11** : 9-12.

Gouvea, C.M.C.P., Souza, J.F., Magalhaes, A.C.N., and Martins, I.S. (1997) NO-releasing substances that induce growth elongation in maize root segments. *Plant Growth Regul.*, **21** : 183-187.

Grubišic, D. and Konjevic, R. (1990) Light and nitrate interaction in phytochrome-controlled germination of *Paulownia tomentosa* seeds. *Planta,* **181** : 239-243.

Grubišic, D., Giba, Z. and Konjevic, R. (1991) Organic nitrates stimulated germination of common chick weed (*Stellaria media* L) seeds. *Arch. Biol. Sci.,* **43** : 7P-8P.

Grubišic, D., Giba, Z. and Konjevic, R. (1992) The effect of organic nitrates in phytochrome-controlled germination of *Paulownia tomentosa* seeds. *Photochem. and Photobiol.,* **56** : 629-632.

Grubišic, D., Giba, Z., and Konjevic, R. (1995) Seed germination of *Gentiana cruciata* L. *Bulletin de l'Institut et du Jardin Botaniques de l' Universite de Belgrade,* **29** : 93-100.

Goel, A., Kumar, G., Payne, G.F. and Dube, S.K. (1997) Plant cell biodegradation of a xenobiotic nitrate ester, nitroglycerin. *Nature Biotech.,* **15** : 174-177.

Guerinot, M.L. and Yi, Y. (1994) Iron: nutritious, noxious and not readily available. *Plant Physiol.,* **104** : 815-820.

Hallett, B. and Bewley, D. (2002) Membranes and seed dormancy: beyond the anaesthetic hypothesis. *Seed Sci. Res.*, **12** : 69-82.

Harrison, R. (2002) Structure and function of xanthine oxidoreductase: Where are we now? *Free Radical Biol. Med.,* **33** : 774-797.

Haas, C.J. and Scheuerlein, R. (1991) Nitrate effect on phytochrome-mediated germination in fern spores: Investigation on the mechanism of nitrate action. *J. Plant Physiol.,* **138** : 350-357.

Hendricks, S.B. and Taylorson, R.B. (1974) Promotion of seed germination by nitrate, nitrite, hydroxylamine and ammonium salts. *Plant Physiol.,* **54** : 304-309.

Henning, L., Stoddart, W.M., Dieterle, M., Whitelam, G.C. and Schäfer, E. (2002) Phytochrome E controls light-induced germination of Arabidopsis. *Plant Physiol.,* **128** : 194-200.

Hilhorst, H.W.M. and Karssen, C.M. (1989) Nitrate reductase independent stimulation of seed germination of *Sysimbrium officinale* L. (hedge mustard). *Ann. Bot.,* **63** : 131-137.

Hilton, J.R. (1985) The influence of light and potassium nitrate on the dormancy and germination of *Avena fatua* L. seed stored buried and under natural conditions. *J. Exp. Bot.,* **36** : 974-979.

Hsiao, A.I. and Quick, W.A. (1996) The roles of inorganic nitrogen salts in maintaining phytocrome- and gibberellin A_3-mediated germination control in skotodormant lettuce seeds. *J. Plant Growth Regul.,* **15** : 159-165.

Huang, X., von Rad, U. and Durner, J. (2002) Nitric oxide induces transcriptional activation of the nitric oxide – tolerant alternative oxidase in *Arabidopsis thaliana* suspension cells. *Planta,* **215** : 914-923.

Ignarro, L.J. (2002) After 130 years the molecular mechanism of action of nitroglycerin is revealed. *Proc. Nat. Acad. Sci. USA,* **99** : 7816-7817.

Karssen, C.M. (1982) Seasonal patterns of dormancy in weed seeds. In : *The Physiology and Biochemistry of Seed Development, Dormancy and Germination.* (Ed. A.A. Khan) Elsevier Biomedical Press. pp 243-270.

Karssen, C.M. and Hilhorst, H.W.M. (1992) *Seeds: The Ecology of Regeneration in Plant Communities.* (Ed. Fenner, M.) CAB International, Wallingford. pp. 327-348.

Keeley, J.E. Morton, B.A., Pedrosa, A. and Trotter, P. (1985) Role of allelopathy, heat, and charred wood in the germination of chaparral herbs and sufrutescens. *J. Ecol.,* **73** : 445-458.

Keeley, J.E. and Pizzorno, M. (1986) Charred wood stimulated germination of two fire-following herbs of the California chaparral and the role of hemicellulose. *American J. Bot.,* **73** : 1289-1297.

Keeley, J.E. (1991) Seed germination and life history syndromes in the California chaparral. *Bot. Rev.,* **57** : 81-116.

Keeley, J.E. (1993) Smoke-induced flowering in the fire-lily *Cyrtanthus ventricosus. South African J. Bot.,* **59** : 638.

Keeley, J.E. and Fotheringham, C.J. (1997) Trace gas emissions and smoke-induced seed germination. *Science,* **276** : 1248-1250.

Keeley, J.E. and Fotheringham, C.J. (1998) Smoke-induced seed germination in California chaparral. *Ecology,* **79** : 2320-2336.

Kozlov, A.V., Staniek, K. and Nohl, H. (1999) Nitrite reductase activity is a novel function of mammalian mitochondria. *FEBS Lett.,* **454** : 127-130.

Lamattina, L., Garcia-Mata, C., Graziano, M. and Pagnussat, G. (2003): Nitric oxide: the versatility of an extensive signal molecule. *Annu. Rev. Plant Biol.,* **54** : 109-136

Lehmann, E. (1909) Zur Keimungphysiologie und -biologie von *Ranunculus sclereatus* L. und einigen anderen Samen. *Berichte der Deutschen Botanischen Gesellschaft,* **27** : 476-494.

Leshem, Y.Y. (1996) Nitric oxide in biological systems. *Plant Growth Regul.,* **18** : 155-159.

Leshem, Y.Y., Wills, R.B.H. and Ku, V.V. (1998) Evidence for the function of the free radical gas - nitric oxide (NO) as an endogenous maturation and senescence regulating factor in highr plants. *Plant Physiol. Biochem.,* **36** : 825-833.

Millar, A.H. and Day, D.A. (1996) Nitric oxide inhibits the cytochrome oxidase but not alternative oxidase of plant mitochondria. *FEBS Lett.,* **398** : 155-158.

Mustilli, A.C. and Bowler, C. (1997) Tuning into the signals controlling photoregulated gene expression in plants. *The EMBO J.,* **16** : 5801-5806.

Nagy, F and Schäfer, E (2000) Control of nuclear import and phytochromes. *Curr. Opin. Plant Biol.,* **3** : 450-454.

Nielsen, T., Pilegaard, K., Egelov, A.H., Granby, K., Hummelshoj, P., Jensen, N.O. and Skov, H. (1996): Atmospheric nitrogen compounds: occurrence, composition and deposition. *Science of Total Environment,* **189/190** : 459-465.

Nikolaeva, M.G., Rasumova, M.V. and Gladkova, V.N. (1985) *Reference Book of Dormant Seed Germination.* (Ed. Danilova, F.) (in Russian) Leningrad, "Nauka" Publishers, Leningrad Branch.

Noritake, T., Kawakita, K., and Doke, N. (1996): Nitric oxide induces phytoalexin accumulation in potato tuber tissues. *Plant Cell Physiol.,* **37** : 113-116.

O'Neil, P. (1985) *Environmental Chemistry. George* Allen and Unwin, London, UK.

Parmeter, J.R. and Uhrenholdt, B. (1975) Some effects of pine needle or grass smoke on fungi. *Phytopathology,* **65** : 28-31.

Palmer, R.M.J., Ashton, D.S. and Moncada, A.S. (1988) Vascular endothelial cells synthesize nitric oxide from L-arginine. *Nature,* **333** : 664-666.

Penson, S., Schuurink, R., Fath, A., Gubler, F., Jacobsen, J. and Jones R. (1996) cGMP is required for gibberellic acid-induced gene expression in barley aleurone. *Plant Cell,* **8** : 2325-2333.

Pfeiffer, S., Janystin, B., Jessner, G., Pichorner, H. and Ebermann, R. (1994) Gaseous nitric oxide stimulates guanosine-3,5-cyclic monophosphate (cGMP) formation in spruce needles. *Phytochemistry,* **36** : 259-262.

Pierce, S.M., Esler, K. and Cowling, R.M. (1995) Smoke-induced germination of succulents (Mesembryanthemaceae) from fire-prone and fire-free habitats in South Africa. *Oecologia,* **102** : 520-522.

Plummer, J.A., Rogers, A.D., Turner, D.W. and Bell, D. T. (2001) Light, nitrogenous compounds, smoke and GA_3 break dormancy and enhance germination in the Australian everlasting daisy *Shoenia filifolia* subsp. *subuilifolia. Seed Sci. Technol.,* **29** : 321-330.

Polglase, P.J., Attiwill, P.M. and Adams, M.A. (1986) Immobilization of soil nitrogen following wildfire in two eucalypt forests of south-eastern Australia. *Acta Oecologica./Oecologia Plantarum,* **7** : 261-271.

Pons, T.L. (1989): Breaking of seed dormancy by nitrate gap detection mechanism. *Ann. Bot.,* **63** : 139-143.

Powlson, D.S., Saffigna, P.G. and Kragt-Cottaar, M. (1988) Denitrification at sub-optimal temperatures in soil from different climatic zones. *Soil Biol. Biochem.,* **20** : 719-723.

Remde, A., Slemr, F. and Conrad, R. (1989): Microbial production and uptake of nitric oxide in soil. *FEMS Microbiol. Ecol.*, **2** : 221-230.

Rockel., P., Strube, F., Rockel., A., Wildt, J. and Kaiser, W.M. (2002) Regulation of nitric oxide (NO) production by plant nitrate reductase *in vivo* and *in vitro. J. Exp. Bot.,* **53** : 103-110.

Ruge, U. (1955) Zur Analyse der Saatguterhitzung von Malvaceen. *Beiträge zur Biologie der Pflanzen,* **31** : 409-417.

Schneider, M., Luxenhofer, O., Dessler, A. and Balschmiter, K. (1998) C_1-C_{15} alkyl nitrates, benzyl nitrate, and bifunctional nitrates: Measurement in California and south Atlantic air and global comparison using C_2Cl_4 and $CHBr_3$ as marker molecules. *Environ. Sci. Technol.,* **32** : 3055-3062.

Shinomura, T., Nagatani, A., Chory, J. and Furuya, M. (1994) The induction of seed germination in *Arabidopsis thaliana* is regulated principally by phytochrome B and secondarily by phytochrome A. *Plant Physiol.,* **104** : 363-371.

Shinomura, T., Nagatani, A., Hanzawa, H., Kubota, M., Watanabe, M. and Furuya, M. (1996) Action spectra for phytochrome A- and B-specific photoinduction of seed germination in *Arabidopsis thaliana. Proc. Natl. Acad. Sci. USA,* **93** : 8129-8133.

Schütz, I. and Furuya, M. (2001) Evidence for type II phytochrome-induced rapid signalling leading to *cab:luciferase* gene expression in tobacco cotyledones. *Planta,* **212** : 759-764.

Skiba, U., Hargreaves, K.J., Fowler, D. and Smith, K.A. (1992) Fluxes of nitric and nitrous oxides from agricultural soils in a cool temperate climate. *Atmospheric Environ.,* **2A** : 2477-2488.

Slemr, F. and Seiler, W. (1991) Field study of environmental variables controlling the NO emission from soil and NO compensation point. *J. Geophysical Res.,* **9** : 13017-13031.

Smith, H.B. (1999) Constructing signal transduction pathways in Arabidopsis. *Plant Cell,* **11** : 1-3.

Stamler, J.S., Singel, D.J. and Loscalzo, J. (1992) Biochemistry of nitric oxide and its redox-related forms. *Science,* **258** : 1898-1902.

Stöhr, C., Strube, F., Marx, G., Ulrich, W. and Rockel, P. (2001) A plasma-membrane bound enzyme of tobacco roots catalyses the formation of nitric oxide from nitrite. *Planta,* **212** : 835-841.

Stushnoff, C. and Hough, L.F. (1968) Response of blueberry seed germination to temperature, light, potassium nitrate and coumarin. *American Soc. Hort. Sci.,* **93** : 260-266.

Roberts, E.H. (1973) Oxidative processes and the control of seed germination. In : *Seed Ecology* (Ed. W. Heydecker) Butterworths. London. pp. 189-231.

Taylor, J.L.S. and Van Staden, J. (1996): Root initiation in *Vigna radiata* (L). Wilczek hypocotyl cuttings is stimulated by smoke-derived extracts. *Plant Growth Regul.,* **18** : 165-168.

Thanos, C.A. and Mitracos, K. (1979): Phytochrome-mediated germination control of maize caryopses. *Planta,* **146** : 415-417.

Thanos, C.A., Gheorgiu, K., Kadis, C. and Pantazi, C. (1992) Cistaceae: a plant family with hard seeds. *Israel J. Bot.,* **41** : 1-263.

Thanos, C.A. and Rundel, P.W. (1995) Fire-followers in chaparral: nitrogenous compounds trigger seed germination. *J. Ecol.,* **83** : 207-216.

Tirlapur, U.K. Dahse, I., Probandt, R. Fischer, W. and Appenroth, K.J. (1999) Phytochrome-induced transmissible signal elicits germination response in turions of *Spirodela polirhyza. Physiol. Plant.,* **105** : 539-545.

Thomas, T.H. and Van Staden, J. (1995) Dormancy break of celery (*Apium graveolens* L.) seeds by plant-derived smoke extract. *Plant Growth Regul.,* **17** : 195-198.

Toole, E.H., Toole, V.K., Bortwick, H.A. and Hendricks, S.B. (1955) Photocontrol of *Lepidium* seed germination. *Plant Physiol.,* **30** : 15-21.

Tornton, F.C. and Valente, R.J. (1996) Soil emissions of nitric oxide and nitrous oxide from no-till corn. *Soil Sci. Soc. American J.,* **60** : 1127-1133.

Torreilles, J. (2001) Nitric oxide: one of the more conserved and widespread signaling molecules. *Frontiers in Bioscience,* **6** : d1161-1172.

Tortoso, A.C. and Hutchinson, C.L. (1990) Contributions of autotrophic and heterotrophic nitrifiers to spill nitrogen oxide and nitrogen dioxide emissions. *Applied Environ. Microbiol.,* **56** : 1799-1805.

Toyomasu, T., Kawaide, H., Mitsuhashi, W., Inoue, Y. and Kamiya, Y. (1998) Phytochrome regulates gibberellin biosynthesis during germination of photoblastic lettuce seeds. *Plant Physiol.,* **118** : 1517-1523.

Tsuchiya, K., Takasugi, M., Minakuchi, K. and Fukuzawa, K. (1996) Sensitive quantitation of nitric oxide by EPR spectroscopy, *Free Radical Biol. and Med.,* **21** : 733-737.

Van Cleemput, O. and Samater, A.H. (1996) Nitrite in soils: accumulation and role in the formation of gaseous N compounds. *Fertilizer Research,* **45** : 81-89.

Van Staden, J., Drewes, F.E. and Jäger, A.K. (1995a) The search for germination stimulants in plant-derived smoke extracts. *South African J. Bot.,* **61** : 264-269.

Van Staden, J., Jäger, A.K. and Strydom, A. (1995b) Interaction between plant-derived smoke extract, light and phytohormones on the germination of light-sensitive lettuce seeds. *Plant Growth Regul.,* **17** : 213-218.

Vujièic, R., Grubišic, D. and Konjevic, R. (1993) Scanning electron microscopy of the seed coat in the genus *Paulownia* (Scrophulariaceae). *Bot. J. Linnean Society,* **111** : 505-511.

Wendehenne, D., Pugin, A., Klessig, D.F., and Durner, J. (2001) Nitric oxide: comparative synthesis and signaling in animal and plant cells. *Trends in Plants Sci.,* **6** : 177-183.

Wentworth, P., Wentworth, A., Zhu, X., Wilson, I., Janda, K., Eschenmoser, A. and Lerner, R. (2003) Evidence for the production of trioxygen species during antibody-catalyzed chemical modification of antigens. *Proc. Nat. Acad. Sci. USA,* **100** : 1490-1493.

Wicklow, D.T. (1977) Germination response on *Emenanthe penduliflora* (Hydrophylaceae). *Ecology,* **58** : 201-205.

Yamasaki, H., Sakihama, Y. and Takahashi, S. (1999) An alternative pathway for nitric oxide production in plants: new features of an old enzyme. *Trends in Plant Sci.,* **4** : 128-129.

Yamulki, S., Harrison, R.M., Goulding, K.W.T. and Webster, C.P. (1997): N_2O, NO and NO_2 fluxes from a grassland: effect of soil pH. *Soil Biol. Biochem.,* **29** : 1199-1208.

Ye, R.W., Averill, B.A. and Tiedje, J.M. (1994) Denitrification: production and consumption of nitric oxide. *Applied Environ. Microbiol.,* **60** : 1053-1058.

Yoshimura, T., Yokoyama, H., Fujii, S., Takayama, F., Oikawa, K. and Kamada, H. (1996). *In vivo* EPR detection and imaging of endogenous nitric oxide in lipopolysaccharide-treated mice. *Nature Biotech.,* **14** : 992-994.

Zhiqiang, C., Zhang, J. and Stamler, J.S. (2002) Identification of the enzymatic mechanism of nitroglycerin bioactivation. *Proc. Natl. Acad. Sci. USA,* **99** : 8306-8311.

Zweier, J.L., Penghai, W., Samuilov, A. and Kuppusamy, P. (1995) Enzyme-independent formation of nitric oxide in biological tissues. *Nature Medicine,* **1** : 804-808.

Chapter 13

NITRIC OXIDE, PEROXIDASE AND LIGNIFICATION IN HIGHER PLANTS

A. Ros Barceló*, C. Gabaldón and F. Pomar

Department of Plant Biology, University of Murcia, E-30100 Murcia, Spain
*Corresponding author : E-mail : rosbarce@um.es

Summary

Lignins are complex cell wall phenolic heteropolymers which result from the peroxidase-mediated oxidative polymerization of three p-hydroxycinnamyl alcohols, p-coumaryl, coniferyl and sinapyl alcohol, giving rise within the lignin polymer to H (p-hydroxyphenyl), G (guaiacyl), and S (syringyl) units, respectively. Lignins are mainly localized in the impermeable water transport conduits of the xylem and phloem fibers, vascular tissues which are also capable for sustaining NO production. In fact, when NO production was estimated in both vascular bundles and tracheary elements from Zinnia elegans *by confocal laser scanning microscopy, using 4,5-diaminofluorescein diacetate as fluorescent probe, NO was mainly located in primary and secondary xylem cell walls, as well as in phloem fibers, regardless of the cell differentiation status. However, there was evidence for an NO gradient related to the degree of xylem differentiation, so that the capability for NO production and cell wall lignification are apparently two inversely related metabolic processes. These results are not surprising since all the branching (also rate limiting) enzymes of the lignin biosynthetic pathway (cinnamate-4-hydroxylase, p-coumarate-3-hydroxylase and*

In : Nitric Oxide Signaling in Higher Plants, 2004
(Eds Jose R. Magalhaes, Rana P. Singh and Leonidas P. Passos)
Studium Press, LLC, Houston, USA, pp 277-308

coniferylaldehyde-5-hydroxylase) and the lignin-assembling enzyme (peroxidase) are hemeproteins and, therefore, possible targets of NO action. This is especially important since any possible metabolic control of NO on these enzymes would enable it to regulate not only the global p-hydroxycinnamyl alcohol pools in lignifying plant cells and the H/G/S ratio for carbon partitioning, but also their rates of polymerization. In the case of peroxidase, there is also evidence for a lot of additional cross-talks with NO, so that these cross-talks should be considered far from being fortuitous. In fact, NO may be regarded as a peroxidase substrate, and so peroxidase could play any role in NO (and $ONOO^-$) detoxification in lignifying plant tissues. That this situation is rather complex is illustrated, finally, by the participation of peroxidase in bio-mimetic NO-synthesizing pathways, which are similar to that catalyzed by the NO-forming enzyme, NO synthase.

Key Words : Cytochrome P450 mono-oxygenases, Lignification, Nitric oxide, Nitric oxide detoxification, Peroxidase.

1. LIGNINS

Lignins (from the Latin *lignum* : wood) are complex cell wall phenolic heteropolymers covalently associated with both polysaccharides and proteins. They are mainly localized in the impermeable water transport conduits of the xylem and other supporting tissues, and result from the oxidative polymerization of three *p*-hydroxycinnamyl alcohols (monolignols) in a reaction mediated by peroxidases (Ros Barceló, 1997), leading to an optically inactive hydrophobic heteropolymer (Ralph *et al.*, 1999). The process of sealing plant cell walls through lignin deposition is known as lignification, and provides mechanical strength to the stems, protecting cellulose fibers from chemical and biological degradation (Grabber *et al.*, 1988). In xylem vessels, lignins contribute to the resistance of the tensile forces of the water columns they contain, also imparting water impermeability. However, the impact of lignin quantity and quality on the chemical, physical, and mechanical properties of the cell wall remains uncertain.

Lignins represent the second most abundant organic compound on the earth's surface after cellulose, and account for about 25% of the plant biomass (Higuchi, 1990). They are found specifically in vascular plants (tracheophyta), and occur selectively in greatest quantity in the secondary cell walls of particular cells, which form parts of woody tissues, such as fibers, xylem vessels, and tracheids. Lignins have been identified in pteridophytes, widely considered to be the first vascular plants, and are likely to have played a key role in the colonization of the terrestrial landscape by plants in the early Devonian period, 400 to 450 million years ago (Lewis and Davin, 1994). Lignins therefore occur in cell walls of true vascular plants, ferns and, probably, in club mosses, although in the last case there is some controversy (Lewis and Yamamoto, 1990). However, they are absent from those mosses which have no tracheids and, of course, from algae (Lewis and Yamamoto, 1990). Thus, and from the botanical standpoint, the phenomenon of lignification is essentially associated with the development of the vascular body in the plant.

Lignins are three-dimensional and amorphous heteropolymers which result from the oxidative coupling of the three *p*-hydroxycinnamyl alcohols (Fig. 1), *p*-coumaryl (1), coniferyl (2), and sinapyl (3) alcohol, giving rise within the lignin polymer to H (*p*-hydroxyphenyl), G (guaiacyl), and S (syringyl) units, respectively. Lignins exhibit a high degree of structural variability, which depends not only on the species and tissue but also on the cell type. This heterogeneity concerns the relative proportion of the three constituent monomers, the different types of interunit linkages, and the occurrence of non-conventional phenolic units within the polymer (Lapierre *et al.*, 1995). These phenylpropane units are interconnected in lignins by a series of ether and carbon-carbon linkages, in various bonding patterns (Higuchi, 1990), leading to the following main sub-structures (Fig. 2): guaiacylglycerol-β-aryl ether (1), phenylcoumaran

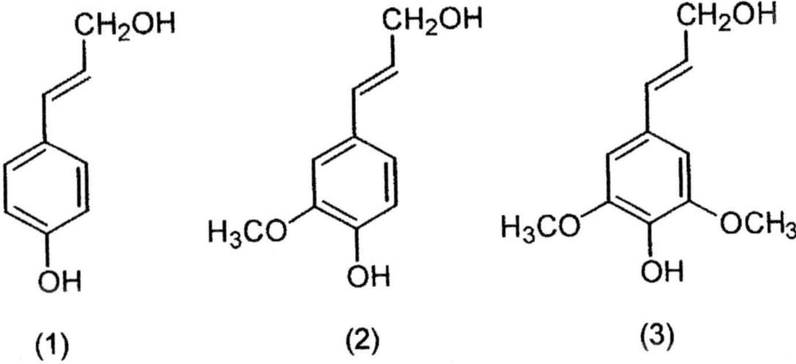

Figure 1 : Structures of *p*-coumaryl (1), coniferyl (2) and sinapyl (3) alcohols.

(2), diarylpropane (3), resinol (4), biphenyl (5), and diphenyl ether (6), as well as other sub-structures of minor importance.

The most frequent β-*O*-4 inter-unit bonds are present in guaiacylglycerol-β-aryl ether substructures (1, Fig. 2), and are the targets of most lignin depolymerization processes. In contrast, other bonds, such as β-5 (in phenylcoumaran), β-1 (in diarylpropane), β-β (in resinol), 5-5 (in biphenyl) and 5-*O*-4 (in diphenyl ether) interunit bonds, are very resistant to degradation. Besides these main inter-unit linkages, there are minor ones, such as the β-6 bonds of phenylisochroman structures and the noncyclic benzyl ether bonds, α-*O*-4 (Lapierre *et al.*, 1995).

The fact that the three monomer constituents of lignins differ as to the extent of their methoxylation (Fig. 1) suggests that a variety of substructures may be formed during polymerization. Thus, at the chemical level, lignins are ill-defined polymers whose monomeric composition is strongly variable, as is the nature of their inter-unit linkages. This means that the expression “lignins” is preferable to

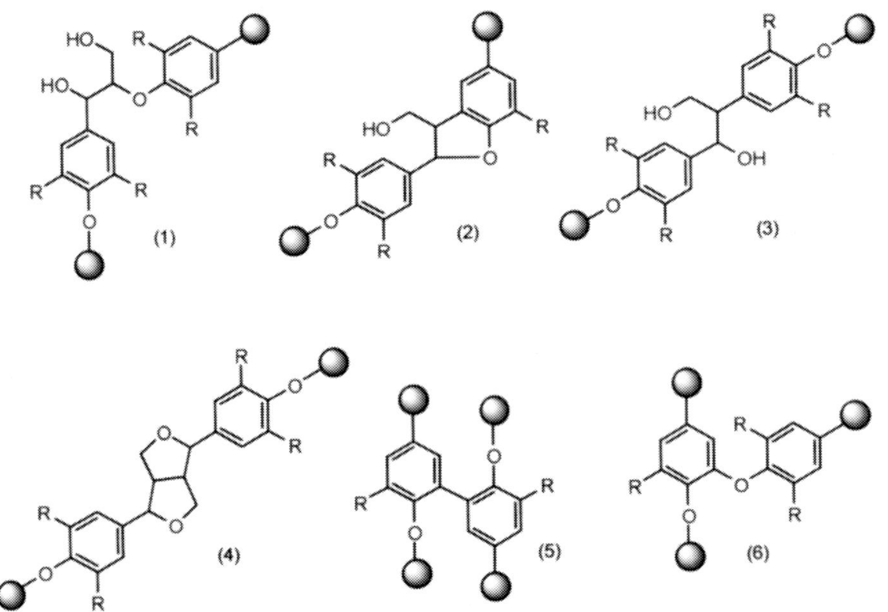

Figure 2 : Principal bonding patterns between phenylpropanoid units in native lignins: guaiacylglycerol-β-aryl ether (1), phenylcoumaran (2), diarylpropane (3), resinol (4), biphenyl (5) and diphenyl ether (6). R = H or OCH_3.

the use of "lignin," since a great diversity of chemical structures probably exists within natural lignins (Boudet *et al.*, 1995).

By far, the greatest proportion of lignins found in vascular plants is deposited in the cell walls of a limited number of cell types, such as tracheids, vessel elements, xylem and phloem fibers, and sclereids, the precise nature of such lignins differing according to cell type. For example, guaiacyl lignins predominate in the xylem of *Arabidopsis* stems, while the adjacent, heavily lignified sclerenchyma cells contain mainly syringyl lignins (Chapple *et al.*, 1992).

Lignins may also vary within a given cell wall. In fact, lignin heterogeneity is regulated during secondary cell wall deposition to form layers of lignins that can differ both in the amount and in the average composition of the monomers. Thus, in xylem cell walls from conifers (Donaldson, 1985), the lignin concentration in the secondary thickening is 16-27%, while the cell corner and middle lamella have a lignin content of 38-88%. Furthermore, lignins deposited in the middle lamella and the cell corners are rich in H units, while G lignins, which are the predominant type, are deposited in both the middle lamella and the secondary thickening (Terashima and Fukushima, 1989).

The most distinctive variation in lignin content and lignin monomer composition in vascular plants is that found between the two main subphyla of spermatophyta. Thus, in gymnosperms, lignins are typically composed of G units, with a minor proportion of H units, while in angiosperms lignins are mainly composed of G-S units (Higuchi, 1990). However, caution should be exercised when attempting to define lignin composition as a function of taxonomy, since there are some gymnosperms in which S moieties predominate, and some angiosperms in which lignins are principally of the G type (Lewis and Yamamoto, 1990).

In grasses, lignins are more complex (Ralph *et al.*, 1994), since they contain additionally significant amounts of esterified *p*-coumaric acid. These data suggest the existence of possible ester and ether linkages of p-coumaric acid in gramineae lignins, and has led some authors (Billa *et al.*, 1996) to propose the occurrence in straw of lignins with a G-S core, analogous to standard angiosperm lignins, onto which H derivative units are grafted in structures which are characterized by their differing susceptibility to degradation. Thus, during the course of evolution, the chemical complexity of lignins has increased from ancient pteridophytes and gymnosperms to the most evolved grasses. All these structural features of native lignins point to a highly controlled polymerization process, which would apparently give rise to a structurally ordered polymer, predominantly composed of lineal β-*O*-4 linked fragments, and with a non-random distribution of the H, G, and S building blocks (Lapierre *et al.*, 1995).

2. NO IS SYNTHESIZED BY THE LIGNIFYING XYLEM

2.1. Lignifying Vascular Bundles

One of the most attractive models for studying the time-course of cell wall lignification is to follow the pattern of vascular differentation in stems (Shininger, 1979). In these organs, vascular development involves, in the first instance, the formation of procambium and, second, the overt cytodifferentiation of these procambial cells into lignifying xylem elements (vessels and fibers) and phloem elements. In axillary shoots, procambium and xylem elements are thought to differentiate acropetally, while the direction is basipetal in adventitious buds (Shininger, 1979). Likewise, in stems, the differentiation of lignifying tracheary elements (TEs) always follows a continuous centripetal sequence, unlike phloem elements, which follow a centrifugal one. Therefore, although not geometrically simple organs, stems show progressive stages of xylem element differentiation and lignin deposition in a true linear sequence. That is, TEs which differ as to the extent of cell wall lignification are both arranged vertically from the meristem to the mature stem region and horizontally from the procambium to the oldest xylem region.

Both xylem and phloem lignifying elements are capable of sustaining NO production, as is revealed by the use of NO-fluorescent indicators. NO-fluorescent indicators, based on the fluorescein chromophore, allow real-time biological imaging of NO. One of these indicators, 4,5-diaminofluorescein (DAF) (Foissner *et al.*, 2000), is in itself non-fluorescent, but reacts with NO in the presence of oxygen to form the highly fluorescent triazolofluorescein (DAF-2T) which has excitation and emission maxima at 495 and 515, respectively. The membrane-permeable form of the dye, its diacetate ester derivative, may be taken up by the cells and hydrolyzed by cellular esterases to again form the membrane-impermeable compound, DAF.

When NO production was estimated by confocal laser scanning microscopy and 4,5-diaminofluorescein diacetate (DAF-DA) as fluorescent probe in mature vascular bundles from stems of *Zinnia elegans*, PTIO [2-phenyl-4,4,5,5-tetramethylimidazol inone-1-oxyl-3-oxide]-sensitive green fluorescence (PSGF) was mainly located in primary and secondary xylem (X) cell walls, as well as in phloem fibers (P), regardless of the cell differentiation status (Fig. 3A). However, there was evidence for an NO gradient related to the degree of differentiation, since cell walls from young xylem cells (large arrows) clearly manifiest a higher capacity for NO production than cell walls from old xylem cells (minor arrows). This gradient for NO production is inversely related to the lignification gradient that shows these xylem cells, as one may expect from their differentiation status, which is determined by their radial position. These results suggest that the capability for NO production and cell wall lignification are two inversely related metabolic processes during xylem differentiation. In fact, a burst of NO production may be observed in differentiating (non-collapsed) young xylem cells (Fig. 3B, arrow), which suggests a strong relationship between NO production and the phenomenon of programmed cell death (PCD), which occurs during xylem differentiation, and which precedes to the phenomenon of cell wall lignification.

2.2. Limits of the Boundary Layer of NO Diffusion (action)

NO is a gaseous and reactive free radical, which has a relatively short half-life of a few seconds. Thus, NO rapidly reacts with O_2 to yield NO_2 ($2\ NO + O_2 \rightarrow 2\ NO_2$) in a complex reaction that is second order in NO ($k\ [NO]^2\ [O_2]$); therefore, the NO half-life depends critically on both the initial NO concentration and the O_2 tension in the air phase (Stamler *et al.*, 1992). NO_2 can further react with NO to yield other nitrogen oxides, such N_2O_3 ($NO + NO_2 \rightarrow N_2O_3$), which rapidly decays in aqueous solutions to lead to nitrite and nitrate. This raises the question of whether the NO produced by lignifying tissues (Fig. 3) is involved in signalling between neighbouring plant organs, neighbouring plant tissues, or neighbouring plant cells. The answer to this question is contained in Fick's second law :

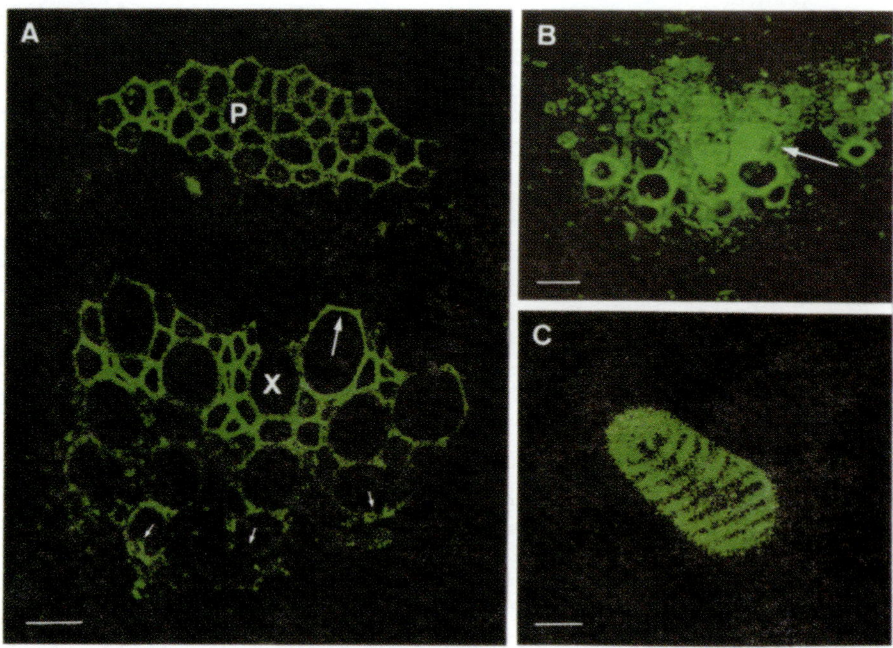

Figure 3 : Image of NO production (green fluorescence) by mature (A) and young (B) *Z. elegans* vascular bundles, and differentiated *Z. elegans* tracheary elements (C) after staining with DAF-DA, as viewed by confocal laser scanning microscopy. P, phloem fibers; X, xylem. Bars = 20 μm (in A), 30 μm (in B), and 15 μm (in C).

$$\chi^2 = 4\ D_{NO}\ \zeta_{1/2}$$

From this equation, it can be established that the length of the boundary layer of NO diffusion (action) (χ) may be as long as 10,000 μm, if we assume that the diffusion coefficient of NO (D_{NO}) is ~ $1 \cdot 10^{-5}\ m^2\ s^{-1}$ (like other gases in air) and its half-life in an aerobic atmosphere ($\varsigma_{1/2}$) ~ 2.5 s. These calculations suggest that NO production by the lignifying xylem, if NO diffuses through the air spaces, may also affect the neighbouring (non lignifying) tissues, and therefore the cell targets of NO may be located beyond the xylem and phloem.

This situation varies if we consider that NO diffusion take place through the aqueous layer (as is probable). In this case, and assuming that the diffusion coefficient of NO (D_{NO}) is ~ $1 \cdot 10^{-9}\ m^2\ s^{-1}$ (like other gases in water), the length of the boundary layer of NO action (χ) would be as short as 100 μm. In this case, the targets for NO action would be exclusively the xylem (or phloem) cells in which NO is generated.

2.3. Lignifying Plant Cell Cultures

The biosynthesis of lignins is developmentally activated in specific tissues and cell types, such as xylem elements during vascular development (see above), but can also be activated in cells that normally do not accumulate lignins, such as leaf mesophyll cells, in response to wounding and hormonal factors (Fukuda, 1997). Thus, lignin genes which are transcriptionally silenced in certain cell types may be activated in response to environmental factors, so that this system constitutes a special and valuable model for studying the nature and time-course of lignin-gene transcription (Demura *et al.*, 2002).

Studies on inducible lignification in plant cell cultures have provided a strong support in the knowledge of lignin biosynthesis. Thus, cultured plant cells can be induced by a variety of stimuli to proceed through temporally controlled metabolic and developmental programs that conclude in the formation of single-cell-derived vascular TEs. The best characterized cell culture system used for studying lignification is that derived from isolated *Z. elegans* mesophyll cells. *Zinnia* mesophyll cells can be induced to differentiate into cells resembling TEs in a medium containing suitable levels of auxin and cytokinin (Fukuda, 1997), and have been considered as a suitable model for studying the sequence of events which take place during xylem differentiation, including PCD and secondary cell wall synthesis (Fukuda, 1997; Milloni *et al.*, 2002). This is due to the observation that the differentiation of mesophyll cells into TEs is synchronously induced in a large number of cells in a relatively homogeneous cell population system. This makes it possible to study the biochemistry and physiology of xylogenesis without the complexity which heterogeneous plant tissues impose. TE differentiation in plant cell cultures, especially in those derived from mesophyll cells, represents a unique and interesting model system, in which secondary cell wall deposition, lignin biosynthesis and PCD may be studied (Fukuda, 1996), and where the interrelationship among these unique cellular events may be explored.

As regards NO production, *trans*-differentiating *Zinnia* mesophyll cells maintain the same property shown by lignifying vascular bundles. Thus, when NO production was again estimated in TEs derived from *Zinnia* mesophyll cell cultures by confocal laser scanning microscopy using DAF-DA as fluorescent probe, PSGF was found mainly confined to differentiating (secondary cell wall forming) TEs (Fig. 3C), although both undifferentiating mesophyll cells and differentiating mesophyll cells undergoing PCD (as tested by plasmolysis) also showed a capacity for NO production. In fact, cytoplasm-disorganised mesophyll cells, which are beginning to experience PCD, show a burst of NO production which is sustained while secondary cell wall synthesis is in progress (manuscript in preparation). These results confirm that NO formation by TEs is not only related to the early processes of PCD, which occurs during xylem differentiation from cambial derivatives, but also to the late process of secondary cell wall formation, which precedes to monolignol assembly.

3. THE LIGNIN BIOSYNTHETIC PATHWAY

The biosynthesis of lignins proceeds through a long sequence of reactions that involve (i) the shikimate pathway, which provides phenylalanine and tyrosine, (ii) the common phenylpropanoid pathway from phenylalanine (and/or tyrosine) to the *p*-hydroxycinnamoyl CoAs, and (iii) the lignin-specific pathway, which channels the *p*-hydroxycinnamoyl CoAs toward the synthesis of *p*-hydroxycinnamyl alcohols, and their later polymerization in cell walls (Whetten and Sederoff, 1995). Plants are the only living organisms capable of channeling carbon from the primary metabolism toward lignin biosynthesis and, today, it is widely accepted that the evolutionary acquisition of the phenylpropanoid pathway has played a key role in the ability of plants to colonize land, not only because one of its products (the lignins) serves to strengthen the aerial organs of the plant, but also because other products of the pathway (the flavonoids) act as protectors against dangerous UV radiation. To integrate this novel evolutionary pathway and the general aromatic amino acid biosynthesis pathway in an efficient metabolic highway in vascular plants, the activities of the enzymes of the shikimate pathway are closely coordinated with the activities of the enzymes of the phenylpropanoid pathway (Whetten and Sederoff, 1995).

The entry point into the shikimate pathway from both the glycolytic (phosphoenol piruvate) and the pentose phosphate pathway (erythrose-

4-phosphate) intermediates is catalyzed by the enzyme, 3-deoxy-D-arabino-heptulosonate-7-phosphate synthase (EC 4.1.2.15). This pathway leads, via chrorismate, to the biosynthesis of the amino acids, phenylalanine, tyrosine, and tryptophan (Herrmann, 1995).

The enzymes involved in the phenylpropanoid pathway (Fig. 4) are phenylalanine ammonia-lyase (PAL, EC 4.3.1.5), cinnamate-4-hydroxylase [C4H, trans-cinnamate, NADH: oxygen oxidoreductase (4-hydroxylating), EC 1.14.13.11], *p*-coumarate-3-hydroxylase (C3H), specific *S*-adenosyl-L-methionine-dependent *O*-methyltransferases, such as the caffeoyl-CoA-*O*-methyl-transferase (CCoAOMT, *S*-adenosyl-L-methionine: caffeoyl-CoA 3-*O*-methyltransferase, EC 2.1.1.104), and *p*-hydroxycinnamate CoA ligase (4CL, *p*-coumarate: CoA ligase, EC 6.2.1.12). The end products of this pathway, the *p*-hydroxycinnamoyl CoAs, are the precursors of lignins but also of other phenolic compounds such as flavonoids and tannins, which accumulate in great amounts in plant tissues.

The enzyme PAL catalyzes the first committed metabolic step from primary metabolism into phenylpropanoid metabolism, which is the deamination of phenylalanine to produce cinnamic acid (Fig. 4). Cinnamic acid is further modified by the consecutive action of the hydroxylases (C4H and C3H) and CCoAOMTs, leading to *p*-coumaroyl and feruloyl CoA (Fig. 4). Although C4H has long been known as a plant cytochrome P-450 monooxygenase (Schuler, 1996), the precise cytochrome P-450 mono-oxygenase nature of C3H has only recently been reported (Franke *et al.* 2002). Following the step catalyzed by C3H, the enzyme 4CL catalyzes the formation of *p*-hydroxycinnamoyl CoA esters, after which these activated intermediates are reduced to *p*-hydroxycinnamyl alcohols via the *p*-hydroxycinnamyl aldehydes (Fig. 4). 4CL is unique in that it also represents the branching point within the general phenylpropanoid metabolism toward the synthesis of flavonoids or lignins. Although this sequence of reactions has recently been proposed to occur in woody angiosperms (Li *et al.*, 2003; Boerjan *et al.*, 2003), such a view is continuously subject to revision as new data become avalable.

The steps which belong to the lignin-specific pathway, channel the *p*-hydroxycinnamoyl CoAs toward the synthesis of *p*-hydroxycinnamyl alcohols (Fig. 4). This segment of the route involves two characteristic reductive steps that convert the *p*-

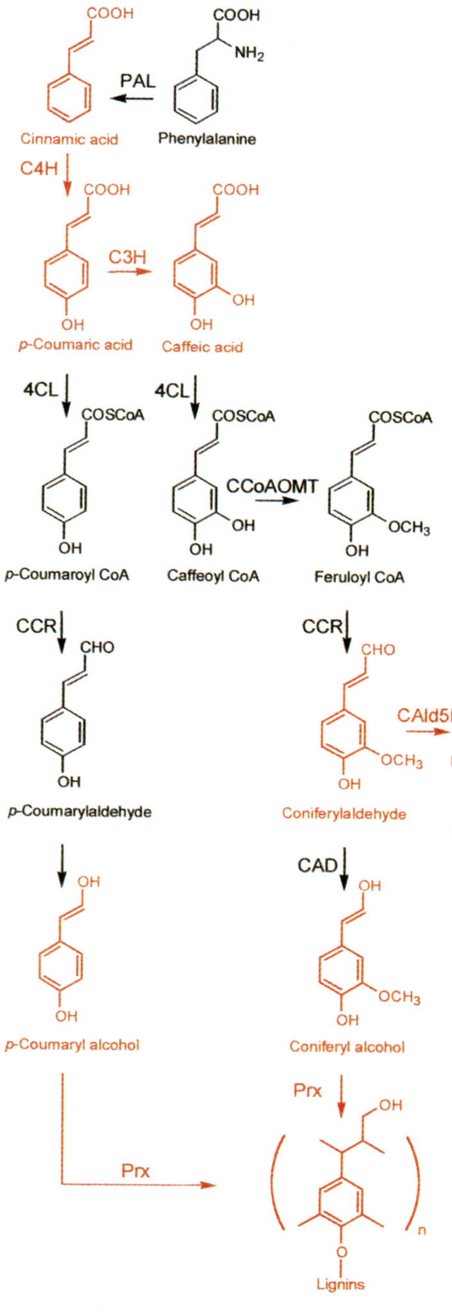

Figure 4 : Metabolic pathway for lignin biosynthesis in angiosperms. Taken from Li *et al.* (2003) with modifications.

hydroxycinnamoyl-CoA esters into *p*-hydroxycinnamyl alcohols. The reductive steps are catalyzed by the enzymes, *p*-hydroxycinnamoyl-CoA reductase (CCR, EC 1.2.1.44) and *p*-hydroxycinnamyl alcohol dehydrogenase (*p*-hydroxycinnamyl aldehyde: NADPH oxidoreductase, EC 1.1.1.195), in its two distinguishable forms, coniferyl alcohol (CAD) and sinapyl alcohol dehydrogenase (SAD), and are often considered to be specific to the lignification pathway. CCR catalyzes the conversion of *p*-hydroxycinnamoyl-CoA esters to their corresponding aldehydes (Fig. 4), this being the first committed step in the lignin specific pathway. Although CCR was thought to be an important control point regulating the flux of phenylpropanoid metabolites toward the biosynthesis of lignins, recent data do not support its role as rate limiting step of the pathway (Anterola *et al.*, 2002; Anterola and Lewis, 2002).

Intercalated in this reductive segment of the route, there is a novel branching reaction catalyzed by coniferylaldehyde-5-hydroxylase (CAld5H) (Fig. 4), which allows the synthesis of syringyl units in angiosperms. This hydroxylase is again a plant cytochrome P-450 mono-oxygenase, and shows certain specificity for coniferylaldehyde and coniferyl alcohol when compared with ferulic acid (Humphreys *et al.*, 1999; Franke *et al.*, 2000). This hydroxylase is coupled to a specific OMT (AldOMT) (Osakabe *et al.*, 1999; Li *et al.*, 2000), which catalyzes the subsequent conversion of 5-hydroxy-coniferylaldehyde into sinapylaldehyde.

The last reaction of the route is the oxidative polymerization of the three p-hydroxycinnamyl alcohols (*p*-coumaryl alcohol, coniferyl alcohol and sinapyl alcohol), to yield lignins in a reaction catalyzed by heme-containing plant peroxidases (Prx, hydrogen donor: H_2O_2 oxidoreductase, EC 1.11.1.7) (Fig. 4). *p*-Hydroxycinnamyl alcohols are thus dimerized and polymerized in xylem cell walls by peroxidases in a reaction dependent on the H_2O_2 delivered by an NADPH-oxidase-like enzyme (Ros Barceló, 1998, Ros Barceló and Pomar, 2001). In the case of coniferyl alcohol (1, Fig. 5), the main products of coniferyl alcohol dimerization by peroxidase are the neolignans, pinoresinol (2), dehydrodiconiferyl alcohol (3) and guaiacylglycerol-β-*O*-coniferyl alcohol ether (5), which are considered as intermediates in macromolecular lignin assembly (Nose *et al.*, 1995).

Cinnamyl alcohol dimers, therefore, are not the end products of the pathway, and they may be oxidized by peroxidases to yield a growing lignin polymer that remains anchored to the cell wall. Lignins can be anchored to cell wall polysaccharides by the nucleophilic addition of hydroxyl groups of polysaccharides to the quinone methide structure (4, Fig. 5) resulting from the β-*O*-4 coupling mode of two

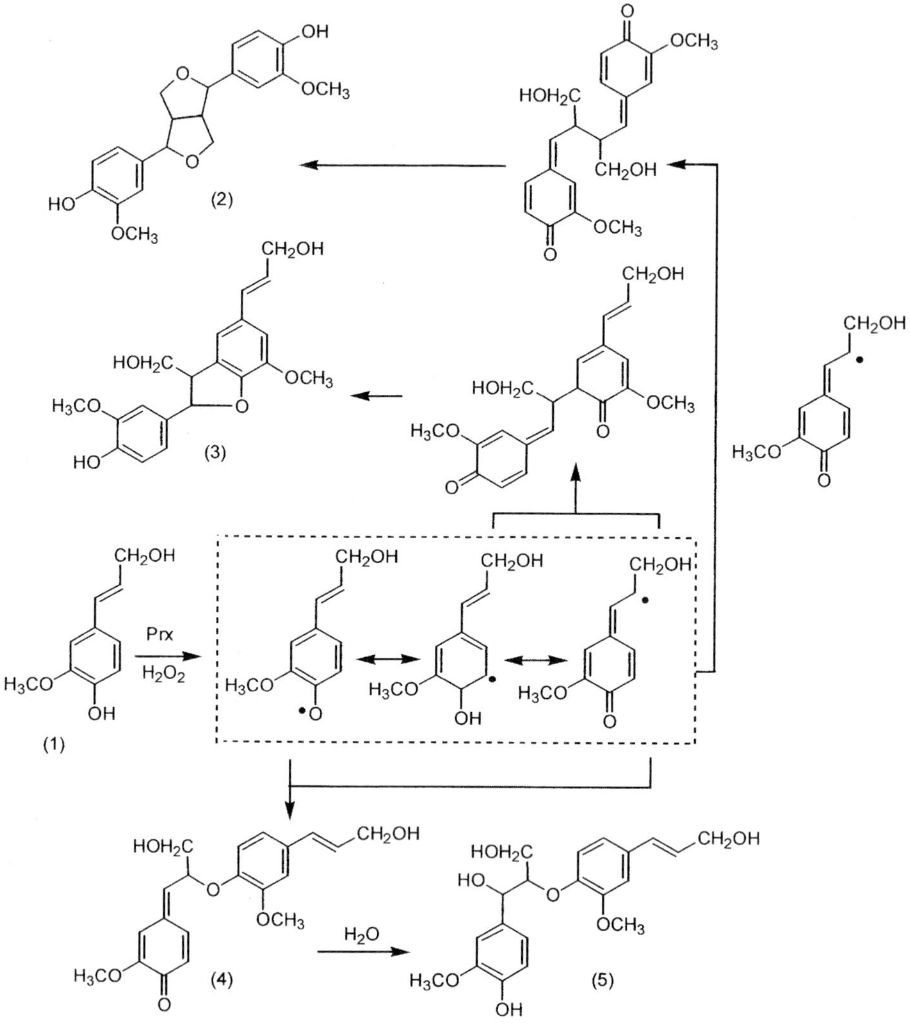

Figure 5 : Phenoxy radicals formed by oxidation of coniferyl alcohol (1) by peroxidase, and formation of the dimers, pinoresinol (2), dehydrodiconiferyl alcohol (3) and guaiacyl-glycerol-β-*O*-coniferyl alcohol ether (5) via quinone methides (4).

cinnamyl alcohol radicals. From a functional point of view, lignins impart strength to plant cell walls, facilitate water transport and impede the degradation of wall polysaccharides, acting as a major line of defense against vertebrate herbivores, insects and fungi.

3.1. Rate Limiting Steps

It has long been proposed by several authors that the metabolic flux (carbon allocation) in the phenylpropanoid pathway is controlled at multiple enzymatic levels. In the absence of an unequivocal experimental design, since initial studies were performed by feeding more or less specific enzyme inhibitors to lignifying plant tissues, rate-limiting steps were largely established at the level of PAL, 4CL and CCR (see Ros Barceló, 1997, and references cited therehein). Most recent studies involve the down-regulation of enzyme expression by silencing specific genes in transgenic plants (Anterola and Lewis, 2002), although in most cases they have been unable to point to any unequivocal stretching or particularly novel feature (Anterola and Lewis, 2002).

Only recently has the question been properly addressed by means of two complementary approximations. Thus, by feeding increasing amounts of phenylalanine or cinnamic acid to lignifying cell cultures of *Pinus taeda*, a gymnosperm, and from the simultaneous study of both the metabolic pools (Anterola *et al.*, 1999) and the transcriptional profile of the enzymes of the phenylpropanoid pathway using quantitative real time PCR analyses (Anterola *et al.*, 2002), it has been clearly established that both carbon allocation to the pathway, and its differential distribution into the two monolignols, *p*-coumaryl and coniferyl alcohol, is controlled by the rate of phenylalanine supply and the differential modulation of C4H and C3H, respectively. In these studies, a coordinated up-regulation of PAL, 4CL, CCoOMT, CCR and CAD in respose to phenylalanine supply was also found, which indicates that these steps are not truly rate-limiting since they are modulated according to metabolic demand.

At this point, it is neccessary to remember that C4H introduces the OH group into C4, yielding *p*-hydroxycinnamyl backbones (e.g. *p*-coumaryl alcohol), susceptible of being polymerized by peroxidase, while C3H introduces the OH group in C3, yielding firstly *o*-catechols, and later guaiacyl backbones (e.g. coniferyl alcohol), which are the precursor of guiacyl lignins (Fig. 4). In angiosperms, there is a novel branching point, the step catalyzed by CAld5H (Fig. 4), that diverts guaiacyl backbones for the synthesis of syringyl (i.e. sinapyl alcohol) moieties, the precursors of syringyl lignins. CAld5H was not studied in *P. taeda* cell culttures since they are derived from a gymnosperm, but one may extrapolate, in the absence of available data, that, like C4H and C3H in gymnosperms, CAld5H might also constitute a rate limiting step in angiosperms.

All the branching (also rate limiting) enzymes of the lignin biosynthetic pathway (C4H, C3H and CAld5H) and the lignin assembling enzyme (peroxidase) are hemeproteins, and therefore susceptible targets of NO action (Tsai, 1994). Any possible metabolic control of NO over these enzymes would lead to the regulation of not only the global *p*-hydroxycinnamyl alcohol pools in lignifying plant cells, and the ratio H/G/S for carbon partitioning, but also of their rates of polymerization.

4. NITRIC OXIDE EFFECTS

NO is a relatively stable paramagnetic free-radical molecule involved in many physiological processes in plants, where it serves as a synchronizing chemical messenger involved in cytotoxicity and programmed cell death (Van Camp *et al.*, 1998, Durner and Klessig, 1999; Neill *et al.*, 2003). In animal cells, most of the biological regulatory properties of NO have been explained on the basis of its capacity to act as an iron ligand in hemeproteins (Tsai, 1994). Dual (activating or inhibitory) effects of NO on hemeproteins have been described (Tsai, 1994), and the nature of the effect seems to depend to a great extent on the resting state (oxidation state) of the hemeprotein, which conditions the ligand properties and the electronic configuration of the heme iron (Tsai, 1994). However, the chemistry of NO means (Stamler *et al.*, 1992) that other transition metallo-proteins, such as those containing copper or zinc, and even thiol-containing proteins, may also be targets of NO action.

In such a scenario, downstream effects of NO on the lignin biosynthetic pathway may be directly induced by interactions with metallo-enzymes (C4H, C3H, CAld5H and peroxidase), or indirectly (subrogated effects) by interaction with signaling proteins such as protein kinases, or second messenger-generating enzymes, such as guanylyl cyclase (GC) (Neill *et al.*, 2003), which coordinate in a one-way highway plant cell gene expression and metabolism.

4.1. Subrogated NO Effects

PAL, which is the first committed enzyme of the phenylpropanoid pathway, channels carbon from primary metabolism to the synthesis of *p*-hydroxycinnamyl alcohols (Fig. 4). PAL expression (as assayed by Northern analyses) is activated in response to NO (Durner *et al.*, 1998; Delledonne *et al.*, 1998) when plant tissues are fed either recombinant NO synthase or NO donors, such as sodium nitroprusside (SNP), *S*-nitroso-*N*-acetyl-DL-penicillamine (SNAP) or *S*-nitroso-L-glutathione (GSNO). This NO effect appears to be mediated by cyclic guanosine monophosphate (cGMP) (Durner *et al.*, 1998). cGMP is a well established second messenger (intracellular signaling) molecule in plant cells, whose levels are transiently modified in response to external stimulus (Wendehenne *et al.*, 2001; Neill *et al.*, 2003). cGMP concentrations are increased by the activity of GC which synthesizes it from GTP, and are returned to resting values by the action of phosphodiesterases (Neill *et al.*, 2003). In plant cells, NO activates cGMP production via direct activation of GC (a hemeprotein), the activation persisting only so long as NO is present (Neill *et al.*, 2003). cGMP-mediated effects on the expression of PAL appears to be mediated by cADP-ribose (cADPR) (Durner *et al.*, 1998), a calcium-mobilising second messenger, whose synthesis is stimulated by cGMP. However, PAL not only participates in the lignin biosynthetic pathway, but also in the synthesis of the whole plethora of phenolics present in plant cells, so that any effect of NO on PAL should be interpreted with caution, since it might not neccesarily reflect modifications in the rate of carbon allocation in the lignin biosynthetic pathway.

Subrogated NO effects on the biosynthesis of lignins could also be manifested at the level of enzyme co-substrates. The most striking cosubstrate used for the synthesis of lignins is H_2O_2 (Ros Barceló, 1998), which is used by peroxidases as an oxidizing agent for polymerizing *p*-hydroxycinnamyl alcohols (Fig. 5). H_2O_2 also works with NO to trigger PCD during the plant hypersensitive response through a finely balanced NO/H_2O_2 cooperation (Delledonne *et al.*, 2001), and it should not be surprising that both species also work together during the PCD which accompanies xylem differentiation.

NO does not affect H_2O_2 production by the lignifying xylem (Ferrer and Ros Barceló, 1999) nor H_2O_2 production by isolated plant mitochondria (one of the organelles involved in H_2O_2 production) (Yamasaki *et al.*, 2001). This is not surprising since the synthesis of both H_2O_2 and NO in plant cells is simultaneously coordinated in response to environmental/hormonal stimuli without negative cross-talks between them (Delledonne *et al.*, 1998).

NO could also control H_2O_2 levels by modulating the enzymes involved in its removal/detoxification. The two enzymes involved in its catabolism are catalase (EC 1.11.1.6), which dismutates H_2O_2 into O_2 and H_2O, and ascorbate peroxidase (EC 1.11.1.11), which reduces H_2O_2 into H_2O at the expense of ascorbic acid. Both enzymes are hemeproteins, and therefore susceptible to direct inhibition by NO, as it is produced (Clark *et al.*, 2000), but any effect of NO on the removal of H_2O_2 should be restricted to the above described enzyme inhibitions, since NO has no effect on the abundance of ascorbate peroxidase transcripts or on transcripts encoding other key enzymes of ascorbate metabolism (Pignocchi and Foyer, 2003).

4.2. Direct NO Effects

It may be expected that the direct NO effects are manifested especially on the metallo-enzymes of the lignin biosynthetic pathway, particularly the hemeproteins, C4H, C3H, CAld5H and peroxidase. However, it seems that such effect are rather complex since, a priori, NO could modulate the activity of hemeproteins through a combination of effects, including transcriptional regulation, altering substrate availability (see above), and direct reaction with enzyme turnover intermediates. At present, there are no molecular data available concerning the interaction of NO with the heme-containing mono-oxygenases (C4H, C3H and CAld5H) of the lignin biosynthetic pathway, and the only effects described concern C4H (Enkhardt and Pommer, 2000). In the case of C4H, NO reacts with the heme group of this cytochrome P450 mono-oxygenase, causing a non-competitive inhibition of the enzyme (Enkhardt and Pommer, 2000). Since the step catalyzed by C4H is a rate limiting step in the lignin biosynthetic pathway (see above), it remains to be determined how and to what extent this inhibition alters the rate of carbon allocation in the pathway.

Unlike plant heme-containing mono-oxygenases, our present knowledge of the effect (and interactions) of NO on (with) plant peroxidases is weightier, and therefore described below.

4.3. NO Effects on Peroxidases

It has long been known that NO reversibly binds to the heme prosthetic group of plant peroxidases (Yonetani *et al.*, 1972; Ascenzi *et al.*, 1989). Optical and electron paramagnetic resonance (EPR) studies suggest that, once the complex is formed (Fig. 6), a one electron transfer between NO and the protoporphyrin IX prosthetic group of peroxidases takes place, which results in the formation of spin-paired complexes (Yonetani *et al.*, 1972). The binding of NO to plant peroxidases results in spectrally distinct complexes (Yonetani *et al.*, 1972; López-Molina *et al.*, 2003), in which the Soret peak at 405 nm shifted to 419 nm, and the α and β-absorption bands at 495 and 636 nm shifted to 533 and 568 nm, respectively. The shapes and positions of these bands (Fig. 7) are typical of low spin ferrous [Fe(II)] peroxidases, confirming the existence of the equilibrium described in Fig. 6. Kinetic analyses using flash photolysis and stopped-flow methods (Kobayashi *et al.* 1989) have revealed a reaction constant (k) of 1.9×10^5 M^{-1} s^{-1}.

The formation of these ferrous nitrosyl complexes [$Fe(II)NO^+$] removes enzyme turnover intermediates such as the resting form, Fe(III):

$$Fe\ (III) + NO \rightarrow Fe\ (III)\ NO \rightarrow Fe\ (II)\ NO^+$$

so that NO is capable of inhibiting peroxidase-catalyzed reactions with K_i in the μM range (Ischiropoulos *et al.*, 1996, Ferrer and Ros Barceló, 1999), especially when peroxidase is assayed with physiological substrates, such as coniferyl alcohol.

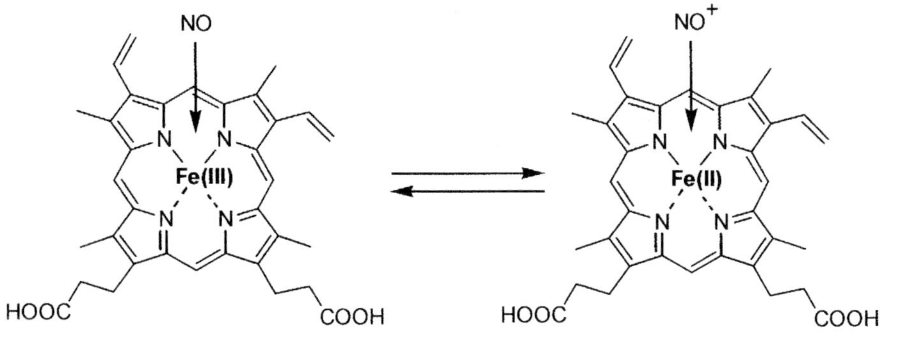

Figure 6 : Complex formed by binding of NO to the protoporphyrin IX prosthestic group of plant peroxidases which results from a one electron-transfer in the formation of a spin-paired complex.

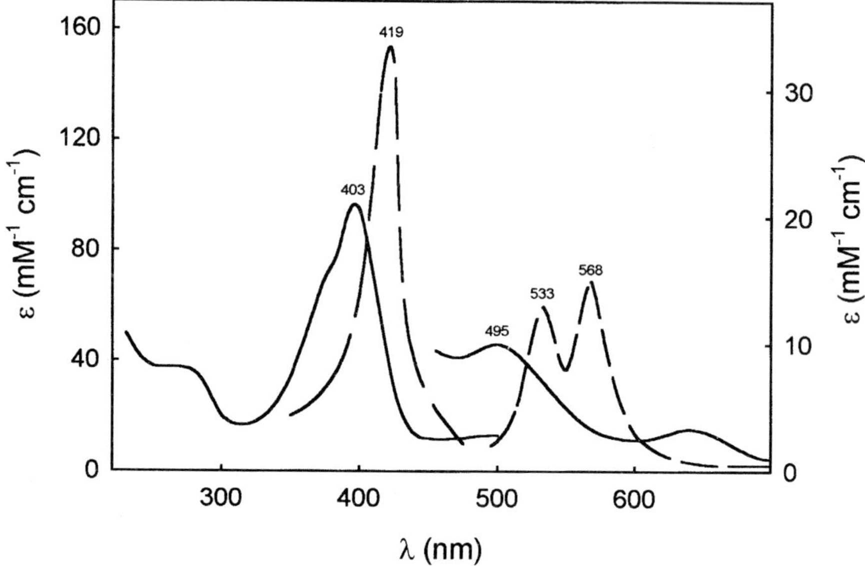

Figure 7 : UV-visible absorption spectrum of a typical plant peroxidase (—) and of its NO complex (— — —).

The inhibitory effect of NO on peroxidase activity may be easily monitored by using NO donors, such as SNAP and SNP. In such studies, it is absolutely necessary to test that the effect of NO donors is reversed by a biologically active NO-scavenger, such as PTIO (Ros Barceló *et al.*, 2002). This is due to the fact that, although these NO-donors are capable of sustaining μM NO concentrations in the aqueous phase (Fig. 8), at least during the first 60 min of their decomposition, the NO chemistry is extremely complex (Stamler *et al.*, 1992), and in aerated aqueous solutions, NO rapidly reacts with O_2 to yield NO_2 ($2\ NO + O_2 \rightarrow 2\ NO_2$). NO_2 can further react with NO to yield other nitrogen oxides, such N_2O_3 ($NO + NO_2 \rightarrow N_2O_3$). Thus, NO-donor solutions aged for several minutes contain a complex mixture of nitrogen oxides ($NO_x = NO + NO_2 + N_2O_3$), in which NO may either constitute the dominant species (at short times) or be present in minor amounts (at long times) (Fig. 8). The biological activity of NO_x is thus easily dissected in the presence of PTIO, since PTIO rapidly reacts with NO to yield NO_2 (Akaike and Maeda, 1996). In other words, any biological effect of aged NO-donor solutions on peroxidases reversed by PTIO can be ascribed to NO itself, while the biological effects of NO-donor solutions which are

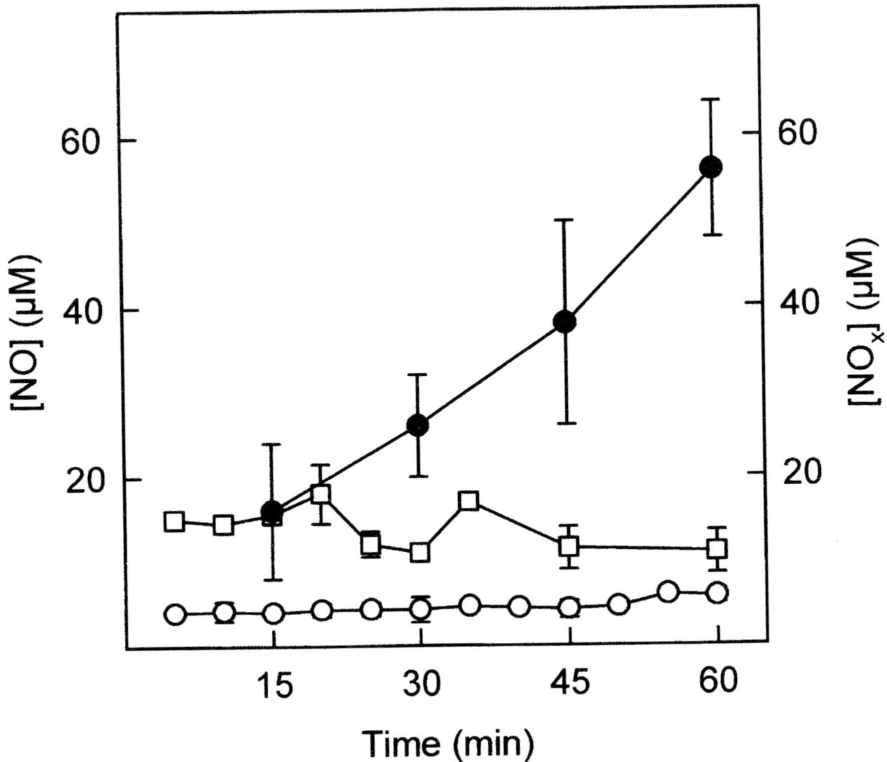

Figure 8 : Time-course of the release of nitrogen oxides (NO_x) (●), monitored with an NO_x electrode, and nitric oxide (NO) (○, □), monitored by the ferrous hemoglobin method, during the decomposition of a 5 mM solution of either SNP (○, ●) or SNAP (□) at 25ºC. Taken from Ros Barceló *et al.* (2002).

not reversed by PTIO must be ascribed to other NO_x, such as NO_2 or N_2O_3. This is especially relevant in the case of peroxidase since reduced nitrogen oxides (NO_x) and especially their decomposition product, NO_2^-, are suitable electron donors in peroxidase-catalyzed reactions (Yamazaki and Yokota, 1973; van der Vliet *et al.*, 1997), and could therefore act as competitive inhibitors of the enzyme.

NO-donors, such as SNP, are capable of inhibiting the peroxidase activity from several sources (Ferrer and Ros Barceló, 1999; Ros Barceló *et al.*, 2002; de Pinto *et al.*, 2002), and this effect is reversed by PTIO (Table 1 and Fig. 9). Histochemical assays confirm that the effect of NO on peroxidase may be extrapolated to those peroxidases involved in the synthesis of lignins. These assays show that

Table 1. Effect of 5 mM SNP on the activity of *Z. elegans* and *Capsicum annuum* peroxidase, assayed with syringaldazine and 3,3',5,5'-tetramethylbenzidine (TMB) as substrates, and reversion of the effect by 150 μM PTIO. nd, not determined.

Treatment	Peroxidase activity (nmol substrate oxidized min^{-1} g^{-1} FW)			
	Syringaldazine		TMB	
	Z. elegans	*C. annuum*	*Z. elegans*	*C. annuum*
Control	56.32 ± 1.80	191.60 ± 4.80	104.2 ± 16.0	810.50 ± 14.70
SNP	37.07 ± 3.61	97.89 ± 1.32	54.6 ± 4.1	372.59 ± 0.45
SNP + PTIO	60.02 ± 2.10	nd	92.0 ± 11.9	961.56 ± 0.33

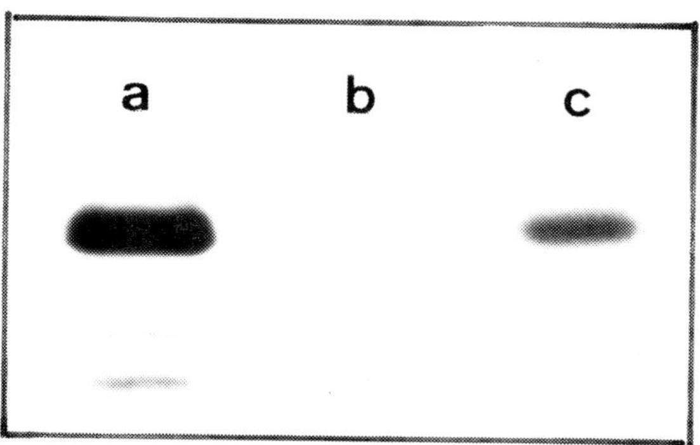

Figure 9 : Inactivation of *Z. elegans* peroxidase isoenzymes by 5 mM SNP (b) and reversion of the inhibitory effect of SNP by 150 μM PTIO (c), compared with control (a). Peroxidase isoenzymes were separated by isoelectric focusing in 3.5-10.5 pH gradients under non-equilibrium conditions. Taken from Ros Barceló *et al.* (2002).

peroxidase located in lignifying tissues is also inhibited by SNP, the effect being again reversed by PTIO (Fig. 10). Histochemical assays depicted in Fig. 10 were performed using 3,3',5,5'-tetramethylbenzidine as peroxidase substrate (Ros Barceló *et al.*, 2002), and using the proper capacity of lignifying tissues to sustain H_2O_2 production (Ros Barceló, 1998), although similar results have been obtained using 3,3'-diaminobenzidine (Jih *et al.*, 2003).

5. NO_X METABOLISM MAY BE DRIVEN BY PEROXIDASES

Peroxidases are heme-containing enzymes which catalyze the one-electron oxidation of several substrates (RH) at the expense of H_2O_2:

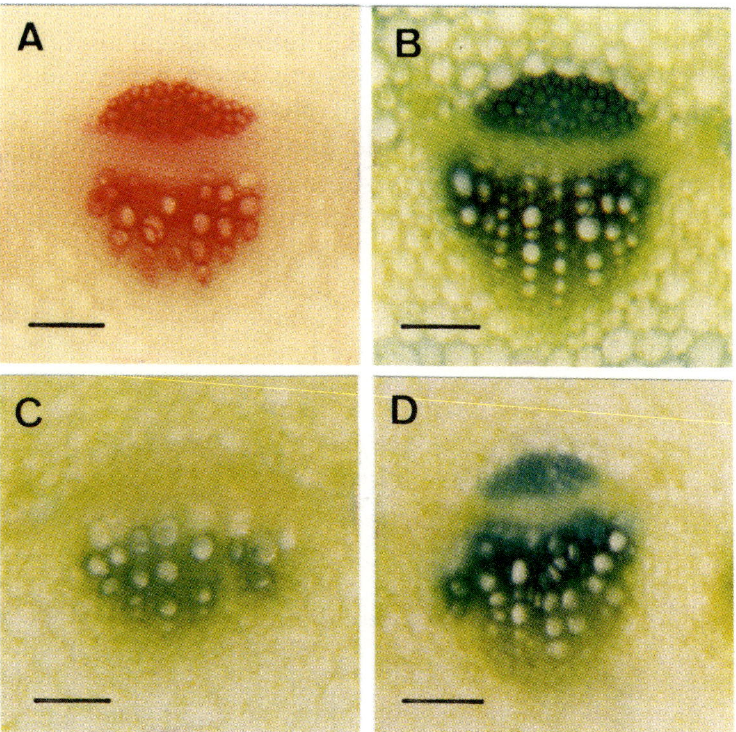

Figure 10 : Effect of the NO-releasing compound, SNP, on peroxidase activity from the *Z. elegans* lignifying xylem. A) Section stained with phloroglucinol to reveal lignins. B) Control section. C) Section pre-incubated for 20 min with 5 mM SNP. D) Section pre-incubated for 20 min with 150 μM PTIO, to which 5 mM SNP was then added. Bars = 100 μm.

$$2\ RH + H_2O_2 \rightarrow 2\ R\cdot + 2\ H_2O$$

including NO (Fig. 11). In their resting state, plant peroxidases contain an iron ion in the oxidation state of +3 (FeIII). The iron is five coordinated to the four pyrrole nitrogens of the heme and to a nitrogen from an axial (proximal) histidine, which has been strongly conserved during evolution, in an aminoacidic sequence motif of the type GgHTiG. The sixth coordination position is free, thus determining a high spin state for the iron (Banci, 1997).

The catalytic cycle for peroxidase (Yamazaki and Yokota, 1973) may be described (Fig. 11) as follows. Hydrogen peroxide oxidizes the ferric form of the enzyme (FeIII) in a two electron oxidation step to yield the enzyme intermediate compound I (CoI). CoI, described as an oxyferryl porphyrin π cation radical containing Fe in the formal oxidation state, Fe(IV) (Banci, 1997), accepts one electron and one proton from NO to yield NO^+ and the oxyferryl heme intermediate known as compound II (CoII). The subsequent one electron reduction of CoII by a second molecule of NO yields the ferric form of peroxidase, FeIII, thus completing the catalytic cycle of the enzyme.

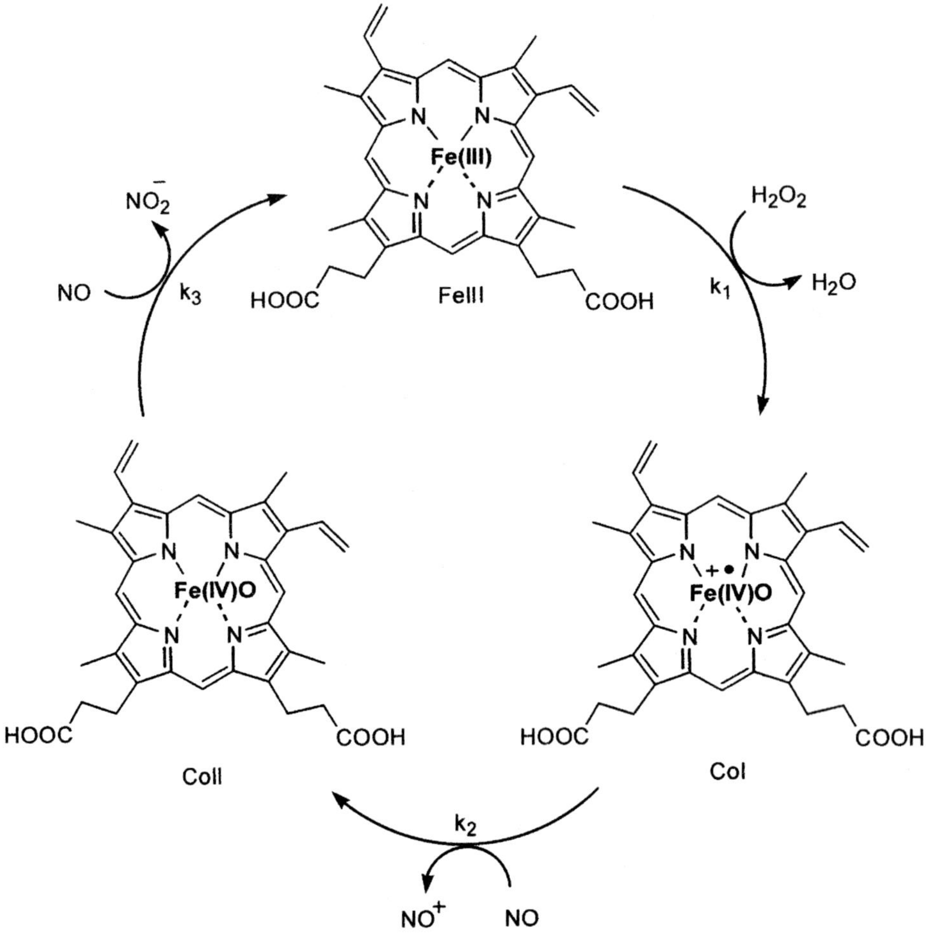

Figure 11 : Catalytic cycle of plant peroxidases in the presence of NO, showing the interconversion between the resting state, Fe(III), and the activated forms, compound I (CoI) and compound II (CoII).

5.1. NO Detoxification

NO is thus regarded as a peroxidase substrate (Glover *et al.*, 1999), and therefore peroxidase could play any role in NO detoxification in plant tissues. In fact, NO reacts with CoI with a rate constant (k_2) of 7.0×10^5 M^{-1} s^{-1} yielding the nitrosyl species, NO^+ (Fig. 11). NO also reacts with CoII with a rate constant (k_3) of 1.3×10^6 M^{-1} s^{-1} to lead probably to the nitrite anion, NO_2^- (Fig. 11). Interestingly, the reaction of CoII with NO is unusually high relative to that of CoI, which is usually the faster reaction. This means that, in the presence of poor electron donors of CoII, such as catechols (Nappi and Vass, 2001) or guaiacol (Uchida *et al.*, 2002, Hung *et al.*, 2002), which show k_3 values ~ 10^2 M^{-1} s^{-1} and ~ 10^5 M^{-1} s^{-1}, respectively (Yamazaki and Yokota, 1973), NO may enhance peroxidase activity by promoting the formation of the native enzyme, FeIII, from CoII. Thus, both activating and inhibitory effects of NO on peroxidase could be present in the lignifying xylem, depending on the reactivity (k_3 value) of the *in situ* substrate of the enzyme.

The product of NO oxidation by CoII of peroxidase is apparently NO_2^-, which can be further oxidized by plant peroxidases to lead to the most likely form, $NO_2\cdot$ (Shibata *et al.*, 1995; van der Vliet *et al.*, 1997). Plant peroxidases are also capable for oxidizing the tyrosine contained in proteins to tyrosine radicals (Tyr·) (1, Fig. 12), which combine with another Tyr· to form dityrosine (Michon *et al.*, 1997). When the oxidation of tyrosine by plant peroxidases is performed in the presence of NO_2^-, a mixture of Tyr· and $NO_2\cdot$ is formed in the reaction media, species which couple yielding 3-nitro-tyrosine (2, Fig. 12) (van der Vliet *et al.*, 1997). 3-Nitro-tyrosine is a well-established marker of oxidative protein damage in mammals (van

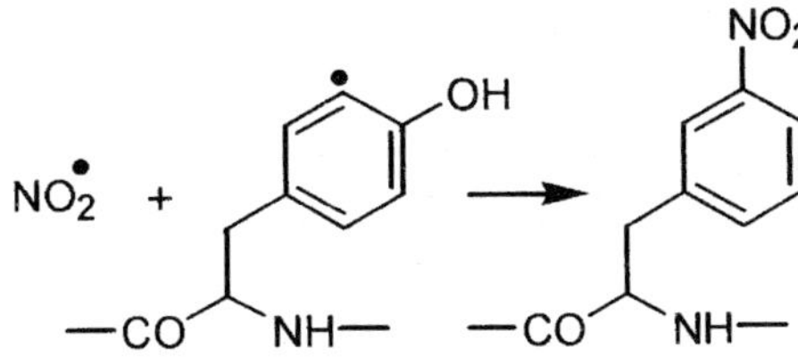

Figure 12 : Coupling of $NO_2\cdot$ and tyrosine radicals (1), formed in peroxidase catalyzed reactions, to lead to 3-nitro-tyrosine (2).

der Vliet *et al.*, 1997) but, to the best of our knowledge, there is no evidence of its presence in lignifying or other plant tissues.

Lignifying plant tissues are capable of sustaining both NO (Fig. 3) and $O_2^{\cdot -}$ (Ogawa *et al.*, 1997; Ros Barceló, 1998 and 1999) production. Therefore, it is likely that in the lignifying xylem both NO and $O_2^{\cdot -}$ will react to lead to the peroxynitrite anion ($ONOO^-$):

$$NO + O_2^{\cdot -} \rightarrow ONOO^-$$

Peroxynitrite is formed at a near difusion controlled rate (k = 6.7 × 10^9 M^{-1} s^{-1}), and it has been postulated that it play a major role in cytotoxicity, including PCD (Wendehenne *et al.*, 2001; Delledonne *et al.*, 2001), although its real role in living cells may be uncertain (Fukuto and Ignarro, 1997). Independently of their real role in plant cells, which remains to be clearly established, peroxynitrite may be regarded as a substrate of plant peroxidases (Floris *et al.*, 1993; Gebicka and Gebicki, 2000). In fact, peroxynitrite reacts (k = 3 × 10^6 M^{-1} s^{-1}) with the resting form of the enzyme to lead to CoII (Floris *et al.*, 1993):

$$Fe(III) + ONOO^- \rightarrow CoII + NO_2^{\cdot}$$

yielding the nitrosylating specie, $NO_2^{\cdot}$. CoII is catalytically inactive towards peroxynitrite, and the decay of CoII to the native enzyme, Fe(III), only takes place in the presence of an external electron donor, such as a phenol. In such a scenario, a role in the scavenging of peroxynitrite has been proposed for the chlorogenic acid/peroxidase system (Grace *et al.*, 1998), which may be functional with any other phenol.

5.2. Bio-mimetic NO-synthesizing Peroxidase-Mediated Pathways

Peroxidase is not only involved in the detoxification of NO species and relatives, but it may also participates in bio-mimetic NO-synthesizing pathways. One of the possible enzymes involved in NO synthesis in plant cells is NO synthase (NOS) (Wendehenne *et al.*, 2001). NOS catalyzes the conversion of L-arginine (1a, Fig. 13) in L-citrulline (1c, Fig. 13) yielding NO. In this reaction, *N*-hydroxy-L-arginine (1b, Fig. 13), a *N*-hydroxyguanidine, acts as key intermediate. *N*-hydroxyguanidines (2b, Fig. 13), including *N*-hydroxy-L-arginine (1b, Fig. 13), may be oxidized by plant peroxidases to release NO, yielding the same products (2c, Fig. 13) as those obtained with NOS (Xian *et al.*, 2001; Cai *et al.*, 2002). The

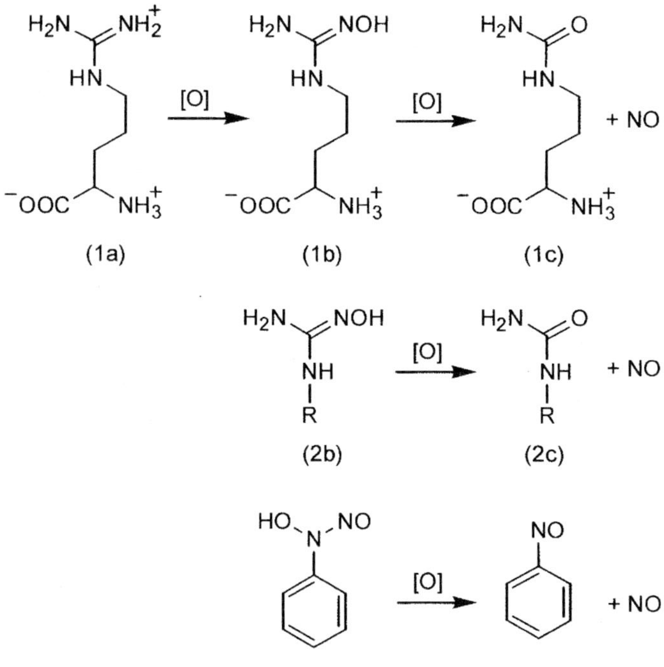

Figure 13 : NOS catalyzed conversion of L-arginine (1a) in L-citrulline (1c) yielding NO. In this reaction, *N*-hydroxy-L-arginine (1b), a *N*-hydroxyguanidine, acts as key intermediate. Peroxidase-catalyzed oxidation of *N*-hydroxyguanidines (2b) to lead to NO. Peroxidase-catalyzed oxidation of *N*-hydroxy-*N*-nitrosamines, such as cupferron (3b), to lead to NO.

ability of plant peroxidases to synthesize NO is not restricted to the course of the oxidation of *N*-hydroxyguanidines, since the oxidation of *N*-hydroxy-*N*-nitrosamines (Alston *et al.*, 1985), such as either cupferron (3b, Fig. 13), a xenobiotic, or alanosine (a natural antineoplastic drug closely related to aspartic acid), also yields NO as collateral product of the reaction.

In such a scenario, it can be expected that all these cross-talks between NO (which is synthesized by the lignifying xylem) and peroxidase (which is contained in the lignifying xylem) should be considered far from fortuitous. Doubtless, once the capability of lignifing tissues for producing NO has been demonstrated and the possible targets in the lignin biosynthesis machinery identified, a challenge for future research in this field will be to identify the exact targets of NO action and to determine (and dissect) the real role played by NO in xylogenesis and cell wall lignification.

ACKNOWLEDGMENTS

This work was supported by grants from the Fundación Séneca (project # PI-70/00615/FS/01) and MCYT (BOS2002-03550). CG holds a fellowship (FPU) from the MECYD (Spain).

REFERENCES

Akaike, T. and Maeda, H. (1996). Quantitation of nitric oxide using 2-phenyl-4,4,5,5-tetramethylimidazoline-1-oxyl-3-oxide (PTIO). *Method. Enzymol.,* **268** : 211-221.

Alston, T. A., Porter, D. J. T. and Bright, H. J. (1985). Generation of nitric oxide by enzymatic oxidation of *N*-hydroxy-*N*-nitrosamines. *J. Biol. Chem.,* **260** : 4069-4074.

Anterola, A. M., Jeon, J., Davin, L. B. and Lewis, N. G. (2002). Transcriptional control of monolignol biosynthesis in *Pinus taeda*. *J. Biol. Chem.,* **227** : 18272-18280.

Anterola, A. M. and Lewis N. G. (2002). Trends in lignin modification: a comprehensive analysis of the effects of genetic manipulations/mutations on lignification and vascular integrity. *Phytochemistry,* **61** : 221-294.

Anterola, A. M., Van Rensburg, H., Van Hereden, P. S., Davin, L. B. and Lewis, N. G. (1999). Multi-site modulation of flux during monolignol formation in loblolly pine (*Pinus taeda*). *Biochem. Biophys. Res. Comm.,* **261** : 652-657.

Ascenzi, P., Brunori, M., Coletta, M. and Desideri, A. (1989). pH effects on the haem iron co-ordination state in the nitric oxide and deoxy derivatives of ferrrous horseradish peroxidase and cytochrome c peroxidase. *Biochem. J.,* **258** : 473-478.

Banci, L. (1997). Structural properties of peroxidases. *J. Biotechnol.,* **53** : 253-263.

Billa, E., Tollier, M. T. and Monties, B. (1996). Characterization of the monomeric composition of in situ wheat straw lignins by alkaline nitrobenzene oxidation: Effect of temperature and reaction time. *J. Sci. Food Agric.,* **72** : 250-256.

Boerjan, W., Ralph, J. and Baucher, M. (2003). Lignin biosynthesis. *Annu. Rev. Plant Biol.,* **54** : 519-546.

Boudet, A. M., Lapierre, C. and Grima-Pettenati, J. (1995). Biochemistry and molecular biology of lignification. *New Phytol.,* **129** : 203-236.

Cai, T., Xian, M. and Wang, P. G. (2002). Electrochemical and peroxidase oxidation study of *N′*-hydroxyguanidine derivatives as NO donors. *Bioorg. Med. Chem. Lett.,* **12** : 1507-1510.

Chapple, C. C. S., Vogt, T., Ellis, B. E. and Somerville, C. R. (1992). An *Arabidopsis* mutant defective in the general phenylpropanoid pathway. *Plant Cell,* **4** : 1413-1424.

Clark, D., Durner, J., Navarre, D. A. and Klessig, D. F. (2000). Nitric oxide inhibition of tobacco catalase and ascorbate peroxidase. *Mol. Plant Microb. Inter.,* **13** : 1380-1384.

Delledonne, M., Xia, Y., Dixon, R. A. and Lamb, C. (1998). Nitric oxide functions as a signal in plant disease resistance. *Nature,* **394** : 585-588.

Delledonne, M., Zeier, J., Marocco, A. and Lamb, C. (2001). Signal interactions between nitric oxide and reactive oxygen intermediates in the plant hypersensitive disease resistance response. *Proc. Natl. Acad. Sci. USA,* **98** : 13454-13459.

Demura, T., Tashiro, G., Horiguchi, G., Kishimoto, N., Kubo, M., Matsuoka, N., Minami, A., Nagata-Hiwatashi, M., Nakamura, K., Okamura, Y., Sassa, N., Suzuki, S., Yazaki, J., Kikuchi, S. and Fukuda, H. (2002). Visualization by comprehensive microarray analysis of gene expression programs during transdifferentiation of mesophyll cells into xylem cells. *Proc. Natl. Acad. Sci. USA,* **99** : 15794-15799.

De Pinto, M. C., Tommasi, F. and De Gara, L. (2002). Ascorbate and ascorbate peroxidase in programmed cell death. In: Plant Peroxidases (Eds. Acosta, M., Rodriguez-López, J. and Pedreño, M. A.) University of Murcia, Murcia pp 237-244.

Donaldson, L. A. (1985). Within- and between-tree variation in lignin concentration in the tracheid cell wall of *Pinus radiata. N.Z. J. For. Sci.,* **15** : 361-369.

Durner, J. and Klessig, D. F. (1999). Nitric oxide as a signal in plants. *Curr. Opin. Plant Biol.,* **2** : 369-374.

Durner, J., Wendehenne, D. and Klessig, D. F. (1998). Defense gene induction in tobacco by nitric oxide, cyclic GMP, and cyclic ADP-ribose. *Proc. Nat. Acad. Sci. USA,* **95** : 10328-10333.

Enkhardt, U. and Pommer, U. (2000). Einflub von stickstoffmonoxid und nitrit auf die aktivität der zimtsäure 4-hydroxylase in *Zea mays* in vitro. *J. Appl. Bot.,* **74** : 151-154.

Ferrer, M. A. and Ros Barceló, A. (1999). Differential effects of nitric oxide on peroxidase and H_2O_2 production by the xylem of *Zinnia elegans. Plant Cell Environm.,* **22** : 891-897.

Floris, R., Piersma, S. R., Yang, G., Jones, P. and Wever, R. (1993). Interaction of myeloperoxidase with peroxynitrite. A comparison with lactoperoxidase, horseradish peroxidase and catalase. *Eur. J. Biochem.,* **215** : 767-775.

Foissner, I., Wendehenne, D., Langebartels, C. and Durner, J. (2000). In vivo imaging of an elicitor-induced nitric oxide burst in tobacco. *Plant J.,* **23** : 817-824.

Franke, R., Hemm, M. R., Denault, J. W., Ruegger, M. O., Humphreys, J. M. and Chapple, C. (2002). Changes in secondary metabolism and deposition of an unusual lignin in the ref8 mutant of *Arabidopsis. Plant J.,* **30** : 47-59.

Franke, R., McMichael, C. M., Meyer, K., Shirley, A. M., Cusumano, J. C. and Chapple, C. (2000). Modified lignin in tobacco and poplar plants over-expressing the *Arabidopsis* gene encoding ferulate 5-hydroxylase. *Plant J.,* **22** : 223-234.

Fukuda, H. (1996). Xylogenesis: Initiation, progression and cell death. *Annu. Rev. Plant Physiol. Plant Mol. Biol.,* **47** : 299-325.

Fukuda, H. (1997). Tracheary element differentiation. *Plant Cell,* **9** : 1147-1156.

Fukuto, J. M. and Ignarro, L. J. (1997). *In Vivo* aspects of nitric oxide (NO) chemistry: does peroxynitrite ($^{-}$OONO) play a major role in cytotoxicity?. *Accounts Chem. Res.,* **30** : 149-152.

Gebicka, L. and Gebicki, J. L. (2000). Reactions of heme peroxidases with peroxynitrite. *IUBMB Life,* **49** : 11-15.

Glover, R., Koshkin, V., Dunford, H. B. and Mason, R. P. (1999). The reaction rates of NO with horseradish peroxidase compounds I and II. *Nitric oxide: Biol. Chem.,* **3** : 439-444.

Grabber, J. H., Ralph, J. and Hatfield, R. D. (1988). Severe inhibition of maize wall degradation by synthetic lignins formed with coniferylaldehyde. *J. Sci. Food Agric.,* **78** : 81-87.

Grace, S. C., Salgo, M. G. and Prior, W. A. (1998). Scavenging of peroxynitrite by a phenolic/peroxidase system prevents oxidative damage to DNA. *FEBS Lett.,* **426** : 24-28.

Herrmann, K. M. (1995). The shikimate pathway: Early steps in the biosynthesis of aromatic compounds. *Plant Cell,* **7** : 907-919.

Higuchi, T. (1990). Lignin biochemistry: Biosynthesis and biodegradation. *Wood Sci. Technol.,* **24** : 23-63.

Humphreys, J. M., Hemm, M. R. and Chapple, C. (1999). New routes for lignin biosynthesis defined by biochemical characterization of recombinant ferulate 5-hydroxylase, a multifunctional cytochrome P450-dependent monooxygenase. *Proc. Natl. Acad. Sci., USA,* **96** : 10045-10050.

Hung, K. T., Chang, C. J. and Kao, C. H. (2002). Paraquat toxicity is reduced by nitric oxide in rice leaves. *J. Plant Physiol.,* **159** : 159-166.

Ischiropoulos, H., Nelson, J., Duran, D. and Al-Mehdi, A. (1996). Reactions of nitric oxide and peroxynitrite with organic molecules and ferrihorseradish peroxidase: interference with the determination of hydrogen peroxide. *Free Rad. Biol. Med.,* **20** : 373-381.

Jih, P., Chen, Y. and Jeng, S. (2003). Involvement of hydrogen peroxide and nitric oxide in expression of the ipomoelin gene from sweet potato. *Plant Physiol.,* **132** : 381-389.

Kobayashi, K., Tamura, M. and Hayashi, K. (1982). Kinetic analyses of the recombination of NO with ferrihemoproteins by the flash photolysis method. *Biochemistry,* **21** : 729-732.

Lapierre, C., Pollet, B. and Rolando, C. (1995). New insights into the molecular architecture of hardwood lignins by chemical degradative methods. *Res. Chem. Intermed.,* **21** : 397-412.

Lewis, N. G. and Davin, L. B. (1994). Evolution of lignan and neolignan biochemical pathways. In: Evolution of Natural Products. (Eds. Nes, D.) ACS Symp. Series, Vol. 562, Am. Chem. Soc., Washington DC., pp 202-246.

Lewis, N. G. and Yamamoto, E. (1990). Lignin: Occurrence, biogenesis and biodegradation. *Annu. Rev. Plant Physiol. Plant Mol. Biol.,* **41** : 455-496.

Li, L., Popko, J. L., Umezawa, T. and Chiang, V. L. (2000). 5-hydroxyconiferyl aldehyde modulates enzymatic methylation for syringyl monolignol formation, a new view of monolignol biosynthesis in angiosperms. *J. Biol. Chem.,* **275** : 6537-6545.

Li, L., Zhou, Y., Cheng, X., Sun, J., Marita, J. M., Ralph, J. and Chiang, V. L. (2003). Combinatorial modification of multiple lignin traits in trees through multigene cotransformation. *Proc. Natl. Acad. Sci. USA,* **100** : 4939-4944.

López-Molina, D., Heering, H. A., Smulevich, G., Tudela, J., Thorneley, R. N. F., García-Cánovas, F. and Rodríguez-López, J. N. (2003). Purification and characterization of a new cationic peroxidase from fresh flowers of *Cynara scolymus* L. *J. Inorg. Biochem.,* **94** : 243-254.

Michon, T., Chenu, M., Kellershon, N., Desmadril, M. and Guéguen, J. (1997). Horseradish peroxidase oxidation of tyrosine-containing peptides and their subsequent polymerization: a kinetic study. *Biochemistry,* **36** : 8505-8513.

Milloni, D., Sado, P. E., Stacey, N. J., Roberts, K. and McCann, M. (2002). Early gene expression associated with the commitment and differentiation of a plant tracheary element is revealed by cDNA-amplified fragment length polymorphism analysis. *Plant Cell,* **14** : 2813-2824.

Nappi, A. J. and Vass, E. (2001). The effects of nitric oxide on the oxidations of L-Dopa and dopamine mediated by tyrosinase and peroxidase. *J. Biol. Chem.,* **276** : 11214-11222.

Neill, S. J., Desikan, R. and Hancock, J. T. (2003). Nitric oxide signaling in plants. *New Phytol.,* **159** : 11-35.

Nose, M., Bernards, M. A., Furlan, M., Zajicek, J., Eberhardt, T. L. and Lewis, N. G. (1995). Towards the specification of consecutive steps in macro-molecular lignin assembly. *Phytochemistry,* **39** : 71-79.

Ogawa, K., Kanematsu, S. and Asada, K. (1997). Generation of superoxide anion and localization of CuZn-superoxide dismutase in the vascular tissue of spinach hypocotyls: their association with lignification. *Plant Cell Physiol.,* **38** : 1118-1126.

Osakabe, K., Tsao, C. C., Li, L., Popko, J. L., Umezawa, T., Carraway, D. T., Smeltzer, R. H., Joshi, C. P. and Chang, V. L. (1999). Coniferyl aldehyde 5-hydroxylation and methylation direct syringyl lignin biosynthesis in angiosperms. *Proc. Natl. Acad. Sci., USA,* **96** : 8955-8960.

Pignocchi, C. and Foyer, C. H. (2003). Apoplastic ascorbate metabolism and its role in the regulation of cell signalling. *Curr. Opin. Plant Biol.,* **6** : 379-389.

Ralph, J., Hatfield, R. D., Quideae, S., Helm, R. F., Grabber, J. H. and Jung, H. J. G. (1994). Pathway of *p*-coumaric acid incorporation into maize lignin as revealed by NMR. *J. Am Chem. Soc.,* **116** : 9448-9456.

Ralph, J., Peng, J., Lu, F., Hatfield, R. D. and Helm, R. F. (1999). Are lignins optically active?. *J. Agric. Food Chem.,* **47** : 2991-2996.

Ros Barceló, A. (1997). Lignification in plant cell walls. *Int. Rev. Cytol.,* **176** : 87-132.

Ros Barceló, A. (1998). The generation of H_2O_2 in the xylem of *Zinnia elegans* is mediated by an NADPH-oxidase-like enzyme. *Planta,* **207** : 207-216.

Ros Barceló, A. (1999). Some properties of the H_2O_2/O_2^- generating system from the lignifying xylem of *Zinnia elegans. Free Rad. Res.,* **31** : S147-154.

Ros Barceló, A. and Pomar, F. (2001). Oxidation of cinnamyl alcohol and aldehydes by a basic peroxidase from lignifying *Z. elegans* hypocotyls. *Phytochemistry,* **57** : 1105-1113.

Ros Barceló, A., Pomar, F., Ferrer, M. A., Martínez, P., Ballesta, M. C. and Pedreño, M. A. (2002). In situ characterization of a NO-sensitive peroxidase in the lignifying xylem of *Zinnia elegans. Physiol. Plant,* **114** : 33-40.

Schuler, M. A. (1996). Plant cytochrome P450 monooxygenases. *Crit. Rev. Plant Sci.,* **15** : 235-284.

Shibata, H., Kono, Y., Yamashita, S., Sawa, Y., Ochiai, H. And Tanaska, K. (1995). Degradation of chlorophyll by nitrogen dioxide generated from nitrite by the peroxidase reaction. *Biochim. Biophys. Acta,* **1230** : 45-50.

Shininger, T. L. (1979). The control of vascular development. *Annu. Rev. Plant Physiol.,* **30** : 313-337.

Stamler, J.S., Singel, D.J. and Loscalzo, J. (1992). Biochemistry of nitric oxide and its redox-activated forms. *Science,* **258** : 1898-1902.

Tsai, A. L. (1994). How does NO activate hemeproteins?. *FEBS Lett,* **341** : 141-145.

Terashima, N. and Fukushima, K. (1989). Biogenesis and structure of macromolecular lignin in cell wall of tree xylem as studied by microautoradiography. In: Plant Cell Wall Polymers: Biogenesis and Biodegradation. (Eds. Lewis, N. G. and Paice, M. G.) ACS Symp. Series, Vol. 399, Am. Chem. Soc., Washington DC., pp 160-168.

Uchida, A., Jagendorf, A. T., Hibino, T., Takabe, T. and Takabe, T. (2002). Effects of hydrogen peroxide and nitric oxide on both salt and heat stress tolerance in rice. *Plant Sci.,* **163** : 515-523.

Van Camp, W., Van Montagu, M. and Inze, D. (1998). H_2O_2 and NO: redox signals in disease resistance. *Trends Plant Sci.,* **3** : 330-334.

Van der Vliet, A., Eiserich, J. P., Halliwell, B. and Cross, C. E. (1997). Formation of reactive nitrogen species during peroxidase-catalized oxidation of nitrite. *J. Biol. Chem.,* **272** : 7617-7625.

Wendehenne, D., Pugin, A., Klessig, D. F. and Durner, J. (2001). Nitric oxide: comparative synthesis and signaling in animal and plant cells. *Trends Plant Sci.,* **6** : 177-183.

Whetten, R. and Sederoff, R. (1995). Lignin biosynthesis. *Plant Cell,* **7** : 1001-1013.

Xian, M., Li, X., Tang, X., Chen, X., Zheng, Z., Galligan, J. J., Kreulen, D. L. and Wang, P. G. (2001). *N*-Hydroxy-derivatives of guanidine based drugs as enzymatic NO donors. *Bioorg. Med. Chem. Lett.,* **11** : 2377-2380.

Yamasaki, H., Shimoji, H., Ohsiro, Y. and Sakihama, Y. (2001). Inhibitory effects of nitric oxide on oxidative phosphorylation in plant mitochondria. *Nitric oxide: Biol. Chem.,* **5** : 261-270.

Yamazaki, I. and Yokota, K. N. (1973). Oxidation states of peroxidase. *Mol. Cell Biochem.,* **2** : 39-52.

Yonetani, T., Yamamoto, H., Erman, J. E., Leigh, J. S., Jr., and Reed, G. H. (1972). Electromagnetic properties of hemoproteins. *J. Biol. Chem.,* **247** : 2447-2455.

Chapter 14

NITRIC OXIDE AS A SIGNALING MOLECULE IN EARLY PLANT DEVELOPMENT AND OXIDATIVE STRESS

Yoki Kwok-Chu Butt, John Hon-Kei Lum and Samuel Chun-Lap Lo⋆

Dept. of Applied Biology and Chemical Technology, The Hong Kong Polytechnic University, Kowloon, Hong Kong SAR

⋆Corresponding author : E-mail : bcsamlo@inet.polyu.edu.hk
Tel. : (852)-27666686; Fax : (852)-23649932

Summary

Nitric oxide (NO) is a highly diffusible and reactive gas with a small molecular size of 30Da. As a small uncharged molecule, NO diffuses and permeates biological membranes freely and rapidly. With its unpaired electron, NO has a strong tendency to interact rapidly with different molecular targets, including oxygen (O_2), superoxide (O_2^-) and transition metals for producing diverse arrays of metabolic effects in cells (Charaborti et al., 1999; Grisham et al., 1999). Such characteristics enable NO to act as a multifunctional signaling molecule in a large number of diverse mammalian physiological processes, including relaxation of smooth muscle, inhibition of platelet aggregation, neural communication and immune regulation (Marletta, 1989).

As a consequence of the physiological importance of NO in mammals, research on its effect in plant has gained considerable attention in recent years. The presence of NO in plant has been known for some

In : Nitric Oxide Signaling in Higher Plants, 2004
(Eds Jose R. Magalhaes, Rana P. Singh and Leonidas P. Passos)
Studium Press, LLC, Houston, USA, pp 309-341

time (Cooney et al., 1994). However, very little is known on the functions of NO in plants. Similar to the case in the animal kingdom, increasing evidence suggested that NO is at heart of many important physiological processes in plants. Acting as a phytohormone, NO was found to function as an endogenous growth regulator in developmental processes in plants (Beligni et al., 2001). Other studies suggested that NO acts as a signaling molecule in plant defense responses against pathogens (Delledonne et al., 1998; Durner et al., 1999; McDowell et al., 2000). Importantly, several downstream key components known to operate in the NO signaling pathway in animals were also found to operate in plants. This suggested that many components of the NO signaling pathway are shared between plants and animals.

In this chapter, we intended to give an account of our current understanding of the role of NO as a signaling molecule in early plant development as well as during oxidative stress. However, we would like to start with a brief summary of our current understanding on how NO is generated in plants.

Keywords : Nitric oxide, plant development, oxidative stress.

1. GENERATION OF NITRIC OXIDE IN PLANTS

Nishimura and coworkers (1986) was first to report that plants not only accumulate but also metabolize atmospheric NO. Subsequent studies (described below) showed that NO could be produced in *vivo* from many possible sources (Figure 1). Amounts of NO produced from these sources are not only more than sufficient to act as signals but also enough to be released into the atmosphere in detectable quantities.

1.1. Non-enzymatic Production of NO

Cooney *et al.* (1994) was first to report results of mechanistic studies on NO generation in plants. They found that NO was generated non-enzymatically from nitrites through a light-mediated conversion of carotenoids. Under acidic conditions (pH 3-6), nitrite is chemically reduced by ascorbic acid to produce NO, dehydroascorbate and a semi-hydroascorbyl radical (Henry *et al.*, 1997). This chemical conversion may occur in the chloroplasts and also in the apoplastic

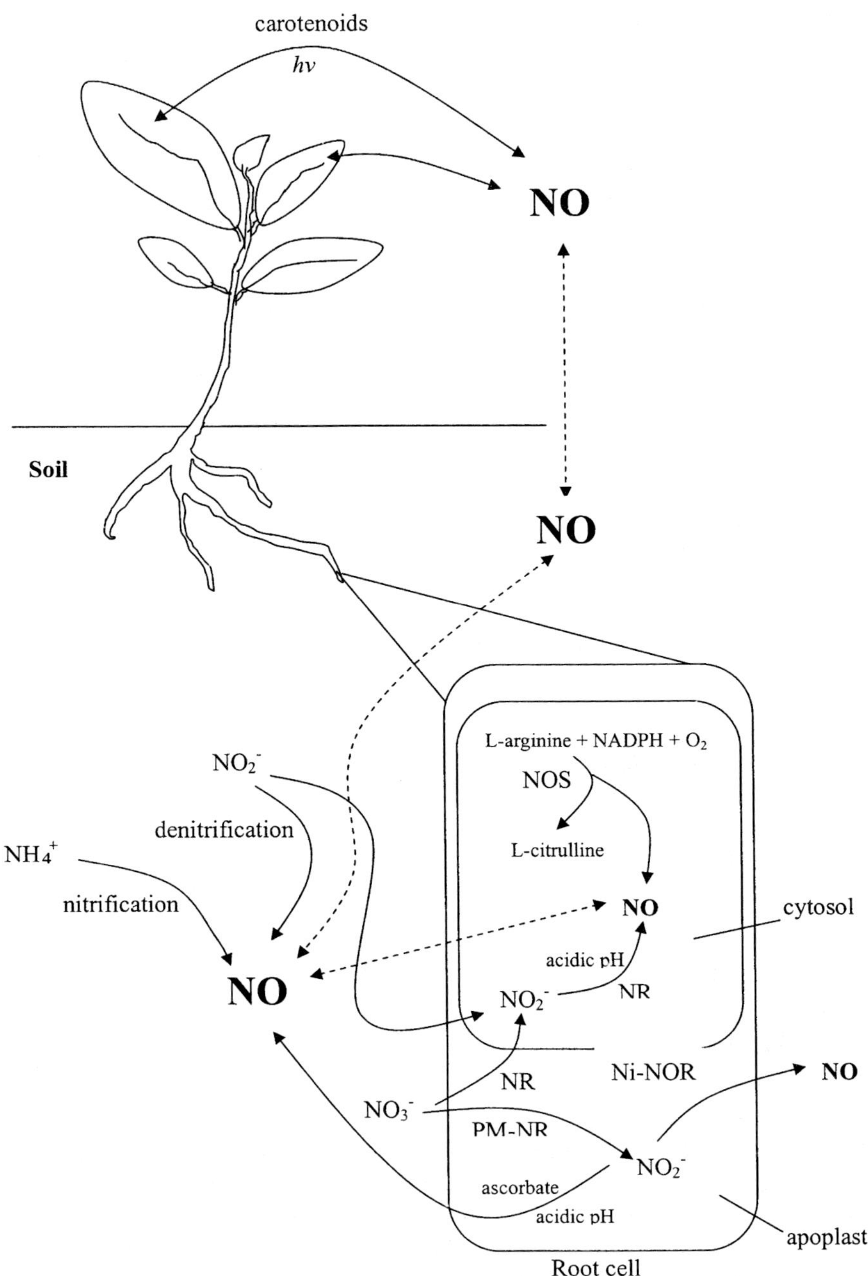

Figure 1 : Possible sources of NO production in plants.

space for which the presence of ascorbic acid had been reported (Horemans *et al.*, 2000). It should be noted that this chemical reaction requires an undissociated acid form of nitrite but such form rarely exists at normal physiological pH. Therefore, this NO formation has probably to occur in tissues and compartments of organelles under acidic conditions (Weitzberg *et al.*, 1998). In addition, an extracellular accumulation of nitrite, which may occur under anaerobic conditions, is also required. Hence, this non-enzymatic generation of NO is probably an uncommon event as intracellular nitrite concentration is always kept extremely low. High concentration of intracellular nitrite is devastating as nitrite as well as its acidic form, nitrous acid, are highly toxic. Nitrite is rapidly reduced to ammonia by nitrite reductase (NiR) in normal plants.

1.2. Enzymatic Production of NO

In plants, NO can be produced by several enzymatic pathways. NO can be released from inorganic nitrogen sources during nitrogen assimilation pathways in the Nitrogen Cycle. NO can also be produced by a putative but yet to be confirmed constitutive plant nitric oxide synthase.

1.2.1. Nitrate reductase

Nitrate reductase (NR) is the key enzyme of nitrogen assimilation. It catalyzes the reduction of nitrate to nitrite using NADPH or NADH as an electron donor. In soybeans, NR produces NO via the reduction of nitrite (Dean *et al.*, 1998; Klepper, 1990). These results were repeated in maize and spinach (Rockel *et al.*, 2002). In addition, Yamasaki *et al.* (1999) showed that the reduction of nitrite to NO could be a side-reaction of NR as it requires the presence of high concentrations of nitrite (Km of NR from nitrite is about 300 μM; Yamasaki *et al.*, 2000). As described earlier, accumulation of high concentrations of nitrite is not favored in healthy plants, therefore, the formation of NO by NR seems unlikely under normal conditions. Only under anaerobic condition does nitrite starts to accumulate *in vivo* (Botrel *et al.*, 1996). The formation of NO by NR and its accumulation was suggested to occur only when cells are in transition to unfavorable anaerobic conditions (Stohr *et al.*, 2002). This situation will occur more often in roots than in leaves. Additionally, the distribution of NR in plant organs seems to depend on nitrate supply. It was shown that NR activity in tobacco roots was induced with a limiting external nitrate concentration. Conversely, higher external nitrate was associated with a lowered NR activity (Stohr, 1999). Being freely exposed to the atmosphere and hence could be aerobic, it is believed that contribution of NO from NR activities in leaves is insignificant. Altogether, it seems that NR might produce some NO in leaves with significant amounts of nitrate, whereas this NR-mediated NO formation might probably occur in conditions when nitrate is limiting or/and under anaerobic conditions in the roots whereas NR is highly activated and nitrite is accumulating.

1.2.2. Nitrite : NO reductase

In roots, it was found that NO can also be produced by a newly found plasma membrane-bound enzyme, nitrite:NO reductase (Ni-NOR) (Stohr *et al.*, 2001). The mechanism of NO formation by Ni-NOR seems to be different from that mediated by NR as it uses reduced cytochrome c and not NAD(P)H as the electron donor. Moreover, this enzymatic NO formation is independent of oxygen and does not require an increased nitrite concentration as a pre-requisite. Meyer and Stohr (2000) showed that Ni-NOR and another plasma membrane-bound NR (PM-NR) might interact to produce NO at the apoplastic surface of plasma membranes of roots.

1.2.3. Nitric oxide synthase

In mammals, NO generation is mainly through a group of isoenzymes called nitric oxide synthase (NOS). NOS catalyzes the NADPH-dependent conversion of L-arginine to NO plus L-citrulline in the presence of FAD, FMN and tetrahydro-L-biopterin (BH_4). In some cases, CaM and calcium are also required (Knowles *et al.*, 1994). Recently, there is increasing evidence suggesting the occurrence of mammalian NOS-like enzyme in plants. Table 1 summarized reports which inferred the presence of NOS in actively growing tissues or cells/tissues after pathogen infection or under stress conditions by indirect methodologies. Based on the formation of L-citrulline from L-arginine, NOS-like activity was directly detected in extracts from leguminous plant *Mucuna hassjoo* (Ninnemann *et al.*, 1996), in *Lupinus albus* roots and its nodules (Cueto *et al.*, 1996), in *Zea mays* root tips and young leaves (Ribeiro *et al.*, 1999) and in isolated peroxisomes from *Pisum sativum* leaves (Barroso *et al.*, 1999).

As in mammals, pathogen infection induces NOS-like activity in plant. Durner *et al.* (1999) revealed that NOS activity was increased

Table 1 : Summary of results from the literature predicting the existence of a putative NOS in plant tissues.

Species	Tissues or cell types	Assays	References
Pisum sativum	Embryonic axes	Immunological analysis	Sen *et al.*, 1995
	Leaves	Gaseous NO emission sensitive to NOS inhibitors	Leshem *et al.*, 1996
	Peroxisomes from leaves	L-citrulline conversion assay, immunological and immunocyto-chemical analysis	Barroso *et al.*, 1999
Lupinus albus	Roots and nodules	L-citrulline conversion assay, NADPH diaphorase activity	Cueto *et al.*, 1996
Mucuna hassjoo	Total extract	L-citrulline conversion assay	Ninnemann *et al.*, 1996
Glycine max	*Pseudomonas syringae*-infected cell suspensions	NO production sensitive to NOS inhibitors	Delledonne *et al.*, 1998
	Embryonic axes	NADPH diaphorase activity	Caro *et al.*, 1999
Nicotiana tabacum	Tobacco mosaic virus-infected leaves	L-citrulline conversion assay, NO production sensitive to NOS inhibitors	Durner *et al.*, 1999
	Leaf epidermal cells treated with a fungal elicitor	NO production sensitive to NOS inhibitors	Foissner *et al.*, 2000
	Cytokinin-treated cell cultures	NO production sensitive to NOS inhibitors	Tun *et al.*, 2001

Table 1. Continued

Species	Tissues or cell types	Assays	References
	Tobacco mosaic virus-infected leaves	oxy-hemoglobin assay, Greiss assay, L-citrulline conversion assay	Chandok *et al.*, 2003
Zea mays	Root tips and young leaves	L-citrulline conversion assay, immunological and immuno-fluorescence analysis	Ribeiro *et al.*, 1999
Taxus brevifolia	Callus	NO production sensitive to NOS inhibitors	Pedroso *et al.*, 2000
Kalachoe daigremcntiana	Callus	NO production sensitive to NOS inhibitors	Pedroso *et al.*, 2000
Arabidopsis thaliana	*UV-B* treated cell suspensions	NO production sensitive to NOS inhibitors	Soheila *et al.*, 2001
Triticum aestivum	Embryonic axes	Immunological analysis	Kuo *et al.*, 1995
Phaseolus aureus	H_2O_2 treated leaves	Chemiluminescence assay	Lum *et al.*, 2002

in a resistant strain of tobacco plant upon infection with tobacco mosaic virus. This phenomenon was also seen in a plant-pathogen interaction between soybean cell suspension cultures and an incompatible *Pseudomonas syringae* strain (Delledonne *et al.*, 1998). Moreover, an increased NO production was detected in tobacco plant treated either with cryptogein, a fungi elicitor from *Phytophthora cryptogea* (Foissner *et al.*, 2000) or with cytokinin, a plant hormone (Tun *et al.*, 2001). Our group is the first to report a functional NOS activity in hydrogen peroxide-treated mature mung bean leaves (Lum *et al.*, 2002). Additionally, NO production was stimulated by various abiotic stresses, such as in the leaves and callus of *Taxus brevifolia* and *Kalachoe daigremontiana* after centrifugation (Pedroso *et al.*, 2000), and in *Arabidopsis thaliana* with *UV-B* treatment (Soheila *et al.*, 2001). All NO production described above can be suppressed by the addition of L-NAME, a known mammalian NOS inhibitor. Therefore, it is likely that a putative plant NOS exist. Moreover, many of the NOS activity reported above are calcium-dependent and some are triggered within minutes, resembling the mammalian constitutive NOS. Whether this yet to be confirmed putative plant NOS is in fact a constitutive enzyme is currently unclear.

Previously, with the use of mammalian NOS antibodies in western blot analysis, several researchers presented data showing the existence of NOS-like protein in plant extracts (Sen *et al.*, 1995; Ribeiro *et al.*, 1999; Barroso *et al.*, 1999). Using a rabbit nNOS antibody, a single band of 105.4 kDa was detected in pea embryonic axes, whereas 2 distinct bands, respectively of 89.7 and 57.5kDa, were observed in wheat embryonic axes (Sen *et al.*, 1995). Ribeiro *et al.* (1999) found a 166 kDa protein using mouse iNOS antibody and several immunoreactive proteins, ranging from 51 to 166 kDa, using rabbit nNOS antibody in maize root tips and young leaves, respectively. Moreover, mammalian NOS antibodies were also used to localize NOS in maize root tips (Ribeiro *et al.*, 1999), pea leaf peroxisomes as well as chloroplasts (Barroso *et al.*, 1999). With an immunohistochemical staining, mammalian iNOS antibody localized immunoreactive protein in the cytosol of the division zone, as well as in the nucleus of the elongation zone of roots (Ribeiro *et al.*, 1999). Such growth-phase dependent subcellular localization is specific to plant.

Although all these studies support the presence of NOS in plant, no plant NOS gene that is homologous to the mammalian one has been cloned to date. Furthermore, no homologue of mammalian-like NOS has been found in the *Arabidopsis* genome (The *Arabidopsis* Genome Initiative, 2000). In those immunological studies described earlier, some of the NOS-immunoreactive proteins are in the same molecular mass range as described for mammalian NOS (Pollock *et al.*, 1991). However, these immunoreactive proteins have never been successfully identified. By our hands, we firstly reported that mammalian NOS antibodies recognize many NOS-unrelated plant proteins (Butt *et al.*, 2003). These results suggested that plant NOS may be functionally similar to mammalian NOS. However, it probably has a very different primary amino acid sequences to known NOS isoforms. Recently in tobacco, Chandok and coworkers (2003) reported a variant of P protein of glycine decarboxylase complex which is functionally similar to mammalian inducible NOS. Primary amino acid sequence of this P protein of glycine decarboxylase complex is very different from that of mammalian iNOS. This reiterated the opinion that plant and mammalian NOS are very different structurally. Nevertheless, although the source of NO in plant is still controversial and indeed NO may arise from more than one sources, it is no doubt that NO is an important signaling molecule in many plant physiological processes. The following sections will focus on the role of NO in early plant development initially and the role of NO as a signaling molecule under oxidative stress will be discussed later.

2. ROLE OF NITRIC OXIDE IN EARLY PLANT DEVELOPMENT

Plant growth and development are regulated by interactions between both external cues (*e.g.* light, gravity and temperature) and internal regulating factors (*e.g.* hormones). Plant senses changes in environmental light conditions with red and far-red absorbing photoreceptors called phytochromes. Bowler *et al.* (1994) reported that activation of phytochromes is mediated via heterotrimeric G proteins, leading to a CaM/Ca^{2+}-dependent or cGMP-dependent pathways. However, details about these mechanisms are largely unknown. In developing plants, many hormones also regulate growth and development. However, it is unknown whether light and hormones act independently in the developmental processes. Whether plant hormones are involved in the events initiated by phytochromes is also unknown (Chory *et al.*, 1994).

In recent years, NO has been demonstrated to act similar to hormones, affecting various developmental processes in plant (Table 2, Gouvea *et al.*, 1997; Leshem *et al.*, 1998; Beligni *et al.*, 2000; Pagnussat *et al.*, 2003). Most of such information were acquired by studies with exogenous application of NO donors and NO scavengers. Depending on the conditions, NO was either promotive or suppressive in plant development. It was shown that NO takes part in light-mediated events, such as induction of seed germination, de-etiolation as well as inhibition of hypocotyls and internodes elongation. Additionally, NO was also found to be involved in hormone-mediated root development (Table 2). Some studies have shown that the effect of NO is more potent than that of traditional plant hormones. Actions of NO were found to overlap with that of some plant hormones, such as indole acidic acid (IAA) (Gouvea *et al.*, 1997) as well as giberellic acid (GA_3) (Beligni *et al.*, 2000). Acting as a growth regulator, NO

Table 2 : Effects of NO seen in different developmental processes. Brackets in the second column are concentrations of reagents used.

Developmental process	NO-mediated effect	Reference
Seed germination	Induction (with 100mm SNP)	Beligni *et al.*, 2000
	Induction (with 1-10mM SNP with single red light)	Grubisic *et al.*, 1990
Programmed cell death of aleurone cells	Delay of aleurone layer (with 100-300mM)	Beligni *et al.*, 2002
Root growth and differentiation	Induction (with 100pM SNP)	Gouvea *et al.*, 1997
	Induction (with 10mM SNP)	Pagnussat *et al.*, 2003
Leaf growth	Induction	Leshem *et al.*, 1996
De-etiolation	Stimulation (with 100mm SNP)	Beligni *et al.*, 2000
Hypocotyl and internode elongation	Inhibition (with 100mm SNP)	Beligni *et al.*, 2000
Senescence and maturation	Delay (with SIN-1 & PBN)	Leshem *et al.*, 1998

can interact with abscisic acid (ABA) (Neill *et al.*, 2002; Garcia-Mata *et al.*, 2002) and counteract with ethylene (Leshem *et al.*, 1998). Importantly, several researchers presented data demonstrating that NO production can be triggered by plant hormones, such as IAA (Pagnussat *et al.*, 2003) or cytokinin (Tun *et al.*, 2001). Several groups also supported that NO regulates developmental processes through the activation of the GC-catalyzed synthesis of cGMP (Pfeiffer *et al.*, 1994; Durner *et al.*, 1999; Pagnussat *et al.*, 2003).

2.1. NO as Effector in Light-Mediated Early Plant Developmental Processes

2.1.1. Stimulation of seed germination

Seed germination is a complex physiological process triggered by imbibition of water and the release from possible dormancy mechanisms by appropriate triggers (e.g. red light via phytochrome). Besides, several plant hormones are also involved in seed dormancy and germination. For instance, abscisic acid (ABA) establishes dormancy during plant embryo development, while gibberellins (GA) breaks dormancy and is required for germination.

Recently, NO and its related nitrogen oxides have been reported to have similar actions as gibberellic acid, one of the gibberellins, as stimulators of seed germination. After supplying seeds of *Paulownia tormentosa* with optimal concentrations (1-10mM) of sodium nitroprusside (SNP) or some other inorganic and organic nitrates, Grubisic and Konjevic (1990) revealed that only a single red light pulse is sufficient to induce germination in these seeds. A longer period of light irradiation is normally required for these seeds. The participation of NO in light-mediated seed germination was also demonstrated by Beligni and Lamattina (2000). The dark-imposed dormancy of lettuce (*Lactuca sativa* L. cv. Grand Rapids) seeds was broken by the addition of two NO donors, 100mM SNP or S-nitroso-N-acetylpenicillamine (SNAP), as in the same extent with a similar concentration of GA_3. The effect with lower concentration of SNP on seed germination in darkness was also more potent than similar concentration of GA_3. Only 2% germination was observed with 10mM GA_3, whereas 50% germination occurred when the seeds were imbibed with 10mM SNP. This SNP-mediated effect on seed germination can be arrested by the addition of carboxy-PTIO, a NO scavenger. Hence, this effect is specific to NO.

During germination, GA from the embryo diffuses to the aleurone layer and stimulates the synthesis and secretion of alpha-amylase and other hydrolytic enzymes. The aleurone layer is made up of aleurone cells and surrounded by the starchy endosperm and embryo. The aleurone cells contain a suite of reactive oxygen species (ROS)-metabolizing enzymes, which are controlled by ABA as well as GA_3. During normal seedling development, these hydrolytic enzymes degrade the storage reserves in the endosperm and provide the growing embryo with nutrients. Following GA-induced enzyme secretion, the aleurone layer is no longer required for seedling development and hence starts to degrade (Bethke *et al.*, 1999; Fath *et al.*, 2002). ROS produced by fatty acyl coenzyme A and glyoxysomes are key mediators for GA-induced programmed cell death (PCD) in barley aleurone layers. Recently, Beligni *et al.* (2002) reported that NO is synthesized by barley aleurone cells and both SNP and SNAP delay GA-induced PCD of aleurone cells. The effect of these NO donors on cell viability can be reversed by NO scavengers and mimicked by antioxidant, indicating that NO acts as an antioxidant to protect the aleurone cell from oxidative stress. The antioxidant role of NO was supported by data from several studies. NO was found to delay senescence (Mallick *et al.*, 2000), prevent ROS-induced cytotoxicity (Beligni *et al.*, 1999), and slow chlorophyll loss resulting from pathogen infection (Laxalt *et al.*, 1997). Besides acting as an antioxidant, it was also suggested that NO acts as a direct signaling molecule in modulating PCD in barley aleurone cells. However, the interaction between NO and GA in seed germination and development is still poorly understood.

On another front, Sen and Cheema (1995) suggested that NO initiates a cascade of events leading to seed germination through GC-mediated cGMP production. cGMP is an important component of NO-mediated signaling pathways (McDonald *et al.*, 1995). As in mammals, NO is suggested to activate GC and hence stimulate cGMP production in plant. Therefore, this NO-induced cGMP production triggers an anthocyanin biosynthesis and initiates a subsequent cascade of events leading to seed germination. This NO-mediated cGMP production was also found to stimulate a modified light requirement of phytochrome-broken dormancy of *Paulownia tomentosa* seeds through participation in a phytochrome phototransduction signaling pathway (Grubisic *et al.*, 1990). However, positive identification of GC in plant is still controversial.

2.1.2. De-etiolation

Plants appeared etiolated when they were grown in the dark or under short-day photoperiods. Once the seedlings become etiolated, they will appear tall and spindly with rudimentary leaves. Additionally, its chloroplasts appeared undeveloped and the amount of chlorophyll these leaves processed will be extremely low or even none Beligni and Lamattina (2000) reported that NO treatment increased chlorophyll levels of wheat seedlings grown in the dark. When compared with controls, chlorophyll contents of NO-treated seedlings could be increased by short-term pulses of white or red light indicating a putative role for NO in de-etiolation. Moreover, the increase of chlorophyll content in NO-treated seedlings will be even higher when the seedlings were subjected to wounding or infected with fungi. It was shown that NO protects chlorophyll levels in stressed potato leaves (Laxalt *et al.*, 1997; Beligni *et al.*, 1999). This suggests a possible connection between NO, light and stress signaling pathways in plant.

In addition to the lack of light as a stimulus, iron deficiency also decreased chlorophyll biosynthesis and chloroplast development, leading to chlorosis. Exposure to NO was reported to completely prevent leaf interveinal chlorosis in maize growing under iron-deficient conditions. With limited iron in supply, mesophyll cells appeared to contain plastids with few photosynthetic lamella and rudimentary grana. In complete contrast, when treated with NO, mesophyll chloroplasts appear completely developed (Graziano *et al.*, 2002). Moreover, it was found that exposure to NO did not increase the endogenous iron concentration in maize plant. Further, NO was even able to revert the chlorotic phenotype of the iron-inefficient maize mutant *yellow stripe* 1 and *yellow stripe* 3, both impaired in the iron uptake mechanisms. Altogether it was suggested that NO might improve or/and facilitate iron availability for chlorophyll synthesis within the plant, details of the mechanisms involved remains to be elucidated.

2.1.3. Inhibition of hypocotyl and internodal elongation

Another process in the life cycle of a plant that is affected by light is inhibition of hypocotyls and internodal elongation. The inhibition seen is caused by reverting the etiolated phenotype which consists of chlorotic and longer organs. In 3 plant species, *Arabidopsis thaliana, Lactuca sativa* and *Solanum tuberosum*, grown in darkness or low light intensity, nanomolar amount of NO produced from 100mM SNP was able to reduce hypocotyl and internode elongation significantly (Beligni *et al.*, 2000). These responses were found to be dose-dependent and can be arrested by addition of carboxy-PTIO. Although mechanistic study on these responses has not been reported, these authors suggested that NO stimulates this type of photomorphogenesis either dependent or independent of light receptors.

2.2. NO as a Signal in Root Development

NO seems to participate in plant cell expansion and differentiation. The notion that NO may participate in signal pathways associated with plant cell growth agrees with experimental results found when the effects of NO donors on leaf and root tissues were studied. Similar to the situation as documented in leaves by Leshem *et al.* (1996), it was found that nanomolar to low micromolar amounts of the NO donors (such as SNP, sodium nitrite (SN) and nitrosoglutathione (NGLU)) induced maize root tip expansion in a dose-dependent manner. However, an inhibitory effect occurred with higher concentrations of NO (Gouvea *et al.*, 1997). In the same manner, micromolar amount of SNP was also found to promote adventitious root formation in cucumber hypocotyl explants (Pagnussat *et al.*, 2003). This effect observed was in the same extent as seen by the induction elicited by the plant hormone IAA (Pagnussat *et al.*, 2003). It was shown that the growth-promoting effect mediated by either IAA or SNP could be reduced by the addition of NO scavengers. Endogenous NO level is found to be lowered in auxin-depleted explants when compared with that in the non-depleted ones. However, this decreased level can be restored by exogenously applied IAA. Therefore, NO might be a downstream component in the auxin-mediated adventitious rooting process. Participation of NO in root growth and development was strengthened by the documentation of NOS-like protein in subcellular locations in maize roots during specific phases of its cell growth (Ribeiro *et al.*, 1999). These workers reported that NO could activate a nuclear transcription factor necessary for rapid growth in the root elongation zone.

As mentioned before, although NOS-like enzyme was found in plant root tissues, NO can also be produced from nitrite by Ni-NOR in the apoplast. Based on the supply of nitrate in the soil, apoplastic PM-NR can reduce nitrate to nitrite, which further reduces to NO via Ni-NOR present in the apoplast. Higher nitrate concentrations in apoplast lead to higher rates of NO formation. The NO produced (intracellular NO) might be released to the soil and eventually to the atmosphere as gaseous NO. Intracellular NO might be consumed either by assimilation to organic N-compounds (Stohr *et al.*, 2001) or lead to the detrimental effects on growth (Stohr, 1999). Kroncke *et al.* (1997) and Durner *et al.* (1999) supported that NO may function as a signaling molecule in plant root systems. Durner *et al.* (1999) explained that due to its apolarity, NO can easily enter the cells via diffusion across membranes and induce secondary reaction in the cytosol, such as activation of GC, and hence increased cGMP production.

NO released from the root system was proposed to function as a signal when it finally reached the shoots. This may be the case for stomata closure response to drought (Garcia-Mata *et al.*, 2001, 2002). Neill *et al.* (2002) also demonstrated that exogenous and endogenous NO contributes in the ABA-dependent stomata closure. Both NO and ABA can function independently or interact synergistically with each other (Neill *et al.*, 2002). It was proposed that drought affects the root system initially and the roots responded by increasing the release of NO. This NO could be released to the atmosphere directly where it may reach the shoot and induce stomatal closure by the same way as ethylene does (Taylor *et al.*, 1986), although in a lesser extent. Ethylene is a well-known long-distance messenger in the regulation of plant development. Recently, its formation was reported to inversely correlate with the NO formation in various plant species (Leshem *et al.*, 1998; Magalhaes *et al.*, 2000).

In mammals, as mentioned before, NO stimulates the activation of GC which leads to the production of cGMP. In plant, a transient increase of cGMP occurred after various stimuli, such as gibberellic acid treatment of barley aleurone (Penson *et al.*, 1996), light stimulation of bean cells (Brown *et al.*, 1989) and NO treatment of spruce needles (Pfeiffer *et al.*, 1994). Furthermore, the effect of NO-releasing compounds on phytochrome-controlled germination of *Paulownia tormentosa* seeds has been attributed to NO-dependent cGMP production (Giba *et al.*, 1998). The participation of cGMP was also demonstrated in NO signaling in auxin-induced adventitious root development in cucumber (Pagnussat *et al.*, 2003). Consistent with cGMP acting as a second messenger in tobacco (Durner *et al.*, 1998), an inhibitor of GC, LY83583, reduced adventitious root formation in both IAA- and SNP-treated cucumber explants. Such inhibition could be relieved by the addition of cell-permeable cGMP analogue, 8-Br-cGMP. Moreover, the treatment with 8-Br-cGMP alone in non-depleted explants did not show morphological differences with that of the SNP- and IAA-treated explants, indicating that cGMP synthesis is required and might be sufficient for adventitious rooting. In addition, cyclic nucleotide phosphodiesterase (PDE) was found to regulate the endogenous cGMP level within cells. A specific and potent inhibitor of PDE, sildenafil citrate, was able to induce adventitious rooting in non-depleted and auxin-depleted cucumber explants (Pagnussat *et al.*, 2003). Similar response occurred in explants treated either with IAA or SNP. However, when compared with those treated with 8-Br-cGMP, only 60% of root numbers were observed in the explants treated with sildenafil citrate. This observation may be explained by the fact that only low level of cGMP was synthesized in IAA-depleted explants. This decreased cGMP production is the result of the lack of NO to stimulate GC. Thus, although the inhibition of PDE activity only increased cGMP level, its level would not reach those of IAA- or SNP-treated explants. On the other hand, light and Ca^{2+}-calmodulin were also identified as plant PDE effectors (Brown *et al.*, 1989). Thus, cGMP production in plants may be stimulated by phytochromes, which constitute part of the molecular events that control plant development.

Collectively, these results indicated that NO operates downstream of IAA promoting adventitious root development through the GC-catalyzed synthesis of cGMP (Pagnussat *et al.*, 2003). Based on the NO-mediated promoting effect on seed germination and root development, a NO signaling pathway in plant was suggested (Figure 2). As a consequence of appropriate triggers, such as hormones and light, NO can be produced enzymatically or non-enzymatically in plant tissues. The NO produced might activate GC and hence the cGMP-dependent transduction pathway. A potential target for cGMP could be a cGMP-dependent protein kinase. However, this plant protein kinase is yet to be isolated or cloned. In animal cells, cGMP acts via cyclic ADP ribose (cADPR). In plant, cADPR was found to

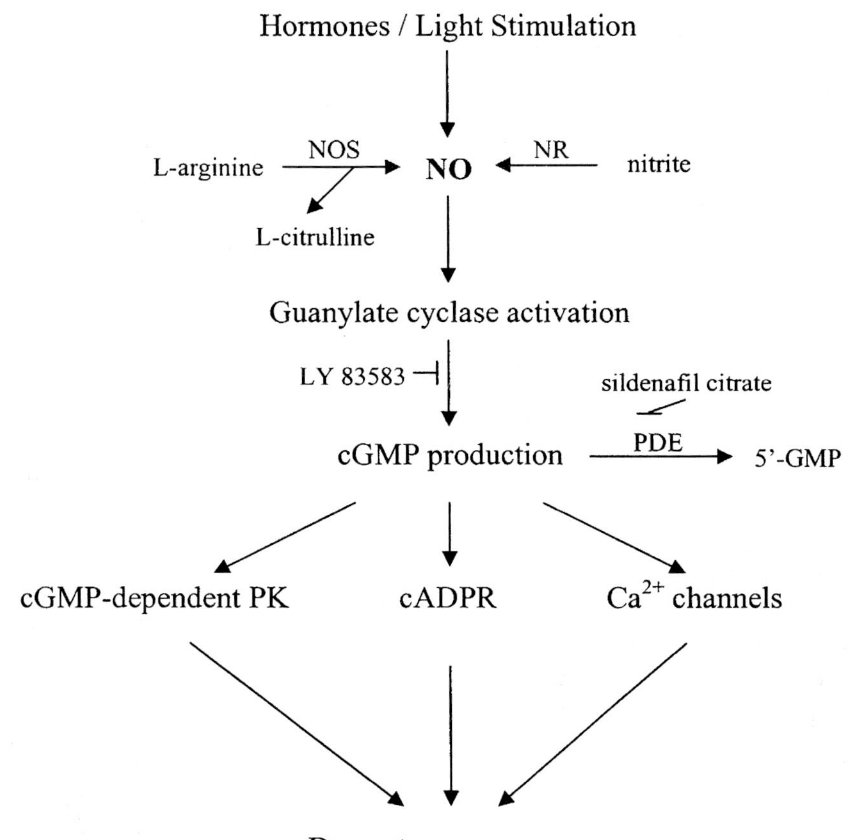

Figure 2 : Proposed NO signal pathway in plants.

be involved in defense genes that are NO-induced (Durner *et al.*, 1998). Moreover, cADPR regulated calcium levels in stomatal guard cells in response to the plant hormone, ABA (Leckie *et al.*, 1998). cADPR was found to elevate cytosolic free calcium in plant, thereby reiterated that it plays an important role in the signal transduction pathways (Navazio *et al.*, 2000). This, cADPR- and cGMP-induced responses can be inhibited by a calcium channel blocker (Durner *et al.*, 1998). Lastly, calcium was also suggested to be involved in a cascade, leading to activation of cell division and elongation (Kiegle *et al.*, 2000).

3. NITRIC OXIDE AS A SIGNALING MOLECULE IN OXIDATIVE STRESS

Oxidative stress is one of the most common abiotic stresses encountered by all aerobic organisms, especially plant. There are numerous different environmental conditions that expose plants to oxidative stress in which reactive oxygen species (ROS) are overproduced. Some ROS are produced exogenously, such as in air pollution or from various pesticides. On the other hand, some ROS can also be generated *in vivo*. For example, singlet oxygen and superoxide are continuously produced in spinach leaves under strong sunlight (Hideg *et al.*, 2002). ROS are also continuously produced during respiration and photosynthesis. Moreover, subsequent to an incompatible infection, strong oxidative bursts were commonly found (Doke *et al.*, 1996).

Indeed, ROS are continuously being generated and removed both enzymatically and non-enzymatically in plant. Under normal conditions, the ROS removing systems balanced or even outperformed the ROS generating system and therefore keep the amount of ROS at low level. However, under specific conditions, such as in plant defense responses, in certain growth/developmental processes and apoptosis, the generation and removal of ROS becomes imbalanced leading to excessive generation of ROS. These ROS are detrimental to plants as they lead to unwanted cellular toxicity. For example, hydroxyl radical (OH^-) can convert fatty acids to toxic lipid peroxides which destroy integrity of biological membranes. Moreover, hydroxyl radicals can cause perpetuating lipid peroxidation and damage nucleic acids as well as proteins. However, it is now very clear that plant uses ROS as important signals to modulate different cellular activities (Neill *et al.*, 2002; Kawano, 2003; Hoeberichts *et al.*, 2003). The production of different ROS is therefore a double-edged sword because their levels must be tightly controlled.

3.1. Enzymatic Generation and Removal of ROS in Plant

There are many types of ROS present in plant. However, the present chapter will only focus on ROS that are closely related to the growth and developmental process as well as the defense response in plants. ROS that play important signaling roles during these processes include hydroxyl radicals (OH^-), superoxide radicals (O_2^-), hydrogen peroxide (H_2O_2) and nitric oxide (NO). The sources of nitric oxide in plant have been discussed thoroughly in the previous sections. Therefore, we will concentrate on the production of other ROS in this section.

ROS can be generated both enzymatically and non-enzymatically in plant. However, the present review will focus on the enzymatic production of ROS in plants. Plants process many different enzymes capable of producing ROS. These include a plasma membrane localized NADH/NADPH oxidase (Groom *et al.*, 1996), a cell-wall localized peroxidase (Wojtaszek *et al.*, 1997), apoplastic amine, diamine and polyamine oxidase (Allan *et al.*, 1997) and protoplastic enzymes within mitochondria, chloroplasts and peroxisomes (del Rio *et al.*, 2002).

Among these enzymes, the NADPH oxidase and the peroxidase systems received the most attention. The NADPH system is a complex consists of an unusual b-type cytochrome with two subunits, the p22phox and gp91phox (Segal *et al.*, 1993). The analogue of the mammalian neutrophils gp91phox has been cloned in rice and *Arabidopsis* (Groom *et al.*, 1996; Keller *et al.*, 1998). However, genes encoding other components of the plant NADPH oxidase have not been discovered. It is therefore very likely that plant and animal regulate this system by different mechanisms. Another system is the various peroxidases that are used to generate H_2O_2 in plants. Peroxidases have been shown to generate H_2O_2 through a superoxide binding intermediate that requires neutral to slightly alkaline pH and in the presence of a suitable reductant (Wojtaszek *et al.*, 1997). Interestingly, a germin-like oxalate oxidase was found in barley that can also produce H_2O_2 during the hypersensitive response (Dumas *et al.*, 1995). It is therefore very common that more than one of these enzymes is involved, either directly or indirectly by interacting with other enzymes, in generating a single ROS in plant.

Besides, plants also evolved an array of enzymes to scavenge the ROS they encountered so as to control their levels. These include catalase, superoxide dismutase (SOD) and the glutathione (GSH/GSSG) system. Catalase is the main enzyme in plant to remove H_2O_2 by converting it into water. SOD catalyzed the dismutation of O_2^-, forming the relatively more stable H_2O_2. Last but not least, the GSH/GSSG system helps to maintain the cellular redox status and prevents oxidative cellular cytotoxicity.

3.2. Involvement of ROS and NO in Plant Physiological Processes

Both ROS and NO play important roles in many physiological processes in plant. In some cases they act independently but in other cases they act together to evolve a specific signal. There are also cases that ROS and NO react chemically to produce another reactive intermediate that serves in signal transduction, such as the formation of peroxynitrite from NO and O_2^-. In this section, we shall discuss some of the important physiological processes that NO and ROS act as indispensable signaling molecules.

3.2.1. Plant defense response

The oxidative burst during plant defense response is one of the most widely studied oxidative stresses in plant. The first step of the plant defense response involved the recognition or interaction between the avirulence molecules (Avr) of pathogens and the corresponding resistance (R) gene product of plant cells. This often leads to a localized region of cell death, the hypersensitive response (HR), around the site of pathogen infiltration (Dangl *et al.*, 1996). One of the most drastic responses in HR is the production of a wide range of ROS, mainly H_2O_2 and O_2^-, at the site of HR (Levine *et al.*, 1994). Moreover, NO has also been discovered to be generated in the incompatible plant-pathogen interactions (Delledonne *et al.*, 1998). Interestingly, the production of H_2O_2 and NO following HR is both a biphasic process. Upon recognition of the pathogens, H_2O_2 and NO accumulate at the site of interaction within 30 minutes (Phase I production). The Phase I production is followed by another much stronger oxidative burst (5 – 10 folds), known as the Phase II production (Dorey *et al.*, 1999). Further investigation shows that the Phase I production is usually non-specific, i.e. occurs in both compatible and incompatible interactions, where the Phase II production of ROS and NO occurs only in incompatible interactions. It was also found that only the Phase II production of H_2O_2 correlates with establishment of disease resistance. Although the oxidative burst occurs at the site of pathogen recognition, the ROS and NO can diffuse to distal part of the plant, interact with other signaling molecules including salicylic acid and jasmonic acid, to develop the systemic acquire resistance (SAR). The downstream signals for the ROS and NO were not well studied in plant defense response. However, it has been shown that ROS and NO induced the expression of several disease-related proteins, including phenylalanine ammonia lyase (PAL), pathogenesis-related protein (PR-1) and chalcone synthase (CHS) (Durner *et al.*, 1998). However, the immediate upstream signals for inducing these genes expression are unclear at present. H_2O_2, NO as well as peroxynitrite have been suggested to be the signals responsible for these genes expression (Dorey *et al.*,1999; Delledonne *et al.*, 1998; Alamillo *et al.*, 2001). It is very likely that the induction mechanisms were very complicated, as NO and ROS can both act independently or interact together, and may also involved other secondary messenger such as cGMP, Ca^{++} and cADPR (Durner *et al.*, 1998). Moreover, it is very difficult to study the effect of a single ROS or NO individually, as most of the ROS scavengers can eliminate more than one ROS.

3.2.2. Programmed cell death (PCD)

Apoptosis is an essential cellular process for all living organisms. It aimed at removing redundant, misplaced and damaged cells and is therefore vital to the maintenance and development of plant. Although the process of PCD is well characterized in animals, it is ill-defined in plant. Recent studies revealed that PCD in plant is similar to that of animal, with characteristics such as DNA laddering (breakage), cytochrome c release from mitochondria and the involvement of caspase-like proteolytic activity (Kim *et al.*, 2003; Hoeberichts *et al.*, 2003). It is now clear that ROS and NO also play important role in PCD. It was found that the individual increase in NO or ROS had different effects on initiating PCD than the simultaneous increase in the two reactive species (de Pinto *et al.*, 2002). NO or H_2O_2 generation alone had little effect on inducing cell death. However, the simultaneous increase in both NO and H_2O_2 concentration activated a process of cell death with the typical cytological and biochemical features of PCD, including cytoplasmic shrinkage and chromatin condensation. Moreover, the cell death process initiated by H_2O_2 and NO together is sensitive to metabolic inhibitors, as the addition of actinomycin prevents cell death significantly (de Pinto *et al.*, 2002). Another critical finding by Delledonne *et al.* (2001) suggested that not only the simultaneous presence of H_2O_2 and NO have effects on inducing PCD, the concentrations of NO and H_2O_2 are also critical (Delledonne *et al.*, 2001). The ratio of NO to H_2O_2 must be in a specific range (~ 0.3) in order to induce PCD. It is therefore evident that the controlling mechanism of ROS induced

PCD is very delicate and certainly requires further investigation before solid conclusions can be draw.

3.2.3. Wounding

It has long been found that wounding leads to the production of ROS, which in turn induced the expression of many wounding responsive genes. However, it is now evident that NO is also involved in the wounding response. The synthesis of proteinase inhibitor I protein is a well studied wounding response. It was found that NO can block the synthesis of proteinase inhibitor I in tomato plant after wounding (Orozco-Cardenas *et al.*, 2002). However, wounding itself did not trigger the production of NO. More importantly, the production of H_2O_2 in response to wounding is also suppressed by more than 50% by NO. Presence of NO was found to interfere the wounding responsive gene expression in the signaling pathway downstream from jasmonic acid, but before the steps that generate H_2O_2. Indeed, the signaling pathway involved is very complicate and may even require the participation of ethylene. Nevertheless, the specific molecular target for NO inhibition of wound signaling remains to be identified. Besides, the expression of the ipomoelin (IPO) gene was used to monitor the wounding response in sweet potato. It was found that the IPO gene expression was induced by H_2O_2 irrespective of whether the plant sustains an injury or not (Jih *et al.*, 2003). Moreover, the signal transduction of H_2O_2 after wounding probably depends on the NADPH oxidase, which produces H_2O_2 endogenously. The same study also demonstrated that the production of H_2O_2 was strongly suppressed by the presence of NO. However, unlike results from the previous study, production of NO was observed upon wounding in potato. Therefore both H_2O_2 and NO were produced after wounding to modulate the plant's defense system and may further stimulate plants to protect themselves from invasions by pathogens and herbivores (Jih *et al.*, 2003).

3.3. Effect of NO on Redox Signaling Enzymes

3.3.1. Catalase

Catalase is a 240 kDa ferric hemoprotein consists of four subunits. It catalyzes the decomposition of H_2O_2 to water. In plant, protein expression level of catalase is so high that exogenously added H_2O_2 is almost degraded in minutes. As catalase is a ferric hemoprotein, its heme group is an obvious target of NO. Indeed, NO moderately inhibits catalase-dependent scavenging of H_2O_2 (Brunelli *et al.*, 2001). The inhibition effect of NO on catalase activity was found to be dose dependent (Clark *et al.*, 2000). However, this inhibition was reversible, as removal of the NO donor by dialysis almost completely restores catalase activity. The effect of peroxynitrite ($ONOO^-$), which is the reactive intermediate generated by NO and O_2^-, was also examined. It was found that peroxynitrite also strongly inhibits the activity of catalase, however, this inhibition was irreversible. The mechanism of the inhibition was also investigated in the same study. The adsorption spectrum of purified tobacco catalase showed a broad peak centered at around 405nm, which indicates the enzyme existing primarily in an active ferricatalase. However, after incubating with NO donor, the absorbance at 420 to 440nm increased and that at 405nm decreased indicating the decrease in amount of ferricatalase present. This spectra shift could be attributed to the binding of NO to the heme group. It is therefore evident that NO binds to the heme group of catalase and inhibits its activity (Clark *et al.*, 2000).

3.3.2. Cytochrome oxidase

Mitochondria are the site of internal respiration in eukaryotes. Plant mitochondria contain two different terminal oxidases, the cytochrome oxidase and alternative oxidase (AOX), for accepting electrons. These two oxidases compete with each other for electrons in mitochondria, with inhibition of one pathway redirecting flux to the other. Unlike the cytochrome pathway, which is coupled to oxidative phosphorylation via protein translocation, electron transferred from ubiquinol to AOX is non-phosphorylating and releases energy as heat (Breidenbach *et al.*, 1997). Therefore, the cellular oxidative state can be regulated by modulating the activities of these two oxidases. Using isolated soybean mitochondria as the investigating system, Millar *et al.* (1996) demonstrated that the oxygen consumption via the cytochrome pathway was inhibited by NO. However, the respiration via the alternative oxidase was not significantly inhibited. Moreover, the inhibition on cytochrome pathway was readily reversed upon the removal of NO. It is therefore tempting to speculate that the alternative oxidase plays important roles in NO tolerance in higher plant. Moreover, NO was found to increase the generation of H_2O_2 in mitochondria prepared from soybean embryonic axes (Caro *et al.*, 1999). Recently, it was found that NO actually up-regulated the expression of the AOX transcript in *Arabidopsis* suspension culture (Huang *et al.*, 2002). In fact, the expression of AOX was found to over-expressed for nearly 10-fold. Taken together, NO could modulate both the oxygen consumption and H_2O_2 production in mitochondria, thus regulate the cellular redox status.

3.3.3. Ascorbate Peroxidase (APX)

Ascorbate peroxidase (APX) is one of the main H_2O_2 scavenging enzymes in plant cells. It cycles electrons through ascorbate to effectively remove hydrogen peroxide and regenerate NAD^+. APX is also a heme containing enzyme and is therefore a potential target for NO modulation. Clark *et al.* (2000) use the tobacco APX to demonstrate that APX activity was inhibited in the presence of NO. The inhibition is both time and dose dependent. However, the activity of APX can be regained upon removal of the NO. These results indicated that APX is inhibited by the formation of an iron-nitrosyl complex between NO and the iron atom at the heme group. The inhibitory effect of NO on the two H_2O_2 scavenging enzymes in plant, the catalase and APX, indicates that NO plays a pivotal role in regulating H_2O_2 levels which in turn act as signals under different oxidative stresses.

3.3.4. Aconitase

Aconitase is an iron-sulfur (4Fe-4S) containing enzyme that catalyzes the conversion of citrate and isocitrate. Two isoforms of aconitase exist in plants. One isoform is located in the cytosol while the other resides in the mitochondria. The mitochondria aconitase is an important constituent of the Krebs cycle and is related to energy metabolism. The activity of aconitase can somehow alter the cellular redox status because its inactivation will lead to a decrease in energy metabolism and thus protect the cell against additional oxidative stress. Inactivation of aconitase also reduced the electron flow through the mitochondrial electron transport chain and thus decreased the generation of ROS, which were natural by-products of respiration. Navarre and coworkers (2000) had found that aconitase of tobacco was inhibited in the presence of NO donors. The mechanism of inhibition is likely to be different from that of catalase and APX. It probably involved the dissociation of the iron-sulfur cluster from the enzyme. Interestingly, the aconitase activity was also sensitive to

H_2O_2. It is therefore very likely that aconitase was performing the role as an oxidative stress sensor in plant cells.

3.3.5. *Alpha-dioxygenase*

Alpha-dioxygenase is an enzyme that catalyzes the primary oxidation of fatty acids into a new identified group of oxylipins. In comparing with normal plants, transgenic plants with altered level of alpha-dioxygenase were found to have difference in response to various oxidative stresses (de Leon *et al.*, 2002). Reduction of alpha-dioxygenase expression level led to an increased damage of plant cells by the generation of more superoxide radicals. On the contrary, increasing activity of alpha-dioxygenase protected plant tissues against damages caused by superoxide radicals. It is therefore very likely that alpha-dioxygenase is involved in maintaining the cellular redox status. Moreover, the expression of the alpha-dioxygenase transcript was upregulated in the presence of various NO donors and other ROS, including H_2O_2 and superoxide radicals (de Leon *et al.*, 2002). However, the over-expression induced by NO was not an intermediate process and reaches its maximum level after 2 days. This result indicates that alpha-dioxygenase may participate in the regulation of long term cellular redox status after an oxidative burst.

3.4. Relationship of NO and ROS

3.4.1. *NO and H_2O_2*

Peroxisome is the cellular compartment where many of the ROS are generated and removed. It has long been discovered that peroxisomes contain many different enzymes involved in ROS metabolisms. These included xanthine oxidase for generating superoxide radicals and catalase for scavenging H_2O_2. Recently, it has been discovered that peroxisomes are also a potential site for NO production (Barroso *et al.*, 1999). Barroso and coworkers (1999) have purified peroxisomes from pea leaves and found that it contains an NO producing enzyme. This enzyme shared many characteristics with the mammalian NOS, including using arginine as substrate and have similar co-factors requirement. Moreover, they had probed a 130kDa protein in the peroxisome protein extract using the mammalian NOS antibody. Although there are doubts on whether mammalian NOS antibody probed immunoreactive proteins in plants are really plant NOS or not (see section 1 above), this discovery reconfirmed the important role of peroxisomes in ROS metabolism. It also raised the interest for investigating how different ROS interact in peroxisomes.

It is well known that H_2O_2 and NO interact to transmit signals in many physiological processes, including plant defense response, apoptosis and wounding. It is also well known that NO can affect the intracellular H_2O_2 level by modulating its production (NADPH oxidase) or scavenging (catalase). However, how about the NO production? Does H_2O_2 or other ROS affect it ? It was found that when H_2O_2 was added to the lower epidermal of mung bean leaves, NO was produced within minutes as monitored by the NO specific fluorescent dye DAF2-DA (Lum *et al.*, 2000). This induced NO production was mediated by H_2O_2, as the addition of catalase completely blocked the induction. Moreover, this induced NO production was mediated by a mammalian NOS-like enzyme, as it was inhibited by the NOS inhibitor L-NAME but was not affected in the presence of EGTA. It was also demonstrated that the H_2O_2 signal was transmitted via calcium ion flux, as the NO production was blocked by the calcium channel blocker, verapamil. These results clearly demonstrated that NO production was modulated, at least in part, by H_2O_2. NO and H_2O_2 not only interact together as signaling molecules, but also regulate the production of each other.

3.4.2. *Superoxide Radicals and NO*

NO reacts with O_2^- to form peroxynitrite ($ONOO^-$) spontaneously. Although NO is reactive towards many other ROS, it will not readily react with bio-molecules. On the contrary, peroxynitrite is often considered to be the active intermediate of NO. It readily reacts with the lipid bilayer of cell membrane, the tyrosine residue of protein (nitration) and cysteine residue of the protein (nitrosylation). In fact, effects or downstream signals of NO may actually be mediated by peroxynitrite. Alamillo *et al.* (2001) have found that urate, which is a natural scavenger of peroxynitrite, can reduce the lesion formation in hypersensitive response. Moreover, urate also decreased the cell death rate in hypersensitive site. The same study also shows that the expression of many disease-related proteins, including, PR-1, PAL and GST, are actually induced by peroxynitrite instead of NO.

Another well known effect of peroxynitrite is that it caused the nitration of tyrosine molecule of proteins. The degree of protein nitration was correlated with many diseases, such as inflammation and Alzheimer's disease in animals (Halliwell *et al.*, 1999). However, the role and consequence of protein nitration in plant are poorly understood and certainly require further investigations. In addition to nitration, peroxynitrite is also capable of nitrosylating cycteine residues of proteins. Protein nitrosylation is an important signaling mechanism in animals, as exemplified by the production of cGMP with soluble guanyl cyclase. Unlike protein nitration, the nitrosylation process is reversible and is therefore a more flexible signaling mechanism. Although the actions of peroxynitrite seem to do more bad than good, recent studies suggested that it may act as a ROS "sink" in plant. Beligni *et al.* (1999) found that NO donor can counteract the toxicity of methyl viologen (a ROS generator), and preserved the chlorophyll content in plants. The same study also revealed that NO also significantly reduces the ion leakage and the number of lesion formed in potato leaves upon infection with pathogens. Moreover, Delledonne *et al* (2001) also suggested that the formation of peroxynitrite is a system that utilizes NO to dispose excessive amount of O_2^-, or using O_2^- to dispose excessive amount of NO. Their results illustrated that peroxynitrite, up to concentration of 1 mM, was not cellular toxic to soybean cells. It was also found that neither H_2O_2 nor NO alone could induce significant cell death in soybean suspension culture. However, when H_2O_2 and NO are present simultaneously in a specific ratio, cell death was resulted. Under normal conditions, before it can react with H_2O_2 , excessive NO is removed by forming peroxynitrite with O_2^-. At the same time, O_2^- is also removed before it can be dismuted to H_2O_2 by SOD. (Delledonne *et al.*, 2001)

4. CONCLUSIONS AND FUTURE PROSPECTS

Although NO is originally regarded as a toxic molecule in plants, recent evidences have shown that it is an important signal during plant development and defense. NO, by itself, is a growth regulator. However, future investigation in the field should focus on how NO interacts with other growth regulators, such as plant hormones and ethylene, to modulate plant growth. On the other hand, it is now clear that the signaling pathway of plant under oxidative stresses is very complicated. ROS, such as O_2^-, H_2O_2 and NO, do not act only as a signaling molecule in oxidative stresses, they also as a regulator to promote or suppress the generation of other ROS. Further investigation focus on the interactions between different ROS will definitely shed new light in the field.

ACKNOWLEDGEMENT

Ms. Yoki Kwok-Chu Butt is a Ph.D. student while Dr. John Hon-Kei Lum is supported by a Postdoctoral Fellowship (account no.: G-YW80) from The Hong Kong Polytechnic University. This work is supported by The Hong Kong Polytechnic University Central Research Grant (RGC account no: PolyU5254/02M, HKPU no: G-T622) awarded to SCL Lo.

REFERENCES

Alamillo J.M. and Garcia-Olmedo F. (2001). Effects of urate, a natural inhibitor of peroxynitrite-mediated toxicity, in the response of Arabidopsis thaliana to the bacterial pathogen Pseudomonas syringae. *Plant J.,* **25** : 529-540.

Allan A.C. and Fluhr R. (1997). Two distinct sources of elicited reactive oxygen species in tobacco epidermal cells. *Plant Cell,* **9** : 1559-1572.

Barroso J.B., Corpas F.J., Carreras A., Sandalio L.M., Valderrama R., Palma J.M., Lupianez J.A. and del Rio L.A. (1999). Localization of nitric-oxide synthase in plant peroxisomes. *J. Biol. Chem.,* **274** : 36729-36733.

Beligni M.V., Fath A., Bethke P.C., Lamattina L. and Jones R.L. (2002). Nitric oxide acts as an antioxidant and delays programmed cell death in barley aleurone layers. *Plant Physiol.,* **129** : 1642-1650.

Beligni M.V. and Lamattina L. (1999). Nitric oxide protects against cellular damage produced by methylviologen herbicides in potato plants. *Nitric Oxide.,* **3** : 199-208.

Beligni M.V. and Lamattina L. (2000). Nitric oxide stimulates seed germination and de-etiolation, and inhibits hypocotyl elongation, three light-induced responses in plants. *Planta,* **210** : 215-221.

Beligni M.V. and Lamattina L. (2001). Nitric oxide: a non-traditional regulator of plant growth. *Trends Plant Sci.,* **6** : 508-509.

Bethke P.C., Lonsdale J.E., Fath A. and Jones R.L. (1999). Hormonally regulated programmed cell death in barley aleurone cells. *Plant Cell,* **11** : 1033-1046.

Botrel A., Magne C. and Kaiser W.M. (1996). Nitrate reduction, nitrite reduction and ammonium assimilation in barley roots in response to anoxia. *Plant Physiol. Biochem.,* **34** : 645-652.

Bowler C., Neuhaus G., Yamagata H. and Chua N.H. (1994). Cyclic GMP and calcium mediate phytochrome phototransduction. *Cell,* **77** : 73-81.

Breidenbach R.W., Saxton M.J., Hansen L.D. and Criddle R.S. (1997). Heat generation and dissipation in plants: can the alternative oxidative phosphorylation pathway serve a thermoregulatory role in plant tissues other than specialized organs? *Plant Physiol.,* **114** : 1137-1140.

Brown E.G., Newton R.P., Evans D.E., Walton T.J., Younis L.M. and Vaughan J.M. (1989). Influence of light on cyclic nucleotide metabolism in plants: effect of dibutyryl cyclic nucleotides on chloroplast components. *Phytochemistry,* **28** : 2559-2563.

Brunelli L., Yermilov V. and Beckman J.S. (2001). Modulation of catalase peroxidatic and catalatic activity by nitric oxide. *Free Radic. Biol. Med.,* **30** : 709-714.

Butt Y.K., Lum J.H. and Lo S.C. (2003). Proteomic identification of plant proteins probed by mammalian nitric oxide synthase antibodies. *Planta,* **216** : 762-771.

Caro A. and Puntarulo S. (1999). Nitric oxide generation by soybean embryonic axes. Possible effect on mitochondrial function. *Free Rad. Res.,* **31** : S205-212.

Chakraborti T., Das S., Mondal M., Roychoudhury S. and Chakraborti S. (1999). Oxidant, mitochondria and calcium: an overview. *Cell Signal.,* **11** : 77-85.

Chandok M.R., Ytterberg A.J., van Wijk K.J. and Klessig D.F. (2003). The pathogen-inducible nitric oxide synthase (iNOS) in plants is a variant of the P protein of the glycine decarboxylase complex. *Cell,* **113** : 469-482.

Chory J., Elich T., Li H.M., Pepper A., Poole D., Reed J., Susek R., Vitart V., Washburn T. and Furuya M. *et al.* (1994). Genes controlling Arabidopsis photomorphogenesis. *Biochem Soc Symp.,* **60** : 257-263.

Clark D., Durner J., Navarre D.A. and Klessig D.F. (2000). Nitric oxide inhibition of tobacco catalase and ascorbate peroxidase. *Mol. Plant Microbe Interact.,* **13** : 1380-1384.

Cooney R.V., Harwood P.J., Custer L.J. and Franke A.A. (1994). Light-mediated conversion of nitrogen dioxide to nitric oxide by carotenoids. *Environ. Health Perspect.,* **102** : 460-462.

Cueto M., Hernandez-Perera O., Martin R., Bentura M.L., Rodrigo J., Lamas S. and Golvano M.P. (1996). Presence of nitric oxide synthase activity in roots and nodules of *Lupinus albus. FEBS Lett.,* **398** : 159-164.

Dangl J.L., Dietrich R.A. and Richberg M.H. (1996). Death Don't Have No Mercy: Cell Death Programs in Plant-Microbe Interactions. *Plant Cell,* **8** : 1793-1807.

de Leon I.P., Sanz A., Hamberg M. and Castresana C. (2002). Involvement of the Arabidopsis alpha-DOX1 fatty acid dioxygenase in protection against oxidative stress and cell death. *Plant J.,* **29** : 61-62.

de Pinto M.C., Tommasi F. and De Gara L. (2002). Changes in the antioxidant systems as part of the signaling pathway responsible for the programmed cell death activated by nitric oxide and reactive oxygen species in tobacco Bright-Yellow 2 cells. *Plant Physiol.,* **130** : 698-708.

Dean J.V. and Harper J.E. (1998). The conversion of nitrite to nitrogen oxide(s) by the constitutive NAD(P)H-nitrate reductase enzyme from soybean. *Plant Physiol.,* **88** : 389-395.

del Rio L.A., Corpas F.J., Sandalio L.M., Palma J.M., Gomez M. and Barroso J.B. (2002). Reactive oxygen species, antioxidant systems and nitric oxide in peroxisomes. *J. Exp. Bot.,* **53** : 1255-1272.

Delledonne M., Xia Y.J., Dixon R.A. and Lamb C. (1998). Nitric oxide functions as a signal in plant disease resistance. *Nature,* **394** : 585-588.

Delledonne M., Zeier J., Marocco A. and Lamb C. (2001). Signal interactions between nitric oxide and reactive oxygen intermediates in the plant hypersensitive disease resistance response. *Proc. Natl. Acad. Sci. USA,* **98** : 13454-13459.

Doke N., Miura Y., Sanchez L.M., Park H.J., Noritake T., Yoshioka H. and Kawakita K. (1996). The oxidative burst protects plants against pathogen attack: mechanism and role as an emergency signal for plant bio-defence. *Gene,* **179** : 45-51.

Dorey S., Kopp M., Geoffroy P., Fritig B. and Kauffmann S. (1999). Hydrogen peroxide from the oxidative burst is neither necessary nor sufficient for hypersensitive cell death induction, phenylalanine ammonia lyase stimulation, salicylic acid accumulation, or scopoletin consumption in cultured tobacco cells treated with elicitin. *Plant Physiol.,* **121** : 163-172.

Dumas B., Freyssinet G. and Pallett K.E. (1995). Tissue-Specific Expression of Germin-Like Oxalate Oxidase during Development and Fungal Infection of Barley Seedlings. *Plant Physiol.,* **107** : 1091-1096.

Durner J. and Klessig D.F. (1998). Nitric oxide as a signal in plants. *Curr. Opin. Plant Biol.,* **2** : 369-374.

Durner J., Wendehenne D. and Klessig D.F. (1999). Defense gene induction in tobacco by nitric oxide, cyclic GMP, and cyclic ADP-ribose. *Proc. Natl. Acad. Sci. USA,* **95** : 10328-10333.

Fath A., Bethke P., Beligni V. and Jones R. (2002). Active oxygen and cell death in cereal aleurone cells. *J. Exp. Bot.,* **53** : 1273-1282.

Foissner I., Wendehenne D., Langebartels C. and Durner J. (2000). In vivo imaging of an elicitor-induced nitric oxide burst in tobacco. *Plant J.,* **23** : 817-824.

Garcia-Mata C. and Lamattina L. (2002). Nitric oxide and abscisic acid cross talk in guard cells. *Plant Physiol.,* **128** : 790-792.

Garcia-Mata C. and Lamattina L. (2001). Nitric oxide induces stomatal closure and enhances the adaptive plant responses against drought stress. *Plant Physiol.,* **126** : 1196-120[illegible].

Giba Z., Grubisic D., Todorovic S., Sajc L., Stojakovic D. and Konjevic T. (1998). Effect of nitric oxide-releasing compounds on phytochrome-controlled germination of Empress tree seeds. *Plant Growth Regul.,* **26** : 175-181.

Gouvea C.M.C.P., Souza J.F., Magalhaes A.C.N. and Martins I.S. (1997). NO-releasing substances that induce growth elongation in maize root segments. *Plant Growth Regul.,* **21** : 183-187.

Graziano M., Beligni M.V. and Lamattina L. (2002). Nitric oxide improves internal iron availability in plants. *Plant Physiol.,* **130** : 1852-1859.

Grisham M.B., Jourd'Heuil D. and Wink D.A. (1999). Nitric oxide. I. Physiological chemistry of nitric oxide and its metabolites: implications in inflammation. *Am. J. Physiol.,* **276** : G315-321.

Groom Q.J., Torres M.A., Fordham-Skelton A.P., Hammond-Kosack K.E., Robinson N.J. and Jones J.D. (1996). rbohA, a rice homologue of the mammalian gp91phox respiratory burst oxidase gene. *Plant J.,* **10** : 515-522.

Grubisic D. and Konjevic R. (1990). Light and nitrate interaction in phytochrome-controlled germination of *Paulownia tomentosa* seed. *Planta,* **181** : 423-426.

Halliwell B., Zhao K. and Whiteman M. (1999). Nitric oxide and peroxynitrite. The ugly, the uglier and the not so good: a personal view of recent controversies. *Free Radic. Res.,* **31** : 651-669.

Henry Y.A., Ducastel B. and Guissani A. (1997). Basic chemistry of nitric oxide and related nitrogen oxides. In: *Nitric Oxide Research from Chemistry to Biology.* (Eds. Henry Y.A., Guissani A., Ducastel B.) Landes Company, 15-46.

Hideg E., Barta C., Kalai T., Vass I., Hideg K. and Asada K. (2002). Detection of singlet oxygen and superoxide with fluorescent sensors in leaves under stress by photoinhibition or UV radiation. *Plant Cell Physiol.,* **43** : 1154-1164.

Hoeberichts F.A. and Woltering E.J. (2003). Multiple mediators of plant programmed cell death: interplay of conserved cell death mechanisms and plant-specific regulators. *Bioessays,* **25** : 47-57.

Horemans N., Foyer C.H. and Asard H. (2000). Transport and action of ascorbate at the plant plasma membrane. *Trends Plant Sci.,* **5** : 263-267.

Huang X., von Rad U. and Durner J. (2002). Nitric oxide induces transcriptional activation of the nitric oxide-tolerant alternative oxidase in Arabidopsis suspension cells. *Planta,* **215** : 914-923.

Jih P.J., Chen Y.C. and Jeng S.T. (2003). Involvement of hydrogen peroxide and nitric oxide in expression of the ipomoelin gene from sweet potato. *Plant Physiol.,* **132** : 381-389.

Kawano T. (2003). Roles of the reactive oxygen species-generating peroxidase reactions in plant defense and growth induction. *Plant Cell Rep.,* **21** : 829-837.

Keller T., Damude H.G., Werner D., Doerner P., Dixon R.A., Lamb C. (1998). A plant homolog of the neutrophil NADPH oxidase gp91phox subunit gene encodes a plasma membrane protein with Ca^{2+} binding motifs. *Plant Cell,* **10** : 255-266.

Kiegle E., Gilliham M., Haseloff J. and Tester M. (2000). Hyperpolarisation-activated calcium currents found only in cells from the elongation zone of *Arabidopsis thaliana* roots. *Plant J.,* **21** : 225-229.

Kim M., Ahn J.W., Jin U.H., Choi D., Paek K.H. and Pai H.S. (2003). Activation of the programmed cell death pathway by inhibition of proteasome function in plants. *J. Biol. Chem.,* **278** : 19406-19415.

Klepper L. (1990). Comparison between NO, evolution mechanisms of wild-type and *nr1* mutant soybean leaves. *Plant Physiol.,* **93** : 26-32.

Knowles RG, Moncada S (1994). Nitric oxide synthases in mammals. *Biochem. J.,* **298** : 249-258.

Kroncke K.D., Fehsel K. and Kolb-Bachofen V. (1997). Nitric oxide cytotoxicity versus cytoprotection - who, why, when and where. *Nitric oxide,* **1** : 107-120.

Kuo W.N., Kuo T.W., Jones D.L. and Baptiste J. (1995). Nitric oxide synthase immunoreactivity in Baker's yeast, lobster and wheat germ. *Biochem. Arch.,* **11** : 73-78.

Laxalt A.M., Beligni M.V. and Lamattina L. (1997). Nitric oxide preserves the level of chlorophyll in potato leaves infected by *Phytophthora infestans. Eur. J. Plant Pathol.,* **103** : 643-651.

Leckie C.P., McAinsh M.R., Allen G.J., Sanders D. and Hetherington A.M. (1998). Abscisic acid-induced stomatal closure mediated by cyclic ADP-ribose. *Proc. Natl. Acad. Sci. USA,* **95** : 15837-15842.

Lesham Y.Y. and Haramaty E. (1996). The characterization and contrasting effects of the nitric oxide free radical in vegetative stress and senescence of *Pisum sativum* L. foliage. *Plant Physiol.,* **148** : 258-263.

Leshem Y.Y., Wills R.B.H. and Veng-Va Ku V. (1998). Evidence for the function of the free radical gas, nitric oxide (NO·), as an endogenous maturation and senescence regulating factor in higher plants. *Plant Physiol. and Biochem.,* **36** : 825-833.

Levine A., Tenhaken R., Dixon R. and Lamb C. (1994). H_2O_2 from the oxidative burst orchestrates the plant hypersensitive disease resistance response. *Cell,* **79** : 583-593.

Lum H.K., Butt Y.K. and Lo S.C. (2002). Hydrogen peroxide induces a rapid production of nitric oxide in mung bean (*Phaseolus aureus*). *Nitric Oxide,* **6** : 205-213.

Magalhaes J.R., Monte D.C. and Durzan D. (2000). Nitric oxide and ethylene emission in *Arabidopsis thaliana. Physiol. Mol. Biol. Plants,* **6** : 117-127.

Mallick N, Mohn FH, Ria LC, Soeder CJ (2000). Evidence for the non-involvement of nitric oxide synthase in nitric oxide production by the green alga *Scenedesmus obliquus. Plant Physiol.,* **156** : 423-426.

Marletta M.A. (1989). Nitric oxide: biosynthesis and biological significance. *Trends Biochem. Sci.,* **14** : 488-492.

McDonald L.J. and Murad F. (1995). Nitric oxide and cGMP signaling. *Adv. Pharmacol.,* **34** : 263-275.

McDowell J.M. and Dangl J.L. (2000). Signal transduction in the plant immune response. *Trends Biochem. Sci.,* **25** : 79-82.

Meyer C. and Stohr C. (2004). Nitrate reductase and nitrite reductase. In: *Advances in Photosynthesis: Photosynthetic Nitrogen Assimilation and Associated Carbon Metabolism.* (Eds Foyer C.H. and Noctor G.) Dordrecht, The Netherlands: Kluwer Academic Publishers (In press).

Millar A.H. and Day D.A. (1996). Nitric oxide inhibits the cytochrome oxidase but not the alternative oxidase of plant mitochondria. *FEBS Lett.,* **398** : 155-158.

Navarre D.A., Wendehenne D., Durner J., Noad R. and Klessig D.F. (2000). Nitric oxide modulates the activity of tobacco aconitase. *Plant Physiol.,* **122** : 573-582.

Navazio L., Bewell M.A., Siddiqua A., Dickinson G.D., Galione A. and Sanders D. (2000). Calcium release from the endoplasmic reticulum of higher plants elicited by the NADP metabolite nicotinic acid adenine dinucleotide phosphate. *Proc. Natl. Acad. Sci. USA,* **97** : 8693-8698.

Neill S., Desikan R. and Hancock J. (2002). Hydrogen peroxide signalling. *Curr. Opin. Plant Biol.,* **5** : 388-395.

Neill S.J., Desikan R., Clarke A. and Hancock J.T. (2002). Nitric oxide is a novel component of abscisic acid signaling in stomatal guard cells. *Plant Physiol.,* **128** : 13-16.

Ninnemann H. and Maier J. (1996). Indications for the occurrence of nitric oxide synthases in fungi and plants and the involvement in photoconidiation of *Neurospora crassa. Photochem. Photobiol.,* **64** : 393-398.

Nishimura H., Hayamizu T. and Yanagisawa Y. (1986). Reduction of NO_2 to NO by rush and other plants. *Environ. Sci. Technol.,* **20** : 413-416.

Orozco-Cardenas M.L. and Ryan C.A. (2002). Nitric oxide negatively modulates wound signaling in tomato plants. *Plant Physiol.,* **130** : 487-493.

Pagnussat G.C., Lanteri M.L. and Lamattina L. (2003). Nitric oxide and cyclic GMP are messengers in the indole acetic acid-induced adventitious rooting process. *Plant Physiol.,* **132** : 1241-1248.

Pedroso M.C. and Durzan D.J. (2000). Effect of different gravity environments on DNA fragmentation and cell death in *Kalanchoe* leaves. *Ann. Bot. (Lond),* **86** : 983-994.

Penson S.P., Schuurink R.C., Fath A., Gubler F., Jacobsen J.V. and Jones R.L. (1996). cGMP is required for gibberellic acid-induced gene expression in barley aleurone. *Plant Cell,* **8** : 2325-2333.

Pfeiffer S., Janistyn B., Jessner G., Pichorner H. and Eberman R.E. (1994). Gaseous nitric oxide stimiulates guanosine 3' 5;-cyclic monophosphate formation in sprune needles. *Phytochemistry,* **36** : 259-262.

Pollock J.S., Forstermann U., Mitchell J.A., Warner T.D., Schmidt H.H., Nakane M. and Murad F. (1991). Purification and characterization of particulate endothelium-derived relaxing factor synthase from cultured and native bovine aortic endothelial cells. *Proc. Natl. Acad. Sci. USA,* **88** : 10480-10484.

Ribeiro E.A. Jr., Cunha F.Q., Tamashiro W.M. and Martins I.S. (1999). Growth phase-dependent subcellular localization of nitric oxide synthase in maize cells. *FEBS Lett.,* **445** : 283-286.

Rockel P., Strube F., Rockel A., Wildt J. and Kaiser W.M. (2002). Regulation of nitric oxide (NO) production by plant nitrate reductase *in vivo* and *in vitro. J. Exp. Bot.,* **53** : 103-110.

Segal A.W. and Abo A. (1993). The biochemical basis of the NADPH oxidase of phagocytes. *Trends Biochem. Sci.,* **18** : 43-47.

Sen S. and Cheema I.R. (1995). Nitric oxide synthase and calmodulin immunoreactivity in plant embryonic tissue. *Biochem. Arch.,* **11** : 221-227.

Soheila A.H.M.C., Fred J., Brian J. and Brian T. (2001). Early signaling components in ultraviolet-B responses: distinct roles for different reactive oxygen species and nitric oxide. *FEBS Lett.,* **489** : 237-242.

Stohr C. and Ullrich W.R. (2002). Generation and possible roles of NO in plant roots and their apoplastic space. *J. Exp. Bot.,* **53** : 2293-2303.

Stohr C. (1999). Relationship of nitrate supply with growth rate, plasma membrane-bound and cytosolic nitrate reductase, and tissue nitrate content in tobacco plant. *Plant, Cell and Environ.,* **22** : 169-177.

Taylor Jr. G.E. and Gunderson C.A. (1986). The response of foliar gas exchange to exogenously applied ethylene. *Plant Physiol.,* **82** : 653-657.

The *Arabidopsis* Genome Initiative (2000). *Nature,* **408** : 796-815.

Tun N.N., Holk A. and Scherer G.F. (2001). Rapid increase of NO release in plant cell cultures induced by cytokinin. *FEBS Lett.,* **509** : 174-176.

Weitzberg E. and Lundberg J.O. (1998). Non-enzymatic nitric oxide production in humans. *Nitric Oxide,* **2** : 1-7.

Wojtaszek P., Trethowan J. and Bolwell G.P. (1997). Reconstitution in vitro of the components and conditions required for the oxidative cross-linking of extracellular proteins in French bean (*Phaseolus vulgaris* L.). *FEBS Lett.*, **405** : 95-98.

Yamasaki H. and Sakihama Y. (2000). Simultaneous production of nitric oxide and peroxynitrite by plant nitrate reductase: in vitro evidence for the NR-dependent formation of active nitrogen species. *FEBS Lett.,* **468** : 89-92.

Yamasaki H., Sakihama Y. and Takahashi S (1999). An alternative pathway for nitric oxide production: new features of an old enzyme. *Trends Plant Sci.,* **4** : 128-129.

SUBJECT INDEX

A

B

C

D

E

F

G

H

I

T

V

W

X

Z